Advancements in Fuzzy Reliability Theory

Akshay Kumar
Graphic Era Hill University, India

Mangey Ram
Department of Mathematics, Graphic Era (Deemed), India

Om Prakash Yadav
North Dakota State University, USA

IGI Global
PUBLISHER of TIMELY KNOWLEDGE

A volume in the Advances in
Systems Analysis, Software
Engineering, and High Performance
Computing (ASASEHPC) Book Series

Published in the United States of America by
IGI Global
Engineering Science Reference (an imprint of IGI Global)
701 E. Chocolate Avenue
Hershey PA, USA 17033
Tel: 717-533-8845
Fax: 717-533-8661
E-mail: cust@igi-global.com
Web site: http://www.igi-global.com

Library of Congress Cataloging-in-Publication Data

Names: Kumar, Akshay, editor. | Ram, Mangey, editor. | Yadav, Om Prakash, 1964- editor.
Title: Advancements in fuzzy reliability theory / Akshay Kumar, Mangey Ram, and Om Prakash Yadav, editors.
Description: Hershey, PA : Engineering Science Reference, [2021] | Includes bibliographical references and index. | Summary: "This book presents original research studies and articles that investigage the latest applications of fuzzy reliability theory applications in engineering areas"-- Provided by publisher.
Identifiers: LCCN 2020048003 (print) | LCCN 2020048004 (ebook) | ISBN 9781799875642 (h/c) | ISBN 9781799875659 (s/c) | ISBN 9781799875666 (eISBN)
Subjects: LCSH: Reliability (Engineering) | Fuzzy systems.
Classification: LCC TA169 .A357 2021 (print) | LCC TA169 (ebook) | DDC 620/.0045201511313--dc23
LC record available at https://lccn.loc.gov/2020048003
LC ebook record available at https://lccn.loc.gov/2020048004

This book is published in the IGI Global book series Advances in Systems Analysis, Software Engineering, and High Performance Computing (ASASEHPC) (ISSN: 2327-3453; eISSN: 2327-3461)

British Cataloguing in Publication Data
A Cataloguing in Publication record for this book is available from the British Library.

Advances in Systems Analysis, Software Engineering, and High Performance Computing (ASASEHPC) Book Series

ISSN:2327-3453
EISSN:2327-3461

Editor-in-Chief: Vijayan Sugumaran, Oakland University, USA

MISSION

The theory and practice of computing applications and distributed systems has emerged as one of the key areas of research driving innovations in business, engineering, and science. The fields of software engineering, systems analysis, and high performance computing offer a wide range of applications and solutions in solving computational problems for any modern organization.

The **Advances in Systems Analysis, Software Engineering, and High Performance Computing (ASASEHPC) Book Series** brings together research in the areas of distributed computing, systems and software engineering, high performance computing, and service science. This collection of publications is useful for academics, researchers, and practitioners seeking the latest practices and knowledge in this field.

COVERAGE

- Distributed Cloud Computing
- Virtual Data Systems
- Enterprise Information Systems
- Human-Computer Interaction
- Computer System Analysis
- Engineering Environments
- Network Management
- Parallel Architectures
- Storage Systems
- Computer Networking

IGI Global is currently accepting manuscripts for publication within this series. To submit a proposal for a volume in this series, please contact our Acquisition Editors at Acquisitions@igi-global.com or visit: http://www.igi-global.com/publish/.

The Advances in Systems Analysis, Software Engineering, and High Performance Computing (ASASEHPC) Book Series (ISSN 2327-3453) is published by IGI Global, 701 E. Chocolate Avenue, Hershey, PA 17033-1240, USA, www.igi-global.com. This series is composed of titles available for purchase individually; each title is edited to be contextually exclusive from any other title within the series. For pricing and ordering information please visit http://www.igi-global.com/book-series/advances-systems-analysis-software-engineering/73689. Postmaster: Send all address changes to above address.

Titles in this Series

For a list of additional titles in this series, please visit:
http://www.igi-global.com/book-series/advances-systems-analysis-software-engineering/73689

Impacts and Challenges of Cloud Business Intelligence
Shadi Aljawarneh (Jordan University of Science and Technology, Jordan) and Manisha Malhotra (Chandigarh University, India)
Business Science Reference • ©2021 • 263pp • H/C (ISBN: 9781799850403) • US $195.00

Handbook of Research on Modeling, Analysis, and Control of Complex Systems
Ahmad Taher Azar (Prince Sultan University, Saudi Arabia) and Nashwa Ahmad Kamal (Cairo University, Egypt)
Engineering Science Reference • ©2021 • 480pp • H/C (ISBN: 9781799857884) • US $295.00

Artificial Intelligence Paradigms for Smart Cyber-Physical Systems
Ashish Kumar Luhach (The PNG University of Technology, Papua New Guinea) and Atilla Elçi (Hasan Kalyoncu University, Turkey)
Engineering Science Reference • ©2021 • 392pp • H/C (ISBN: 9781799851011) • US $225.00

Advancements in Model-Driven Architecture in Software Engineering
Yassine Rhazali (Moulay Ismail University of Meknes, Morocco)
Engineering Science Reference • ©2021 • 287pp • H/C (ISBN: 9781799836612) • US $215.00

Cloud-Based Big Data Analytics in Vehicular Ad-Hoc Networks
Ram Shringar Rao (Ambedkar Institute of Advanced Communication Technologies and Research, India) Nanhay Singh (Ambedkar Institute of Advanced Communication Technologies and Research, India) Omprakash Kaiwartya (School of Science and Technology, Nottingham Trent University, UK) and Sanjoy Das (Indira Gandhi National Tribal University, India)
Engineering Science Reference • ©2021 • 312pp • H/C (ISBN: 9781799827641) • US $245.00

For an entire list of titles in this series, please visit:
http://www.igi-global.com/book-series/advances-systems-analysis-software-engineering/73689

IGI Global
PUBLISHER of TIMELY KNOWLEDGE

701 East Chocolate Avenue, Hershey, PA 17033, USA
Tel: 717-533-8845 x100 • Fax: 717-533-8661
E-Mail: cust@igi-global.com • www.igi-global.com

Table of Contents

Detailed Table of Contents

The fuzzy cognitive map is considered as a tool for the ranking of factors affecting reliability. The rank of a factor is defined as an equivalent of the Birnbaum importance index in the classical reliability theory. Models and algorithms are proposed for calculation of the importance indexes of single factors and their joint effects on the reliability. The method is exemplified by the reliability and safety of an automobile in the "driver-automobile-road" system subject to the driver's qualification, traffic situation, unit costs of operation, operating conditions, maintenance scheduling, quality of maintenance and repair, quality of automobile design, quality of operational materials and spare parts, as well as storage conditions. The advantages of the method include 1) use of available expert information with no collection and processing statistical data and 2) capability to take into account any quantitative and qualitative factors associated with people, technology, software, quality of service, operating conditions, etc.

In this chapter, the authors study a weighted-((f / (r, s)), k)/ (m, n): G system. The system consists of mn components arranged in a matrix form and the system works if all the sub matrices of order (r, s), the total weight of the working components is greater than f and the total weight of the working components in the system is at least k. This chapter deals with the evaluation of fuzzy reliability and fuzzy mean time to failure of the considered system with the application of fuzzy universal generating function and fuzzy Rayleigh distribution. In this study, the authors formed some prepositions to understand the behaviour of the considered system with respect to different varying parameters and also present an illustrative example to understand them.

Service products are mostly produced and consumed simultaneously through interaction between customer and service providers. To prevent external failures in service operations, it is important to identify potential risks and take relevant actions to eliminate or reduce the occurrence. Therefore, risk assessment is vital to customer satisfaction in any service organization. Failure mode and effects analysis (FMEA) is an effective and useful tool for risk assessment. Although FMEA has been extensively studied in the manufacturing literature, there are a limited number of studies considering the application of FMEA in the hospitality industry. In traditional FMEA, the risk priority of failure modes is determined by generating a crisp risk priority number (RPN). However, it has been claimed in the literature that crisp RPN doesn't have a good performance in reflecting real-life situations. To overcome this shortcoming, a fuzzy hybrid FMEA method is developed. The proposed method has been tested on a case study in a five-star hotel to assess its applicability and benefits.

In the process of dealing with various problems that are related to the control and management of complex systems, uncertainty is always the issue. Most of the decisions to be made by engineers and governors are subject to a lack of data that causes uncertainty. Some information is not always accessible and insufficient at the time of decision-making (DM). It is important to use the concept of a human being's expert knowledge, so developing models to estimate and imitate human decision-making systems has been becoming necessary. This chapter will critically assess the fuzzy reliability theory and systems to demonstrate how fuzzy logic represents human behavior and non-linearity. The authors created a fuzzy inference system to model two different complex systems. Fuzzy approaches to real problems in DM are effective alternatives to traditional approaches. Fuzzy integrations improve DM models in five principal features: (1) expert knowledge, (2) uncertainty handling, (3) human and government behavior modeling, (4) flexible modeling, and (5) simpler representations.

Chapter 5

Burak Efe, Necmettin Erbakan University, Turkey
Mustafa Kurt, Gazi University, Turkey

This chapter aims basically to ensure two benefits, which focus on method and application. Firstly, this chapter proposes a new approach, which is TOPSIS (technique for order preference by similarity to ideal solution) based on interval type-2 trapezoidal fuzzy (IT2TrF) numbers. Secondly, the proposed method is applied to failure mode and effects analysis (FMEA) problem to overcome the drawbacks of traditional FMEA, which determines possible failure modes. In traditional FMEA, the risk priority number method, which is obtained as the product of occurrence, severity, and detection risk factors, has been debated to have several drawbacks. The implementation of risk assessment is realized to indicate the validity of the suggested approach in a construction building. The results of the proposed approach and some other methods are compared. The aim and resources are optimized by using linear programming.

Chapter 6

Kamal Kumar, Amity University, Haryana, India
Naveen Mani, Sandip University, Nashik, India
Amit Sharma, Amity University, Haryana, India
Reeta Bhardwaj, Amity University, Haryana, India

The failure mode and effect analysis (FMEA) is widely used an effective pre-accident risk assessment tool to identify, eliminate, and assess potential failure modes in different industries for enhancing the safety and reliability of systems, process, services, and products. Therefore, this chapter presents a new approach to rank the failure modes under the interval-valued intuitionistic fuzzy set (IVIFS). For this, a novel measure to measure the fuzziness known as entropy measure is proposed. Some properties and axiom definition of the proposed entropy measure have been presented to show the validity of it. Afterwards, the proposed entropy measure is utilized to obtain the weight of risk factor and developed an approach under the

IVIFS environment to determine the risk priority order of failure modes. Finally, a real-life case of FMEA has been discussed to manifest the developed approach, and obtained results are compared with the results obtained by the existing methods for showing the feasibility and validity of the proposed approach.

A new type-2 fuzzy reliability-redundancy allocation problem (IT2FRRAP) is introduced in this study, which maximizes system reliability underneath the type-2 fuzzy uncertainty. Throughout the design process, estimation of system parameters with type-2 fuzzy sets is very rational, as many experts may have varying views on the numerous risks involved with the multi-state framework. The authors consider fuzzy membership functions of interval type-2 to fuzzy set to reflect the reliability of the sub-system. By using the extension theory, they evaluate the total system reliability as per respective configurations (i.e., series, parallel, non-parallel, and bridge). They developed a novel solution method focused on particle swarm optimization to address the IT2FRRAP. They carried out the computational simulations on various numerical databases to show the gravity of the proposed formulations. To compare its productiveness, a detailed analysis of the proposed solution was made with the genetic algorithm (GA).

In a system where it cannot be reduced to a pure series and parallel form (i.e., complex), reliability is obtained by universal generating function (UGF) technique. The intuitionistic fuzzy sets (IFS) concept with triangular fuzzy number (TFN) and Weibull lifetime distribution is introduced to find the fuzzy reliability of the same system. Also, averaging operator with given equal weights is used with the set of three triangular intuitionistic fuzzy number. A numerical example is solved for demonstration.

This chapter considers a consecutive r-out of-k-from n: G system taken for the study. The considered system works if at least r elements out of k consecutive elements within n elements work. The consecutive r-out of-k-from n: G system is special case of linear consecutive k-out-of-r-from-n:G system. A technique is applied for analyzing fuzzy reliability of the system using universal generating function and fuzzy set theory. At the end of the chapter, the proposed technique is demonstrated from a numerical example.

The present study proposes to determine the fuzzy reliability of different systems in which the lifetime of components are following fuzzy exponential distribution where fuzzy reliability function and its α-cut set are presented. The fuzzy reliability of different systems is defined on the basis of octagonal intuitionistic fuzzy numbers. The fuzzy reliability functions of k-out-of-m system, series system, parallel systems, and their fuzzy mean time to failure are discussed respectively using the concept of α-cut of octagonal intuitionistic fuzzy numbers. Finally, some numerical examples are discussed to illustrate how to calculate the fuzzy system reliability and α-cut of fuzzy mean time to failure (FMTTF).

The present work introduces a new technique for analyzing the fuzzy reliability of a system stem under hesitant fuzzy environment where the reliability of a component/ unit of a system is represented by hesitant fuzzy element (HFE). In this study, the authors evaluate the fuzzy reliability for some complex systems (series configuration, parallel configuration, and bridge configuration) using score function which is very useful in reliability modelling (especially in decision making, risk analysis,

Preface

In recent years, substantial efforts are being made in the development of reliability theory including fuzzy reliability theories and their applications to various areas of real-life problems. Fuzzy set theory widely used in decision making and multi criteria such as management and engineering etc. There are several types of fuzzy set theory such as intuitionistic fuzzy, hesitant fuzzy, dual hesitant fuzzy, bifuzzy, picture fuzzy etc. based on triangular and trapezoidal fuzzy number for evaluate the uncertainty of real life systems such as series, parallel and complex systems. In the decade fuzzy reliability having effective tools and techniques based on real set theory for proposed models.

The book *Advancements in Fuzzy Reliability Theory* engrossed on a comprehensive range of fuzzy techniques applied in various fields of engineering problems and real life systems like integrated intuitionistic fuzzy AHP & TOPSIS approach, entropy measure for interval-valued intuitionistic fuzzy set, fuzzy reliability theory, fuzzy reliability in decision making, interval type-2 fuzzy reliability-redundancy, fuzzy PSO algorithm, fuzzy lifetime distribution emphasizing octagonal intuitionistic fuzzy numbers, hesitant fuzzy and score function, fuzzy universal generating function, neuro fuzzy system, adaptive neuro fuzzy inference system, hybrid fuzzy logic genetic algorithms belief networks, etc.

This research book is be used as a support book for the undergraduate, postgraduate students of engineering and allied branches; engineers, research scientists/academicians involved in fuzzy logic applications in engineering areas.

Akshay Kumar
Graphic Era Hill University, India

Mangey Ram
Department of Mathematics, Graphic Era (Deemed), India

Om Prakash Yadav
North Dakota State University, USA

Chapter 1
Fuzzy Cognitive Maps in Reliability Modeling

Alexander Rotshtein
Lev Academic Center, Jerusalem College of Technology, Israel & Vasyl Stus Donetsk National University, Vinnitsa, Ukraine

Ludmila Pustylnik
Afeka Tel-Aviv Academic College of Engineering, Israel

Denys Katielnikov
Vinnytsia National Technical University, Ukraine

ABSTRACT

The fuzzy cognitive map is considered as a tool for the ranking of factors affecting reliability. The rank of a factor is defined as an equivalent of the Birnbaum importance index in the classical reliability theory. Models and algorithms are proposed for calculation of the importance indexes of single factors and their joint effects on the reliability. The method is exemplified by the reliability and safety of an automobile in the "driver-automobile-road" system subject to the driver's qualification, traffic situation, unit costs of operation, operating conditions, maintenance scheduling, quality of maintenance and repair, quality of automobile design, quality of operational materials and spare parts, as well as storage conditions. The advantages of the method include 1) use of available expert information with no collection and processing statistical data and 2) capability to take into account any quantitative and qualitative factors associated with people, technology, software, quality of service, operating conditions, etc.

DOI: 10.4018/978-1-7998-7564-2.ch001

INTRODUCTION

Success in applied modeling problem solving largely depends on the chosen mathematical tools. Reliability theory is like medicine; medicine deals with human health, while reliability theory deals with "the health of man-made systems". The range of problems that may harm the reliability of system at the stages of design, production and operation makes reliability science an endless polygon, and presents opportunities for testing different mathematical theories and new methods for predicting, diagnosing and decision making.

Classical reliability theory [1, 2] arose in the middle of the 20th century when probability theory was the only way to formalize uncertainty. The models of classical reliability theory use the source data about failure probabilities and the laws of random value distribution of time between failures and repairs. It is assumed that these inputs are known from reference books or can be obtained experimentally. In practice, though, obtaining the source data and addressing all the influencing factors is a difficult and often impossible task. These difficulties often force us to substitute approximate source data into the complex probabilistic models, and make it impossible to predict accurately the probability of failure. As a result, engineers do not believe in their own reliability calculations (Barnard, 2012).

The appearance in 1965 an article by L. Zadeh (Zadeh, 1965) laid the foundation for a new method of describing uncertainty in the form of fuzzy logic. This aroused natural interest among specialists in reliability, and as a result, led to the emergence of an independent direction, which can be called Fuzzy Reliability. The first specialized collection of articles in this area is (Onisawa & Kasprzyk, 1995), and the first monographs include (Cai, 1996; Rotshtein & Shtovba, 1997; Utkin & Shubinsky, 1998).

Our experience in using fuzzy mathematics in reliability modeling shows that the following fundamental concepts and principles have the greatest application:

- *Fuzzy numbers* (Dubois & Prade, 1978) are a convenient means of presenting expert information about uncertain source data that are used in probabilistic reliability models (Onisawa et al., 1995; Rotshtein & Shtovba, 1997; Rotshtein & Shtovba, 1998).
- *The extension principle* (Zadeh, 1973) allows converting probabilistic reliability models to the form necessary for working with uncertain source data. It is important to emphasize that calculations with fuzzy numbers are much simpler than similar calculations with probabilistic distribution functions (Onisawa et al., 1995; Rotshtein & Shtovba, 1997; Rotshtein & Shtovba, 1998).

- *Fuzzy if-then rules* (Zadeh, 1973) make it possible to approximate non-linear dependences of the reliability level on influencing factors. Compared with regression models, fuzzy rules require less experimentation due to the use of expert knowledge in a natural language (Onisawa & Kasprzyk, 1995; Rotshtein & Shtovba, 1998; Rotshtein & Shtovba, 2009).
- *The membership function* (Zadeh, 1965; Zadeh, 1973) is an indicator of the level of correctness of an element of the system depending on the value of some parameter. The structure of the system allows one to obtain the function of the correct operation of the system depending on the influencing parameters (Rotshtein, 2010; Rotshtein, 2011) and solve the problems of optimal redundancy without failure statistics (Katelnikov & Rotshtein, 2010; Rotshtein et al., 2012).
- *Fuzzy relation and compositional rule of inference* (Zadeh, 1973) model cause-effect relationships in the form of a system of fuzzy logical equations. This allows you to solve the diagnostic problem, i.e. Failure Mode Effect Analysis - FMEA (International Electrotechnical Commission (IEC), 2006) as the restoration of causes from the observed effects (Rotshtein et al., 2006; Rotshtein & Rakytyanska, 2001; Rotshtein & Rakytyanska, 2008).
- *The transitive closure* (Zadeh, 1973) of the fuzzy similarity relationship allows you to classify the elements of the system and determine the importance of their influence on the reliability of the system as a whole (Rotshtein & Dreyfuss, 2018; Rotshtein, 2017).
- *The principle of fuzzy criteria intersection* by Bellmann – Zadeh (Bellman & Zadeh, 1970) allows to choose the values of factors affecting the system's reliability without complicated calculation of the failure probability (Rotshtein, 2018; Rotshtein, 2019).

Note that the above references reflect only the tastes and experiences of the authors and do not claim to be an exhaustive review of the works on the use of fuzzy mathematics in reliability modeling.

In 1986, fuzzy cognitive maps (FCM) appeared (Kosko, 1986), which are of interest as a new tool of reliability modeling. Currently there are a large number of works on the application of FCM in various fields (Papageorgiu, 2014). FCM applications in reliability problems, as a rule, come down to modeling scenarios of the influence of various causes on failure types, which corresponds to the FMCA method (Palaez & Bowles, 1996; Salmeron & Gutierrez, 2012). It should be noted that the analogue of FMCA in the probabilistic theory of reliability (Barlow & Proschan, 1975) is the Birnbaum importance index (Birnbaum, 1969), which indicates the rank of a system element in the sense of influence on reliability.

In this article, FCM is considered as a model for the interaction of factors affecting system reliability. A method for ranking FCM elements is proposed, which allows one to calculate the ranks of influencing factors by analogy with the Birnbaum importance index (Birnbaum, 1969). The method is illustrated by the example of modeling the reliability and safety of a car, taking into account technical, ergatic, environmental and organizational factors.

BASIC NOTIONS AND MODELS

General Information

FCM were introduced by B. Kosko (Kosko, 1986) as an extension of R. Axelrod's binary cognitive maps (Axelrod, 1976), intended for the dynamics modeling of the causal relationships in socio-political systems. FCM is a directed graph with weighted edges, of which an example is shown in Figure 1. Graph nodes C_i called *concepts* correspond to the input and output variables that are taken into consideration in the model. Weighted edges of the graph reflect the *magnitude of the effect* w_{ij} .of the changes of one variable C_i on the changes of another variable C_j.

Figure 1. An example of fuzzy cognitive map

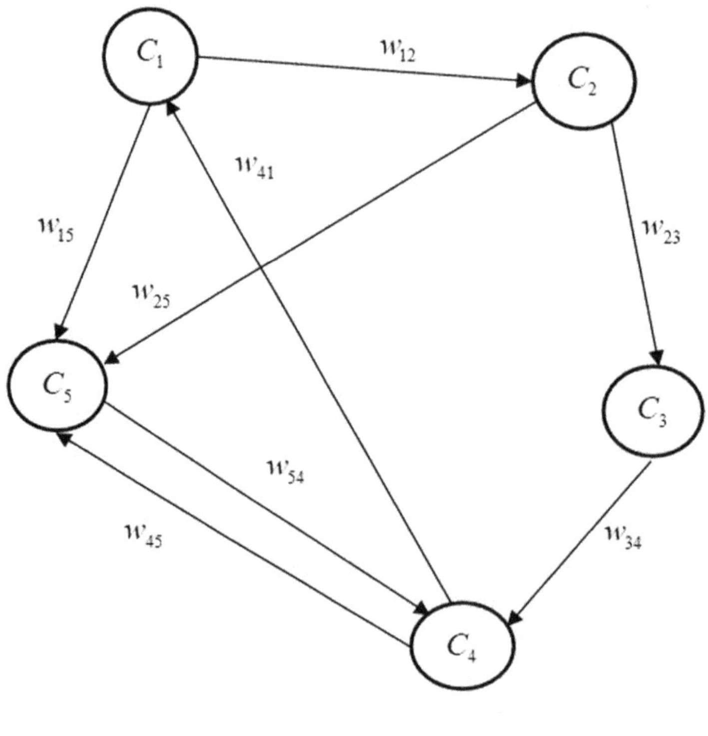

The term "cognitive" implies that the initial data for modeling consists of subjective opinions of an expert expressed as, e.g. "increases" or "decreases", for instance: "increasing C_i causes the decrease of C_j". In binary cognitive maps (Axelrod, 1976), an "increase" is estimated as "+1", while a "decrease" is estimated as "–1".

The term "fuzzy" implies that FCM (Kosko, 1986) use various levels of "increase" and "decrease". They are defined by numbers from the intervals (Barlow & Proschan, 1975) and [–1, 0], which corresponds to the terms "weak", "average", "strong", etc. from the fuzzy set theory (Zadeh, 1965; Zadeh, 1973).

As any approximator, e.g. regression, fuzzy rules, neural network, etc., FCM contains configurable parameters that are to be estimated through minimization of the distance between the model and experimental output values. If the experimental data "inputs–outputs" is not available, the quality of the whole model depends on the expert's qualification. The art of modeling consists in compensating for the missing experimental data through high quality of expert estimates.

It would be relevant comparing FCM and Markovian chains (processes) familiar to the dependability experts. Both types of models are weighted directed graphs. The basic difference between FCM and Markovian reliability models consists in the fundamental difference between the fuzzy logic (causes) and probability theory (effects) shown in Figure 2: the Markovian models reflect the dynamics of system states probabilities accounting for failures and restorations; FCM simulate the level dynamics of interrelated factors that cause failures and affect their probability.

Figure 2. Interrelation of fuzzy and probabilistic tools in reliability modeling

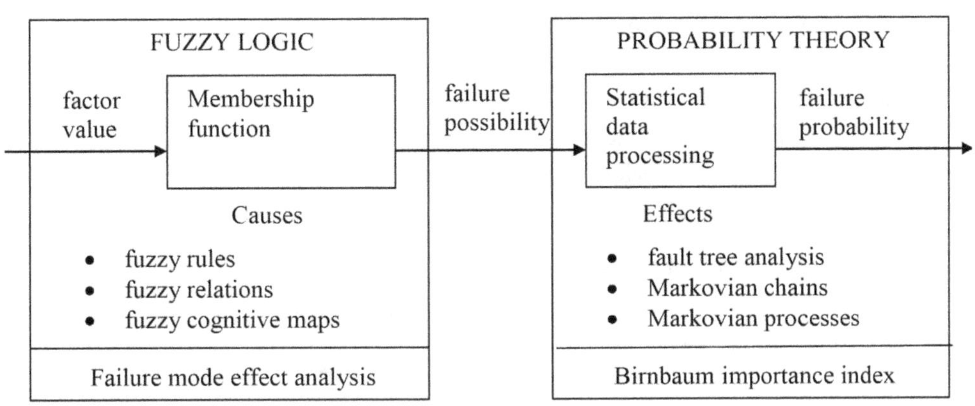

The Figure 2 also explains the place of FCM between other mathematical tools in reliability modeling.

Concepts

Let $C = \{C_1, C_2, \ldots, C_n\}$ .be a known set of concepts, i.e. variables used in the model. According to (Kosko, 1986), each concept $C_i \in C$ .is evaluated with value $A_i \in [0,1]$ .that defines the level of the concept and is based on expert opinion. Value A_i is to be obtained as follows. We will assume each concept $C_i \in C$ .to be a linguistic variable (Zadeh, 1973), that is estimated with value x_i on a universal set, i.e. interval $\left[\underline{x_i}, \overline{x_i}\right]$. where, $\underline{x_i}$.($\overline{x_i}$. is the lower (upper) boundary. We will estimate concept $C_j \in C$ .with the use of the fuzzy terms "perfection of concept C_i .", that is denoted as PC_i .and is a fuzzy set,

$$PC_i = \int_{\left[\underline{x_i}, \overline{x_i}\right]} \pi(x_i)/(x_i).$$

where $\pi(x_i)$ is the membership function of variable x_i in the notion of "perfection of concept C_i". Using this function, each absolute estimate $x_i \in \left[\underline{x_i}, \overline{x_i}\right]$. is associated with number

$A_i = \pi(x_i) \in [0,1]$. that characterizes the proximity of the value of concept $C_i \in C$ .to a certain ideal: 0 is the lowest perfection, 1 is the highest perfection. "Fuzzy perfection" is synonymous with "fuzzy correctness", for which the membership functions were considered in (Rotshtein, 2018). Possible fuzzy boundaries between perfect and non-perfect values of variable x are shown in Figure 3, where, as the value of x grows, the following transitions take place:

1. "non–perfect" (1) – "perfect" (0),
2. "perfect" (1) – "non–perfect" (0),
3. "non–perfect" (0) –"perfect" (1) – "non–perfect" (0).

Figure 3. Membership functions for fuzzy perfection

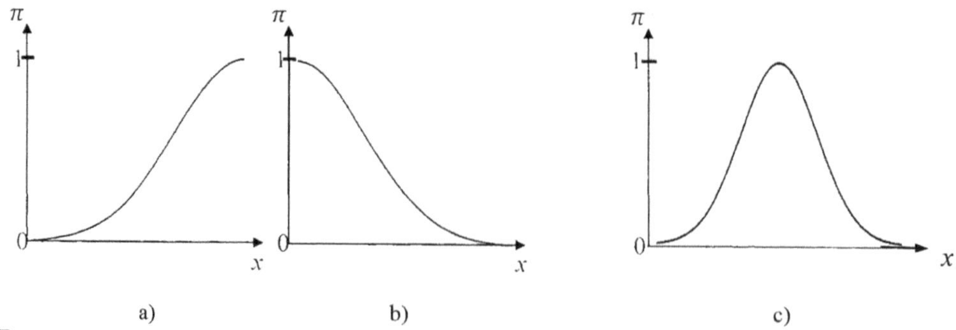

Connections Between Concepts

The weight w_{ij} .of the edge that connects concepts C_i and C_j indicates the magnitude of the effect of C_i on C_j. Let concepts C_i and C_j be characterized by variables x_i and x_j, while – as the result of the experiment – dependence $x_j = \varphi\ (x_i)$ was achieved. Then, the weight w_{ij} is defined as the derivative $w_{ij} = dx_i\ /dx_j$ that can have three forms (Figure 4):

$w_{ij} > 0$, if the increase (decrease) of value x_i causes the increase (decrease) of value x_j (positive effect of C_i on C_j);

$w_{ij} > 0$ if the increase (decrease) of value x_i causes the decrease (increase) of value x_j (negative effect of C_i on C_j);

$w_{ij} = 0$ if value x_j does not depend on value x_i (no effect C_i on C_j).

Figure 4. Types of effect between concepts

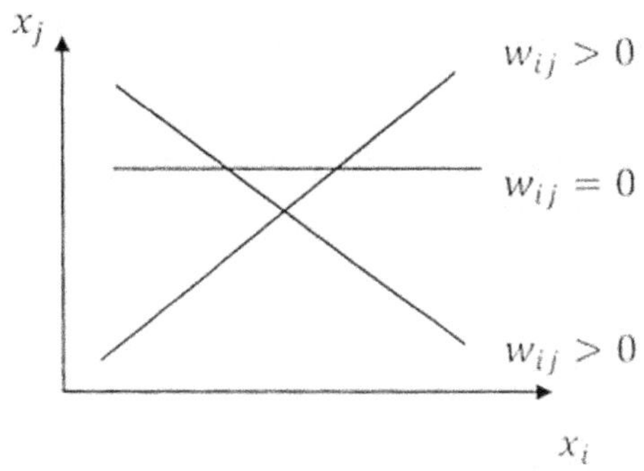

The magnitude of effect (w_{ij}) is estimated expertly by means of linguistic terms and thermometer scale (Table 1).

If several expert opinions are taken into consideration, the value w_{ij} is estimated as the weighted average:

$$w_{ij} = \frac{\alpha_1 w_{ij}^1 + \alpha_2 w_{ij}^2 + \ldots + \alpha_m w_{ij}^m}{\alpha_1 + \alpha_2 + \ldots + \alpha_n}.$$

where w_{ij}^p .s the estimate of the magnitude of the effect of the p-th expert;

α_p is the weight of the p-th expert, $p = 1, 2, \ldots, m$;

m is the number of experts.

Table 1. Methods of estimating the magnitude of an effect

Thermometer scale	Linguistic estimations	Quantitative estimations
 1 0 -1	Positive maximum Positive above average Positive average Positive under average Not available Negative under average Negative average Negative above average Negative maximum	1 0.75 0.5 0.25 0 -0.25 -0.5 -0.75 -1

In order to reduce the subjectivity of expert estimates, the method of the least effect proposed in (Rotshtein & Dreyfuss, 2018) can be used.

Recurrence Equation

According to (Stylios & Groumpos, 2004), the dynamics of concept values variation in FCM are defined by formula

$$A_i^{k+1} = \mathrm{f}\left(\sum_{\substack{j=1 \\ j \neq i}}^{n} \mathrm{A}_j^k \mathrm{w}_{ji} + \mathrm{c}\mathrm{A}_i^{k+1} \right), \qquad \mathrm{k} = 0, 1, 2, \ldots \tag{1}$$

where A_i^{k+1} .is the value of concept C_i at step $k + 1$; A_i^k .nd A_j^k .the values of concept C_i and C_j at step k respectively, w_{ij} is the magnitude of the effect of concept C_j on concept C_i; c is the parameter that takes into consideration the history, i.e. the contribution of the concept's value at the preceding step, $c\in$(Barlow & Proschan, 1975); f is the threshold function, due to which the value of the concept does not exceed one.

In this paper, it is assumed that $c = 1$, while for the threshold function is used the positive part of the hyperbolic tangent (Figure 5):

$$f(x) = \begin{cases} \tanh(x) \; for & x \geq 0 \\ 0 \qquad for & x < 0 \end{cases} \quad \tanh(x) = \frac{e^x - e^{-x}}{e^x + e^{-x}}.$$

Figure 5. Threshold function

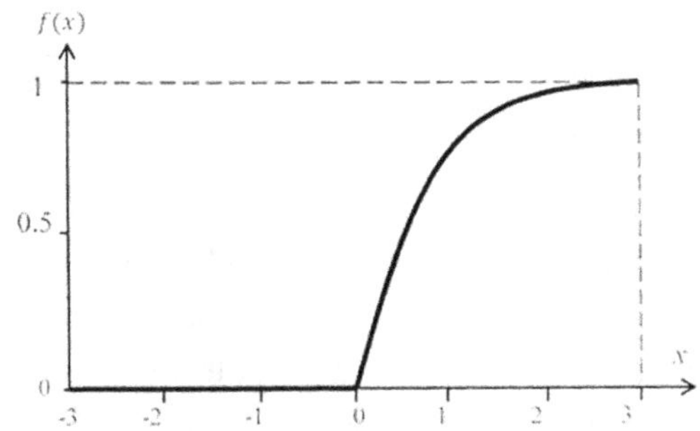

Matrix Model

The recurrence equation (1) can be represented in matrix form Figure 4 – Types of effects between concepts.

$$\boldsymbol{A}^k = f\left(A^k W_0 + cA^k\right)., k = 0, 1, 2, \ldots, \tag{2}$$

where $\boldsymbol{A}^{k+1}$, $\boldsymbol{A}^k$, k = 0, 1, 2, … are (1×n) state vectors of FCM, whose elements define the values of concept at steps k+1 and k respectively;

$$\boldsymbol{W}_0 = \begin{bmatrix} 0 & w_{12} & \cdots & w_{1n} \\ w_{21} & 0 & \cdots & w_{2n} \\ \cdot & \cdot & \cdots & \cdot \\ w_{n1} & w_{n2} & \cdots & 0 \end{bmatrix}. \tag{3}$$

is the (n×n) matrix of the magnitude of mutual effects of concepts C_i, in which diagonal elements are equal to zero. If instead of matrix (3) an (n×n) matrix (4) is used,

$$\boldsymbol{W} = \begin{bmatrix} c & w_{12} & \cdots & w_{1n} \\ w_{21} & c & \cdots & w_{2n} \\ \cdot & \cdot & \cdots & \cdot \\ w_{n1} & w_{n2} & \cdots & c \end{bmatrix}. \tag{4}$$

in which all elements on the main diagonal are equal to parameter $c\in$(Barlow & Proschan, 1975), then we will write formula (2) as

$$A^k = f\left(A^k W\right)., k = 0, 1, 2, \ldots, \tag{5}$$

that is similar to the recurrence equation for a Markovian chain, if we take $f(x) = x$. The fundamental difference consists in the fact that a Markovian chain simulates the dynamics of event probability variation, while FCM simulates the dynamics of the level of causes, i.e. factors that lead to such states or events (see Figure 2).

The initial state of an FCM is defined by vector

$$A^0 = \left[A_1^0, A_2^0, \ldots, A_n^0\right]., \tag{6}$$

whose elements reflect the values of concepts at step $k = 0$. As the result of interaction between concepts FCM enters the steady mode that corresponds with one of the types of stability.

RANKING OF CONCEPT

The allocation of system's reliability resources is based on quantitative estimates (ranks) of its elements importance. In the statistical reliability theory, Birnbaum's importance index (Birnbaum, 1969) of an element is the most widely used (Barlow & Proschan, 1975). It is defined based on the system's reliability function

$$P_s = f_s\left(P_1, \ldots, P_i, \ldots\right)., \tag{7}$$

where P_S and P_i are the system's probability of no-failure and its i-the element respectively.

The first derivative in (7) is the importance index of the system's i-th element according to Birnbaum that is calculated as follows (Barlow & Proschan, 1975):

$$I_i = \frac{\partial P_s}{\partial P_i} = P_s\left(P_1, \ldots, P_{i-1}, i, P_{i+1}, \ldots P_n\right) - P_s\left(P_1, \ldots, P_{i-1}, 0, P_{i+1}, \ldots P_n\right). \tag{8}$$

The second derivative in (7) is the importance index of the joint effect of the i-th and j-th elements (joint reliability importance), that was introduced in (Hong & Lie, 1993).

In our case, the elements of the model include the input concepts, i.e. the factors that affect the output level of system reliability. That explains the requirement to calculate the importance indices of FCM concepts.

Definition of Importance Indices

In the set of concepts $C=\{C_1, C_2, \ldots, C_n\}$ we will assume the following:

C_n is the output concept that defines the level of system reliability and is estimated with number $A_n \in$ (Barlow & Proschan, 1975);

$C_1, C_2, \ldots, C_{n-1}$ are the input concepts that correspond with the interconnected factors affecting system reliability and estimated by levels $A_i \in$ (Barlow & Proschan, 1975), $i=1, \ldots, n-1$.

The value of concept C_n at the l-th step is the function of the elements of vector (6), i.e.

$$A_n^l = F\left(A_1^0, A_2^0, \ldots, A_n^0\right). \tag{9}$$

It is assumed that A_n^l .is the value of concept C_n in the steady state, i.e. at such step l, when A_n^l .is close to A_n^{l-1}. Formula (9) is equivalent to (7), which allows proceeding to the definition of concept ranks based on derivatives.

Let $I(C_j)$ be the importance index of concept C_j, while $I(C_j, C_k)$ be the index of join importance of concepts C_j and C_k. Following (8) and (Birnbaum, 1969; Hong & Lie, 1993), let us identify such importance indices as:

$$I\left(C_j\right) = \frac{\partial A_n^l}{\partial A_j} = \frac{F\left(1_j, O\right) - F(0)}{1-0} = F\left(1_j, 0\right). \tag{10}$$

$$I\left(C_j, C_k\right) = \frac{\partial^2 A_n^l}{\partial A_j \partial A_k} = \frac{F\left(1_j, 1_k, 0\right) - F(0)}{(1-0)(1-0)} = F\left(1_j, 1_k, 0\right)., \tag{11}$$

where $F(1_j, 0)$ is the value of function (9), when $A_j^0 = 1$., while all the other arguments are equal to zero; $F(0)$ is the value of function (9), when all arguments are equal to zero (it is assumed that $F(0) = 0$); $F(1_j, 1_k, 0)$ is the value of function (9), when $A_j^0 = A_k^0 = 1$., while all the other arguments are equal to zero.

Note. The zero values of input concepts (except one in (10) and two in (11) that are equal to one) are selected in order to eliminate the possibility of them having an effect on the output concept through transitive connections.

Algorithm of Importance Index Calculation

Step 1. Specifying the initial vector (6). For importance index $I(C_j)$, the initial vector is specified as follows

$$A^0 = \left[A_j^0 = 1, A_i^0 = 0, i = 1, 2, \ldots, n, i \neq j \right]., \tag{12}$$

while for importance index $I(C_j, C_k)$ it is specified as

$$A^0 = \left[A_j^0 = A_k^0 = 1, A_i^0 = 0, i = 1, 2, \ldots, n, i \neq j, k \right].. \tag{13}$$

Step 2. Using recurrence equation (5), finding the FCM state vector

$$A^l = \left[A_1^l, A_2^l, \ldots, A_n^l \right].. \tag{14}$$

in steady-state operating conditions, i.e. at such step l, whereas $\left| A_i^l - A_i^{l-1} \right| < \varepsilon$. where ε is a small positive number, $i = 1, 2, \ldots, n$.

Step 3. Elements A_n^l .f vector (14) obtained under initial vectors (12) and (13) respectively shall be considered to be importance indices $I(C_j)$ and $I(C_j, C_k)$.

AN EXAMPLE

Concepts and Effects

Let us examine the automobile reliability and safety model in the "driver-automobile-road" system. The fuzzy cognitive map of the system is shown in Figure 6, where the concepts have the following contents: C_1 is the driver's qualification, C_2 is the road conditions, C_3 is the unit costs of operation, C_4 is the operating conditions, C_5 is the frequency of maintenance operations, C_6 is the quality of service and repair, C_7 is the quality of automobile's design, C_8 is the quality of operational materials and spare parts, C_9 is the storage conditions, C_{10} is the reliability and safety of automobile in the "driver-automobile-road" system.

Figure 6. Fuzzy cognitive map for reliability and safety estimation

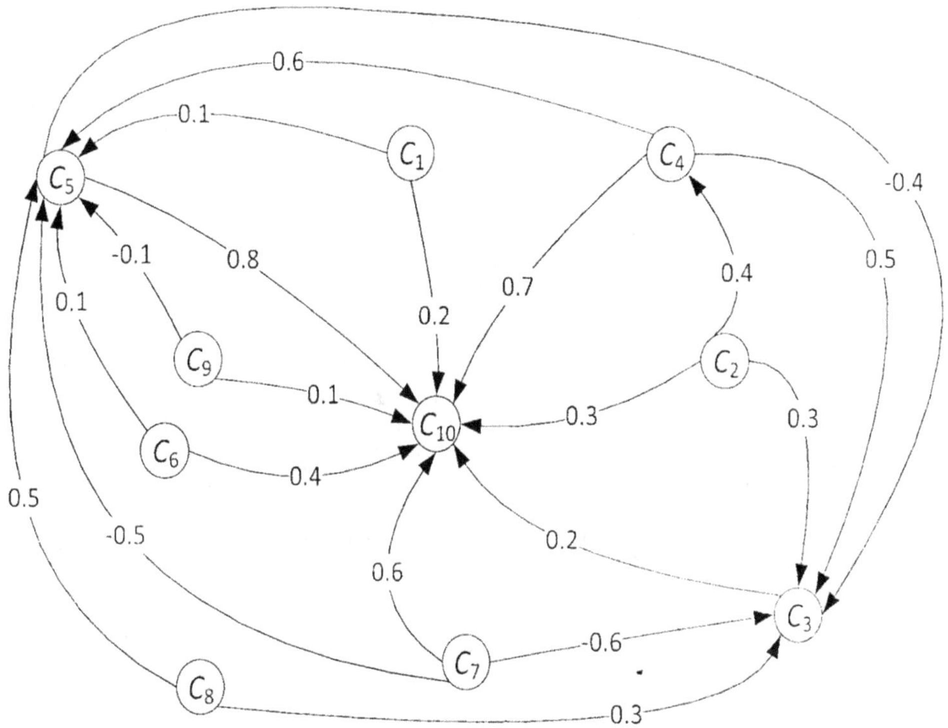

Matrix $\boldsymbol{W}$ (4) with expert estimates of the magnitude of effect, that assumes that $c = 1$, is as follows

$$W = \begin{bmatrix} 1 & 0 & 0 & 0 & 0.1 & 0 & 0 & 0 & 0 & 0.5 \\ 0 & 1 & 0.3 & 0.4 & 0 & 0 & 0 & 0 & 0 & 0.3 \\ 0 & 0 & 1 & 0 & 0 & 0 & 0 & 0 & 0 & 0.2 \\ 0 & 0 & 0.5 & 1 & 0.6 & 0 & 0 & 0 & 0 & 0.7 \\ 0 & 0 & -0.4 & 0 & 1 & 0 & 0 & 0 & 0 & 0.8 \\ 0 & 0 & 0 & 0 & 0.1 & 1 & 0 & 0 & 0 & 0.4 \\ 0 & 0 & -0.6 & 0 & -0.5 & 0 & 1 & 0 & 0 & 0.6 \\ 0 & 0 & 0.3 & 0 & 0.5 & 0 & 0 & 1 & 0 & 0 \\ 0 & 0 & 0 & 0 & -0.1 & 0 & 0 & 0 & 1 & 0.1 \\ 0 & 0 & 0 & 0 & 0 & 0 & 0 & 0 & 0 & 1 \end{bmatrix}.$$

Importance Indices of Concepts

Table 2 contains nine pairs of vectors associated with the calculation of the importance indices of concepts C_1, …, C_9.

Table 2. Values of concepts in steady state for various initial vectors

Step	A_1	A_2	A_3	A_4	A_5	A_6	A_7	A_8	A_9	A_{10}
1 … 3040	1 … 0.022	0 … 0.000	0 … 0.000	0 … 0.000	0 … 0.187	0 … 0.000	0 … 0.000	0 … 0.000	0 … 0.000	0 … 0.68579
1 … 774	0 … 0.000	1 … 0.044	0 … 0.000	0 … 0.365	0 … 0.747	0 … 0.000	0 … 0.000	0 … 0.000	0 … 0.000	0 … 0.94834
1 … 3717	0 … 0.000	0 … 0.000	1 … 0.020	0 … 0.000	0 … 0.000	0 … 0.000	0 … 0.000	0 … 0.000	0 … 0.000	0 … 0.22707
1 … 3014	0 … 0.000	0 … 0.000	0 … 0.000	1 … 0.022	0 … 0.335	0 … 0.000	0 … 0.000	0 … 0.000	0 … 0.000	0 … 0.79115
1 … 5324	0 … 0.000	0 … 0.000	0 … 0.000	0 … 0.000	1 … 0.017	0 … 0.000	0 … 0.000	0 … 0.000	0 … 0.000	0 … 0.33491
1 … 3196	0 … 0.000	0 … 0.000	0 … 0.000	0 … 0.000	0 … 0.186	1 … 0.022	0 … 0.000	0 … 0.000	0 … 0.000	0 … 0.68912
1 … 4953	0 … 0.000	0 … 0.000	0 … 0.000	0 … 0.000	0 … 0.000	0 … 0.000	1 … 0.017	0 … 0.000	0 … 0.000	0 … 0.30912
1 … 2742	0 … 0.000	0 … 0.000	0 … 0.000	0 … 0.000	0 … 0.000	0 … 0.321	0 … 0.000	1 … 0.023	0 … 0.000	0 … 0.77418
1 … 3086	0 … 0.000	0 … 0.000	0 … 0.000	0 … 0.000	0 … 0.000	0 … 0.000	0 … 0.000	0 … 0.000	1 … 0.022	0 … 0.18667

Each pair contains the initial vector (12) and vector (14) in steady-state operating conditions. The last element of the second vector in each pair corresponds to the importance index of the concept, i.e. $I(C_1) = 0.686$. The last column in Table 2 shows the step-by-step change of the level of reliability and safety of an automobile (A_{10}) in case of activation of one of the factors (A_i, $i = 1, …, 9$). The diagram of the importance indices of concepts is shown in Figure 7. The results of calculation of the importance indices of the join effect of concepts are shown in Table 3, i.e. *I*

$(C_1, C_2) = 0.949$. It should be noted that concept C_7 can be detailed subjects to the classical reliability theory (Barlow & Proschan, 1975; Rybinin, 1976).

Figure 7. Diagram of importance indices of factors

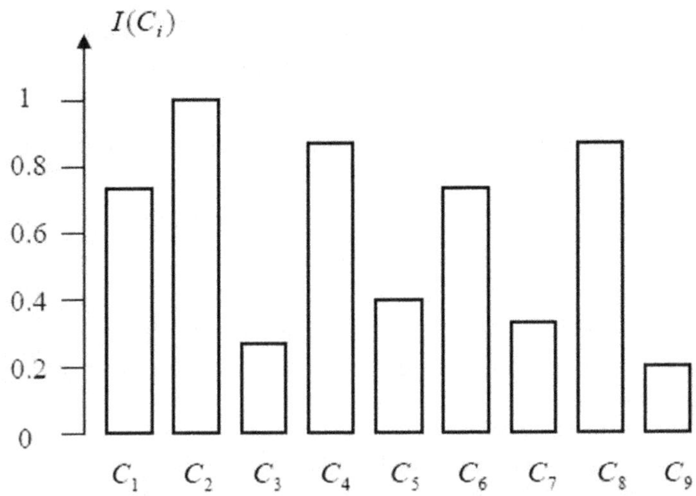

Table 3. Importance indices of combined effect of factors

Concepts	C_2	C_3	C_4	C_5	C_6	C_7	C_8	C_9
C_1 C_2 C_3 C_4 C_5 C_6 C_7 C_8	0.949	0.686 0.948	0.801 0.948 0.791	0.686 0.948 0.335 0.791	0.730 0.950 0.689 0.803 0.689	0.335 0.949 0.309 0.703 0.309 0.356	0.786 0.950 0.774 0.823 0.774 0.788 0.309	0.255 0.948 0.254 0.782 0.187 0.294 0.323 0.763

CONCLUSION

The paper proposes and demonstrates with an example of a man-machine system a method of ranking of factors that affect its reliability. The method is based on the formalization of causal relationships between the contributing factors and the reliability in the form of a fuzzy cognitive map, i.e. directed graph, whose nodes correspond to the system reliability and contributing factors, while the weighted edges indicate the magnitude of the factors' effect on each other and the system's dependability.

The proposed method may be regarded as an equivalent to Birnbaum's ranking of system components in the probabilistic reliability theory. The advantages of the method include:

- Use of available expert information with no collection and processing of statistical data ;
- capability to take into consideration any qualitative and qualitative factors associated with people, Technology, software, quality of service, operating conditions, etc.. In particular, individual concepts can characterize various types of redundancy (structural, algorithmic, etc.), that are used to improve reliability;
- easy to change the number of considered factors through the introduction of new nodes and edges of a Directed graph.

The method can be applied to complex systems with fuzzy structures, whose reliability strongly depends on interrelated factors that are measured by means of expert information.

REFERENCES

Axelrod, R. (1976). *Structure of Decision: The Cognitive Maps of Political Elites*. Princeton University Press.

Barlow, R., & Proschan, F. (1975). *Statistical Theory of Reliability and Life Testing*. Holt, Rinehart and Winston.

Barnard, A. (2012). *Why you cannot predict electronic product reliability*. ARS.

Bellman, R. E., & Zadeh, L. A. (1970). Decision Making in a Fuzzy Environment. *Management Science*, *17*(4), 141–164. doi:10.1287/mnsc.17.4.B141

Birnbaum, Z. W. (1969). On the Importance of Different Components in the Multicomponent System. In Multivariate Analysis (2nd ed., pp. 581-592). New York: Academic.

Cai, K. Y. (1996). *Introduction on Fuzzy Reliability*. Kluwer Academic Publisher. doi:10.1007/978-1-4613-1403-5

Dubois, D., & Prade, H. (1978). Operations on Fuzzy Numbers. *International Journal of Systems Science*, *9*(6), 613–623. doi:10.1080/00207727808941724

Hong, J. S., & Lie, C. H. (1993). Joint Reliability Importance of Two Edges in Undirected Network. *IEEE Transactions on Reliability*, *42*(1), 17–23. doi:10.1109/24.210266

International Electrotechnical Commission (IEC). (2006). *Analysis Techniques for System Reliability: Procedure for Failure Mode Effect Analysis (FMEA)*. Available from: https://webstore.iec.ch/preview/info

Katelnikov, D. I., & Rotshtein, A. P. (2010). Fuzzy Algorithmic Simulation of Reliability: Control and Correction Resources Optimization. *Journal of Computer and Systems Sciences International*, *49*(6), 967–971. doi:10.1134/S1064230710060146

Kosko, B. (1986). Fuzzy cognitive maps. *International Journal of Man-Machine Studies*, *24*(1), 65–75. doi:10.1016/S0020-7373(86)80040-2

Onisawa, T., & Kasprzyk, J. (Eds.). (1995). *Reliability and Safety Analysis under Fuzzyness*. Physica - Verlag. doi:10.1007/978-3-7908-1898-7

Palaez, C. E., & Bowles, J. B. (1996). Using Fuzzy Cognitive Maps as a System Model for Failure Modes and Effect Analysis. *Intelligent Systems*, *88*, 177–199.

Papageorgiu, E. I. (Ed.). (2014). *Fuzzy Cognitive Maps for Applied Sciences and Engineering From Fundamentals to Extensions and Learning Algorithms*. Springer. doi:10.1007/978-3-642-39739-4

Rotshtein, A., & Dreyfuss, M. (2018). Ranking of Elements in System Reliability Modeling: The Least Influence Method. *Systems Engineering*, *21*(5).

Rotshtein, A., Katelnikov, D., & Pustylnik, L. (2012). Reliability Analysis and Optimization using Fuzzy Logic and Chaos Theory. *International Journal of Quality, Statistics and Reliability*. 9.

Rotshtein, A., Posner, M., & Rakytyanska, H. (2006). Cause and Effective Analysis based on Fuzzy Relation. *Reliability and Safety Engineering*, *91*(9), 1095–1101. doi:10.1016/j.ress.2005.11.041

Rotshtein, A., & Rakytyanska, H. (2012). *Fuzzy Evidence in Identification, Forecasting and Diagnosis*. Springer. doi:10.1007/978-3-642-25786-5

Rotshtein, A. P. (1995). Fuzzy Reliability Analysis of Labour (Man –Machine) Systems. In T. Onisawa & J. Kasprzyk (Eds.), *Reliability and Safety Analyses under Fuzzyness* (pp. 245–273). Physica-Verlag. doi:10.1007/978-3-7908-1898-7_17

Rotshtein, A. P. (2010). Algebra of Algorithms and Fuzzy Logic in System Reliability Analysis. *Journal of Computer and Systems Sciences International*, *49*(2), 254–264. doi:10.1134/S1064230710020115

Rotshtein, A. P. (2011). Fuzzy Algorithmic Reliability Analysis of Complex Systems. *Cybernetics and Systems Analysis*, *49*(6), 545–552. doi:10.100710559-011-9371-x

Rotshtein, A. P. (2017). Ranking of System's Elements Based on Fuzzy Relations and Transitive Closure. *Cybernetics and Systems Analysis*, (1), 68–75.

Rotshtein, A. P. (2018). Selection of Human Performance Conditions Based on Fuzzy Perfection. *Journal of Computer and Systems Sciences International*, *49*(6), 927–971. doi:10.1134/S1064230718060096

Rotshtein, A. P. (2019). Reliability – Based Design of Human Performance Conditions using Fuzzy Perfection. *Cybernetics and Systems Analysis*, *55*(2), 68–79. doi:10.100710559-019-00128-5

Rotshtein, A. P., & Rakytyanska, H. B. (2001). Solution of a Diagnostics Problem on the Basis of Fuzzy Relation and Genetic Algorithm. *Cybernetics and Systems Analysis*, *37*(6), 918–925. doi:10.1023/A:1014594201670

Rotshtein, A. P., & Rakytyanska, H. B. (2008). Diagnoses Problem Solving using Fuzzy Relation. *IEEE Transactions on Fuzzy Systems*, *16*(3), 664–676. doi:10.1109/TFUZZ.2007.905908

Rotshtein, A. P., & Shtovba, S. D. (1997). *Fuzzy Reliability of Algorithmic Processes*. Vinnitsa, Continent –Prim. (in Russian)

Rotshtein, A. P., & Shtovba, S. D. (1998). Prediction of Reliability of Algorithmic Processes with Fuzzy Input Data. *Cybernetics and Systems Analysis*, *34*(4), 545–552. doi:10.1007/BF02666999

Rotshtein, A. P., & Shtovba, S. D. (2009). Modeling of the Human Operator Reliability with Aid of the Sugeno Fuzzy Knowledge Base. *Automation and Remote Control*, *70*(1), 163–178. doi:10.1134/S0005117909010123

Rybinin, I.A. (1976). *Reliability of Engineering Systems, Principles and Analysis*. Moscow: Mir.

Salmeron, J., & Gutierrez, E. (2012). Fuzzy Grey Cognitive Maps in Reliability Engineering. *Applied Soft Computing*, *12*(12), 3818–3824. doi:10.1016/j.asoc.2012.02.003

Stylios, C. D., & Groumpos, P. P. (2004). Modeling Complex Systems using Fuzzy Cognitive Maps, *IEEE Transaction on Systems, Man and Cybernetics. Systems and Humans*, *34*(1), 155–162. doi:10.1109/TSMCA.2003.818878

Utkin, L. V., & Shubinsky, I. B. (1998). *Non-conventional Methods of Reliability Analysis of Information Systems*. Lubavic. (in Russian)

Zadeh, L. A. (1965). Fuzzy Sets. *Information and Control*, *8*(3), 338–353. doi:10.1016/S0019-9958(65)90241-X

Zadeh, L. A. (1973). Outline of a New Approach to the Analysis of Complex Systems and Decision Process. *IEEE Transactions on Systems, Man, and Cybernetics*, *SMS-3*(1), 28–44. doi:10.1109/TSMC.1973.5408575

Chapter 2
Fuzzy Reliability of Weighted-((f / (r, s)), k)/ (m, n) G System

Seema Negi
S. V. Govt. P. G. College, Lohaghat, India

S. B. Singh
G. B. Pant University of Agriculture and Technology, Pantnagar, India

Kamlesh Bisht
https://orcid.org/0000-0002-5052-5222
Uttarakhand Open University, Haldwani, India

ABSTRACT

In this chapter, the authors study a weighted-((f / (r, s)), k)/ (m, n): G system. The system consists of mn components arranged in a matrix form and the system works if all the sub matrices of order (r, s), the total weight of the working components is greater than f and the total weight of the working components in the system is at least k. This chapter deals with the evaluation of fuzzy reliability and fuzzy mean time to failure of the considered system with the application of fuzzy universal generating function and fuzzy Rayleigh distribution. In this study, the authors formed some prepositions to understand the behaviour of the considered system with respect to different varying parameters and also present an illustrative example to understand them.

DOI: 10.4018/978-1-7998-7564-2.ch002

INTRODUCTION

Every real life engineering system consists of elements connected in different configurations such as series, parallel, series-parallel, k-out-of-n etc. Of course they often have to deal with the reliability evaluation of these systems. Needless to say, recently especial attention has been paid to k-out-of-n systems due to its wide applications in engineering systems. A number of researchers like (Chung; 1988 and Bueno & Carmo; 2007) and lot of others carried out research on reliability assessment of k-out-of-n: G (F) system. It needs to be pointed out here that in this system if each element is having its own weight within the system, then it is renamed as weighted-k-out-of-n: G (F) system. Quite a few authors like Wu and Chen (1994), Li and Zuo (2008) and Li et al. (2008), Wang et al. (2012) and Negi and Singh (2015) have done incredible works in the field of reliability assessment of weighted-k-out-of-n: G (F) system. There are different extensions of k-out-of-n: G (F) system, consecutive k-out-of-n: G (F) system is one of them. Lam and Ng (2001) and Bollinger (1982) contributed significantly in this extension. The two-dimensional version of consecutive- k-out-of-n: F system is known as a consecutive-(R, S)-out-of-(M, N): F system denoted by $(R, S)/(M, N)$: F. This version is generated by Salvia and Lasher (1990) and used in a sensor system, a pattern system, an x-ray diagnostic system etc. This system consists of M x N components arranged like an (M, N) matrix and fails iff the system has an (R, S) sub matrix that contains all failed components. Habib et al. (2010) evaluated the reliability of this system by putting conditions on the number of failed components. Noguchi et al. (1996), Yamamoto et al. (1995) and Yuge et al. (2000) discussed the reliability of connected-(R, S)-out-of-(M, N): F system. The reliability bounds of connected -(R, S)-out-of-(M, N): F system is derived by Malinowski and preuss (1996). Another system termed as x-within-consecutive-(R, S)-out-of-(M, N): F system is introduced by Makri et al. (1996) and obtained the lower and upper bounds for its reliability. This system fails iff x components in an arbitrary (R, S) sub lattice fail. The system can be designated as x-$(R, S)/(M, N)$: F system. Clearly if $x = RS$, then x-$(R, S)/(M, N)$: F system becomes $(R, S)/(M, N)$: F system.

Zuo et al. (2000) proposed a combined model of a k-out-of-M x N: F and an $(R,S)/(M,N)$: F system which is described as an $((R, S), k)/(M,N)$: F system. This system fails if at least k components fail in the system, or there is at least one cluster (size $RS > k$) of failed components. If $k = MN$, it clearly looks like the $(R, S)/(M,N)$: F system. Though they used recursive equation method to assess the reliability of the system but it cannot be applied to all realistic situations. Furthermore, there are some engineering systems in which each component has its own importance i.e., weight in the system. Correspondingly as stated earlier this system is renamed as weighted-$((f / (r, s)), k)/(m, n)$: F system. Further, sometimes, it is very hard to get precise data of system parameters which leads to the uncertainty in the system parameters. It is a hard fact that reliability of the system depends upon failure data. Therefore if uncertainties are present in the system parameters, it becomes difficult to find out exact reliability characteristics of the system and so we move towards the fuzzy reliability theory.

The fuzzy set theory was introduced by Zadeh (1965) and when it combines with reliability theory it becomes fuzzy reliability theory. Cai et al. (1993) introduced this theory and later the authors Chen (1994) and Kumar et al. (2006) have done noteworthy works to broaden the concept of this theory. Considerable attention was devoted to the fuzzy reliability of the system under uncertainty by Bohra and Singh (2015), Pak et al. (2014), Karpisek et al. (2010) and BALOUI (2011) incorporating fuzzy Rayleigh distribution, fuzzy Weibull distribution and fuzzy exponential distribution respectively. The hazard function of fuzzy Rayleigh distribution is an increasing function of time. Whenever we have this type of situations fuzzy Rayleigh distribution is more suitable than any other distributions. In the frame of this concept if failure times are distributed according to the fuzzy Rayleigh distribution then, an intense aging of the component is incorporated.

There are various methods for reliability assessment of the engineering systems. Out of them universal generating function (UGF) is one of the very useful tools due to its complexity and time reducing capability and nature of straightforwardness. It is worth mentioning here that UGF approach includes the contribution of all the system's components to the entire system performance. It combines two components of the system by defining composition operators to reduce the complexity, which is not the case in other methods. Several authors such as Ushakov (1986), Levitin and Lisniansky (2005) and Levitin (2004) have extended the UGF to evaluate the reliability of the system. UGF technique is not efficient to tackle the uncertainties present in the system therefore it is reformulated as fuzzy UGF. Ding and Lisniansky (2008) and Liu and Huang (2010) used fuzzy UGF in the reliability assessment of the systems.

Keeping above discussion in mind we focus our present study to evaluate the fuzzy reliability characteristics such as fuzzy reliability and fuzzy mean time to failure (FMTTF) of a weighted-$((f/(r,s)),k)/(m,n)$: G system. This study is different from the others in the sense that here we combine the fuzzy universal generating function (FUGF) and fuzzy Rayleigh distribution approaches for the fuzzy reliability and FMTTF evaluation of the target system unlike in the past. With the help of proposed method some prepositions are developed and the variation of fuzzy reliability and FMTTF by incorporating changes on different system parameters has taken into account. We also present numerical example to demonstrate the application of system under study.

NOTATIONS

Notations used in this chapter are given in Table 1.

Table 1. Notations

mn	:	number of components in the system.
g_i	:	states of component i in the system.
$p(q)$	:	probability of success (failure) of the system at time t.
$p_i\,(q_i)$	:	probability of success (failure) of component i in the system at time t.
λ_i	:	failure rate of component i of the system.
w_i	:	weight of component i of the system.
$\tilde{\lambda}_i$	:	fuzzy failure rate of component i of the system
$\tilde{\lambda}_i^L$	:	lower bound of fuzzy failure rate of component i of the system
$\tilde{\lambda}_i^U$	:	upper bound of fuzzy failure rate of component i of the system
$\tilde{R}_w(((f\,/\,(r,s)),k)\,/\,(m,n):\mathrm{G})$	:	fuzzy reliability of weighted-((f / (r, s)), k)/ (m, n): G system
$\tilde{T}_m\left\{(((f\,/\,(r,s)),k)\,/\,(m,n):\mathrm{G})\right\}(\alpha)$	:	fuzzy mean time to failure of weighted-((f / (r, s)), k)/ (m, n): G system
$\tilde{R}_{wL}(((f\,/\,(r,s)),k)\,/\,(m,n):\mathrm{G})(t_1,\alpha)$	:	lower bound of fuzzy reliability of weighted- ((f / (r, s)), k)/ (m, n): G system
$\tilde{R}_{wU}(((f\,/\,(r,s)),k)\,/\,(m,n):\mathrm{G})(t_1,\alpha)$	:	upper bound of fuzzy reliability of weighted-((f / (r, s)), k)/ (m, n): G system
$\tilde{T}_{mL}\left\{(((f\,/\,(r,s)),k)\,/\,(m,n):\mathrm{G})\right\}(\alpha)$	:	lower bound of FMTTF of weighted-((f / (r, s)), k)/ (m, n): G system
$\tilde{T}_{mU}\left\{(((f\,/\,(r,s)),k)\,/\,(m,n):\mathrm{G})\right\}(\alpha)$	:	upper bound of FMTTF of weighted-((f / (r, s)), k)/ (m, n): G system

DEFINITIONS

In this section some of the definitions used in the study are presented.

Weighted-((f / (r, s)), k)/ (m, n): G System

A weighted-((f / (r, s)), k)/ (m, n): G system works if:

1. In all the sub matrices of order (r, s), the total weight of the working components is greater than f.
2. The total weight of the working components in the system is at least k.

Generalised Trapezoidal Fuzzy Number (GTrFN)

Chen (1985) represented a GTrFN Ã as Ã= (a_1, a_2, a_3, a_4; μ), $0 < \mu \leq 1$ where a_1, a_2, a_3, a_4 are real numbers. Its membership function $\xi_{\tilde{A}}^{w}(x): R \rightarrow [0, \mu]$ is defined as

$$\xi_{\tilde{A}}^{w}(x) = \begin{cases} \mu\left(\dfrac{x - a_1}{a_2 - a_1}\right), & a_1 \leq x \leq a_2 \\ \mu, & a_2 \leq x \leq a_3 \\ \mu\left(\dfrac{a_4 - x}{a_4 - a_3}\right), & a_3 \leq x \leq a_4 \\ 0, & \text{otherwise} \end{cases} \tag{1}$$

The operations associated with GTrFN can be seen in Shaw & Roy (2011).

Fuzzy Rayleigh Distribution

The Rayleigh distribution is a continuous probability distribution which is very useful in statistics as it provides population model. The probability density function of a continuous random variable X which follows the Rayleigh distribution is defined as

$$f(x) = \begin{cases} \dfrac{2x}{\theta} exp - \left(\dfrac{x}{\theta}\right)^2, & 0 \leq x < \infty \\ 0 \quad , & otherwise \end{cases} . \tag{2}$$

where $\theta \geq 0$ .is a scale parameter and is reciprocal of the rate parameter (λ). Consider a situation where the scale parameter is not crisp i.e. it takes fuzzy value.

To incorporate this uncertainty let us replace θ .by $\tilde{\theta}$. Now the fuzzy probability of getting a value in the interval [*c*, *d*] and its α-cut can be obtained as

$$\tilde{p}[c \le x \le d][\alpha] = \{\int_c^d \frac{2x}{\theta^2} exp - \left(\frac{x}{\theta}\right)^2 dx \mid \theta \in \tilde{\theta}[\alpha]\} \quad .$$

$$= [p^L[\alpha], p^U .\alpha]] \quad (3)$$

where lower $p^L[\alpha]$ .and upper $p^U[\alpha]$ .bounds of fuzzy probability are given by

$$p^L[\alpha] = \max\{\int_c^d \frac{2x}{\theta^2} exp - \left(\frac{x}{\theta}\right)^2 dx \mid \theta \in \tilde{\theta}[\alpha]\}.$$

$$p^U[\alpha] = \min\{\int_c^d \frac{2x}{\theta^2} exp - \left(\frac{x}{\theta}\right)^2 dx \mid \theta \in \tilde{\theta}[\alpha]\}.$$

Fuzzy Reliability Function

Fuzzy reliability function is the collection of those probabilities that a unit survives beyond a certain time say *t*. As stated earlier in this chapter reliability estimation requires its lifetime data. If the random variable *X* denotes the lifetime of the components of the system, then the fuzzy survival function $\tilde{S}(t)$ .can be calculated as

$$\tilde{S}(t) = P(X > t) = 1 - \tilde{F}_X(t), t > 0. \quad (4)$$

where $\tilde{F}_X(t) = P(X \le t)$ .is the fuzzy cumulative distribution function.

$$\tilde{S}.t) = \{ \int_t^\infty \frac{2x}{\theta^2} exp - \left(\frac{x}{\theta}\right)^2 dx \mid \theta \in \tilde{\theta}[\alpha].\}$$

$$=\{exp\text{-}\left(\frac{t}{\theta}\right)^2 .\theta \in \tilde{\theta}[\alpha]. \tag{5}$$

If $\tilde{\theta}$. $\{ a_1, a_2, a_3, a_4; \mu\}$ .hen its α-cut is obtained as follows

$$\tilde{\theta}[\alpha] = \left\{\left[a_1 + \frac{\alpha}{\mu}(a_2 - a_1), a_4 - \frac{\alpha}{\mu}(a_4 - a_3)\right]\right\}. \tag{6}$$

Applying the above α- cut, the fuzzy survival function (equation (5)) will reduce to the form

$$\tilde{S}(\mathrm{t}) = \left\{\left[\exp - \left\{\frac{\mathrm{t}}{a_1 + \frac{\alpha}{\mu}(a_2 - a_1)}\right\}^2, exp - \left\{\frac{\mathrm{t}}{a_4 - \frac{\alpha}{\mu}(a_4 - a_3)}\right\}^2\right], . \tag{7}$$

From equation (7) it is clear that fuzzy reliability function depends upon t and α.

Fuzzy Mean Time to Failure (*FMTTF*)

It is the expected mean time to failure. *FMTTF* of any system can be calculated as follows:

$$FMTTF[\alpha] = \{\int_0^\infty xf(x)dx \mid S_\alpha\}. = \{\int_0^\infty s(t)dt .\mid S_\alpha\}.$$

$$FMTTF[\alpha] \;=\; \{\int_0^\infty exp - \left(\frac{t}{\theta}\right)^2 dt \mid S_\alpha\}. \;=\; \{\frac{\theta\sqrt{\pi}}{2} \mid \theta \in \tilde{\theta}[\alpha]\}. \;= \{[$$

$$\left(a_1 + \alpha\left(a_2 - a_1\right)\right)\frac{\sqrt{\pi}}{2} .(a_4 - \alpha\left(a_4 - a_3\right))\frac{\sqrt{\pi}}{2} .\} \tag{8}$$

FUZZY UNIVERSAL GENERATING FUNCTION OF WEIGHTED-((*F* / (*R*, *S*)), *K*)/ (*M*, *N*): G SYSTEM

Consider a weighted-((*f* / (*r*, *s*)), *k*)/ (*m*, *n*): G system consisting of *mn* independent and non-identical components arranged in the form of matrix and having some integer weight say w_i, i=1, 2,...., *mn* respectively. The FUGF of proposed system is calculated as

The FUGF (fu- function) for component 1 is given as:

$$\tilde{U}_1(z) = \sum_{g_1=0}^{1} \tilde{p}_1^{g_1} (1 - \tilde{p}_1)^{1-g_1} z^{g_1, w_1} \tag{9}$$

The fu- function for component 1 and 2 is given as:

$$\tilde{U}_{1,2}(z) = \sum_{g_2=0}^{1} \sum_{g_1=0}^{1} \left[\prod_{i=1}^{2} \tilde{p}_i^{g_i} (1 - \tilde{p}_i)^{1-g_i} \right] z^{(g_1, g_2), [w_1, w_2]} \tag{10}$$

....

The fu- function for components 1, 2, 3.......and *mn* is given as:

$$\tilde{U}_{1,2,\dots\dots,MN}(z) = \sum_{g_{mn}=0}^{1} \sum_{g_{mn-1}=0}^{1} \cdots\cdots \sum_{g_1=0}^{1} \left[\prod_{i=1}^{mn} \tilde{p}_i^{g_i} (1 - \tilde{p}_i)^{1-g_i} \right] z^{(g_1, g_2, \dots\dots, g_{MN}), \left[\begin{smallmatrix} w_1, \dots \\ \dots\dots, w_{mn} \end{smallmatrix}\right]} \tag{11}$$

Algorithim:

1. Take a weight matrix say A of order (*m*, *n*) from equation (11).
2. Add the components of matrix A and if the addition is greater than a pre assumed value *k* then proceed further otherwise go back to another matrix, say B and repeat the steps (i) and (ii).
3. Find out all the sub matrices of matrix A of order (*r*, *s*).
4. Now check out each sub matrix for the addition of its components.
5. Add the components of each submatrix of A and if this addition has value greater than a pre assumed value *f* then step ahead otherwise stop and go back to the next matrix B and repeat the steps (i)-(v).
6. If steps (i)- (v) are satisfied by matrix A then, probability corresponding to the matrix A is considered as a part of the fuzzy reliability of the considered system otherwise go back to the step (i) for the next matrix B.

(vii)Repeat this process for all 2^{mn} weight matrices obtained from equation (11).

FUZZY RELIABILITY AND FMTTF ASSESSMENT OF WEIGHTED-((*F* / (*R*, *S*)), *K*)/ (*M*, *N*): G SYSTEM BY USING FUGF

Preposition 1: Let a weighted-((*f* / (*r*, *s*)), *k*)/ (*m*, *n*): G system. Fuzzy reliability of weighted-((*f* / (*r*, *s*)), *k*)/ (*m*, *n*): G system with independent and non-identical components can be expressed as

$$\tilde{R}_w(((f\ /\ (r,s)),k)\ /\ m,n)\ :\ \mathrm{G}) = \Pr\left\{A \geq k;\ B \geq f\right\}$$

where $A = \sum_{i=1}^{mn} X_i$ and $B = \sum_{i=1}^{rs} X_i$ are total weight of working components in $r \times s$ submatrix and $m \times n$ matrix respectively.

Proof: Let fuzzy reliability of weighted-((*f*/ (*r*, *s*)), *k*)/ (*m*, *n*): G system is denoted by $\tilde{R}_w(((f\ /\ (r,s)),k)\ /\ m,n)\ :\ \mathrm{G})$. Let collection of probabilities of those pair of components for which total weight of working components in $m \times n$ matrix ($\sum_{i=1}^{mn} X_i$) be greater than *k* be given by $\Pr\left\{\sum_{i=1}^{mn} X_i \geq k\right\}$ and the pool of probabilities of those pair of components for which total weight of working components in $r \times s$ submatrix ($\sum_{i=1}^{rs} X_i$) be greater than *f* is denoted by $\Pr\left\{\sum_{i=1}^{rs} X_i \geq f\right\}$. From definition of weighted-((*f* / (*r*, *s*)), *k*)/ (*m*, *n*): G system, we can get

$$\tilde{R}_w(((f\ /\ (r,s)),k)\ /\ m,n)\ :\ \mathrm{G}) = \Pr\left\{\sum_{i=1}^{mn} X_i \geq k\right\} \cap \Pr\left\{\sum_{i=1}^{rs} X_i \geq f\right\} \tag{12}$$

Right hand side of the equation (12) can collectively be written as $\Pr\left\{\sum_{i=1}^{mn} X_i \geq k;\ \sum_{i=1}^{rs} X_i \geq f\right\}$ and so equation (12) becomes

$$\tilde{R}_w(((f\ /\ (r,s)),k)\ /\ m,n)\ :\ \mathrm{G}) = \Pr\left\{\sum_{i=1}^{mn} X_i \geq k;\ \sum_{i=1}^{rs} X_i \geq f\right\} \tag{13}$$

Designating A= $\sum_{i=1}^{mn} X_i$ and B= $\sum_{i=1}^{rs} X_i$, equation (13) takes the form

$$\tilde{R}_w(((f\ /\ (r,s)),k)\ /\ m,n)\ :\ \mathrm{G}) = \Pr\left\{A \geq k;\ B \geq f\right\} \tag{14}$$

Preposition 2: Let $\tilde{R}_w(((f\ /\ (r,s)),k)\ /\ m,n) : \mathrm{G})(t,\alpha)$ and $\tilde{T}_m\left\{((f\ /\ (r,s)),k)\ /\ m,n) : \mathrm{G}\right\}(\alpha)$ be fuzzy reliability and FMTTF of weighted-((*f* / (*r*, *s*)), *k*)/ (*m*, *n*): G system with independent and non-identical components respectively, then FMTTF of weighted-((*f* / (*r*, *s*)), *k*)/ (*m*, *n*): G system is obtained as

$$\tilde{T}_m\left\{((f\ /\ (r,s)),k)\ /\ m,n)\ :\ \mathrm{G}\right\}(\alpha) = \int_0^{\infty} \tilde{R}_w(((f\ /\ (r,s)),k)\ /\ m,n)\ :\ \mathrm{G})(t,\alpha)dt$$

Proof: If $\tilde{R}_w(((f\ /\ (r,s)),k)\ /\ m,n) : \mathrm{G})(t,\alpha)$ and $\tilde{T}_m\left\{((f\ /\ (r,s)),k)\ /\ m,n) : \mathrm{G}\right\}(\alpha)$ are fuzzy reliability and FMTTF of weighted-((*f* / (*r*, *s*)), *k*)/ (*m*, *n*): G system respectively then from the classical expression of mean time to failure, we have

$$\mathrm{MTTF} = \int_0^{\infty} R(t)dt, \text{ where } R(t) \text{ is the reliability of the system.}$$

So we can express FMTTF as

$$\tilde{T}_m\left\{((f\ /\ (r,s)),k)\ /\ m,n)\ :\ \mathrm{G}\right\}(\alpha) = \int_0^{\infty} \tilde{R}_w(((f\ /\ (r,s)),k)\ /\ m,n)\ :\ \mathrm{G})(t,\alpha)dt \tag{15}$$

VARIATIONS ON FUZZY RELIABILITY AND FMTTF OF CONSIDERED SYSTEM WITH RESPECT TO TIME AND ALPHA

Preposition 3: If probabilities of success of the components of weighted-((*f* / (*r*, *s*)), *k*)/ (*m*, *n*): G system follow fuzzy Rayleigh distribution having symmetric trapezoidal fuzzy failure rates and

$\tilde{R}_{wL}(((f\,/\,(r,s)),k)\,/\,m,n):\mathrm{G})(t,\alpha)$ and $\tilde{R}_{wU}(((f\,/\,(r,s)),k)\,/\,m,n):\mathrm{G})(t,\alpha)$ are lower and upper bounds of fuzzy reliability of weighted-((*f* / (*r*, *s*)), *k*)/ (*m*, *n*): G system respectively, then the variations on lower and upper bounds of fuzzy reliability of weighted-((*f* / (*r*, *s*)), *k*)/ (*m*, *n*): G system with respect to time (for fixed alpha) are expressed as

$$\tilde{R}_{wL}(((f\,/\,(r,s)),k)\,/\,m,n):\mathrm{G})(t_1,\alpha)\le\tilde{R}_{wL}(((f\,/\,(r,s)),k)\,/\,m,n):\mathrm{G})(t_2,\alpha),t_1\le t_2$$

$$\tilde{R}_{wU}(((f\,/\,(r,s)),k)\,/\,m,n):\mathrm{G})(t_1,\alpha)\le\tilde{R}_{wU}(((f\,/\,(r,s)),k)\,/\,m,n):\mathrm{G})(t_2,\alpha),t_1\le t_2$$

Proof: It is given that probabilities of success of the components follow fuzzy Rayleigh distribution having symmetric trapezoidal fuzzy failure rates so the reliability of each component can be obtained as

$$\tilde{p}(t,\alpha)=\left\{\exp-\left\{\left(a_4-\alpha(a_4-a_3\right)t\right\}^2,\exp-\left\{\left(a_1+\alpha(a_2-a_1\right)t\right\}^2\right\}$$
$$=\left[\tilde{p}_L(t,\alpha),\tilde{p}_U(t,\alpha)\right] \tag{16}$$

It is well known that as *t* increases (fixed alpha), negative exponential of $\left(a_4-\alpha(a_4-a_3\right)t$ and $\left(a_1+\alpha(a_2-a_1\right)t$ decreases hence we must have $\tilde{p}_L(t_1,\alpha)\ge\tilde{p}_L(t_2,\alpha),t_1\le t_2$ and $\tilde{p}_U(t_1,\alpha)\ge\tilde{p}_U(t_2,\alpha),t_1\le t_2$. Since fuzzy reliability of weighted-((*f* / (*r*, *s*)), *k*)/ (*m*, *n*): G system is the collection of probabilities of success of its components so it contains the terms like $\tilde{p}(t,\alpha)$ and its behaviour is similar to $\tilde{p}(t,\alpha)$. From above discussion it can be concluded that

$$\tilde{R}_{wL}(((f\,/\,(r,s)),k)\,/\,m,n):\mathrm{G})(t_1,\alpha)\le\tilde{R}_{wL}(((f\,/\,(r,s)),k)\,/\,m,n):\mathrm{G})(t_2,\alpha),t_1\le t_2$$

$$\tilde{R}_{wU}(((f\,/\,(r,s)),k)\,/\,m,n):\mathrm{G})(t_1,\alpha)\le\tilde{R}_{wU}(((f\,/\,(r,s)),k)\,/\,m,n):\mathrm{G})(t_2,\alpha),t_1\le t_2$$

Preposition 4: By using preposition 3, the changes on lower and upper bounds of fuzzy reliability of weighted-((*f* / (*r*, *s*)), *k*)/ (*m*, *n*): G system with respect to time (for fixed alpha) are obtained as

$$\tilde{R}_{wL}(((f\,/\,(r,s)),k)\,/\,m,n):\mathrm{G})(t,\alpha_1)\le\tilde{R}_{wL}(((f\,/\,(r,s)),k)\,/\,m,n):\mathrm{G})(t,\alpha_2),\alpha_1\le\alpha_2$$

$$\tilde{R}_{wU}(((f\,/\,(r,s)),k)\,/\,m,n):\mathrm{G})(t,\alpha_1)\ge\tilde{R}_{wU}(((f\,/\,(r,s)),k)\,/\,m,n):\mathrm{G})(t,\alpha_2),\alpha_1\le\alpha_2$$

Proof: From equation (16), it can clearly be seen that $p_L(t,\alpha_1)\le p_L(t,\alpha_2),\alpha_1\le\alpha_2$ and $p_U(t,\alpha_1)\ge p_U(t,\alpha_2),\alpha_1\le\alpha_2$. And definition of weighted-((*f* / (*r*, *s*)), *k*)/ (*m*, *n*): G system provides that the lower and upper bounds of fuzzy reliability can be expressed in the terms of $p_L(t,\alpha)$ and $p_U(t,\alpha)$ respectively and of course it will be dependent on the nature of $p_L(t,\alpha)$ and $p_L(t,\alpha)$ with respect to α. Hence

$$\tilde{R}_{wL}(((f\,/\,(r,s)),k)\,/\,m,n):\mathrm{G})(t,\alpha_1)\le\tilde{R}_{wL}(((f\,/\,(r,s)),k)\,/\,m,n):\mathrm{G})(t,\alpha_2),\alpha_1\le\alpha_2$$

$$\tilde{R}_{wU}(((f\,/\,(r,s)),k)\,/\,m,n):\mathrm{G})(t,\alpha_1)\ge\tilde{R}_{wU}(((f\,/\,(r,s)),k)\,/\,m,n):\mathrm{G})(t,\alpha_2),\alpha_1\le\alpha_2$$

Preposition 5: The variations on lower and upper bounds of FMTTF of weighted-((*f* / (*r*, *s*)), *k*)/ (*m*, *n*): G system with respect to alpha are given by

$$\tilde{T}_{mL}\left\{((f\,/\,(r,s)),k)\,/\,m,n):\mathrm{G}\right\}(\alpha_1)\le\tilde{T}_{mL}\left\{((f\,/\,(r,s)),k)\,/\,m,n):\mathrm{G}\right\}(\alpha_2),\alpha_1\le\alpha_2$$

$$\tilde{T}_{mU}\left\{((f\,/\,(r,s)),k)\,/\,m,n):\mathrm{G}\right\}(\alpha_1)\ge\tilde{T}_{mU}\left\{((f\,/\,(r,s)),k)\,/\,m,n):\mathrm{G}\right\}(\alpha_2),\alpha_1\le\alpha_2$$

where $\tilde{T}_{mL}\left\{((f\,/\,(r,s)),k)\,/\,m,n):\mathrm{G}\right\}(\alpha)$ and $\tilde{T}_{mL}\left\{((f\,/\,(r,s)),k)\,/\,m,n):\mathrm{G}\right\}(\alpha)$ are lower and upper bounds of FMTTF of weighted-((*f* / (*r*, *s*)), *k*)/ (*m*, *n*): G system respectively.

Proof: From the definition of FMTTF, the lower and upper bounds of FMTTF can be expressed as

$$\tilde{T}_{mL}\left\{((f\,/\,(r,s)),k)\,/\,m,n):\mathrm{G}\right\}(\alpha)=\int_0^{\infty}\tilde{R}_{wL}(((f\,/\,(r,s)),k)\,/\,m,n):\mathrm{G})(t,\alpha)dt$$
$$=\left\{\frac{\sqrt{\pi}}{2\left(a_4-\alpha(a_4-a_3)\right)}\middle|a_3,a_4\in R\text{ and }a_3<a_4\right\} \tag{17}$$

$$\tilde{T}_{mU}\left\{((f\,/\,(r,s)),k)\,/\,m,n):\mathrm{G}\right\}(\alpha)=\int_0^{\infty}\tilde{R}_{wU}(((f\,/\,(r,s)),k)\,/\,m,n):\mathrm{G})(t,\alpha)dt$$
$$=\left\{\frac{\sqrt{\pi}}{2\left(a_2+\alpha(a_2-a_1)\right)}\middle|a_1,a_2\in R\text{ and }a_1<a_2\right\} \tag{18}$$

Frompreposition4itisclearthatas α increases, $\tilde{R}_{wL}(((f\,/\,(r,s)),k)\,/\,m,n):\mathrm{G})(t,\alpha)$ and $\tilde{R}_{wU}(((f\,/\,(r,s)),k)\,/\,m,n):\mathrm{G})(t,\alpha)$ increases and decreases respectively. Hence from equations (17) and (18) we can have

$$\tilde{T}_{mL}\left\{((f\,/\,(r,s)),k)\,/\,m,n):\mathrm{G}\right\}(\alpha_1)\le\tilde{T}_{mL}\left\{((f\,/\,(r,s)),k)\,/\,m,n):\mathrm{G}\right\}(\alpha_2),\alpha_1\le\alpha_2$$

$$\tilde{T}_{mU}\left\{((f\,/\,(r,s)),k)\,/\,m,n):\mathrm{G}\right\}(\alpha_1)\ge\tilde{T}_{mU}\left\{((f\,/\,(r,s)),k)\,/\,m,n):\mathrm{G}\right\}(\alpha_2),\alpha_1\le\alpha_2$$

EFFECT OF *K* AND *F* ON FUZZY RELIABILITY AND FMTTF OF WEIGHTED-((*F* / (*R*, *S*)), *K*)/ (*M*, *N*): G SYSTEM

Preposition 6: If fuzzy reliability of weighted-((*f* / (*r*, *s*)), *k*)/ (*m*, *n*): G system is $\tilde{R}_{w}(((f\,/\,(r,s)),k)\,/\,m,n):\mathrm{G})$ then the changes on fuzzy reliability of the system with respect to *f* and *k* are obtained as

(a)
$$\tilde{R}_{w}(((f_2\,/\,(r,s)),k)\,/\,m,n):\mathrm{G})(t,\alpha)\le\tilde{R}_{w}(((f_1\,/\,(r,s)),k)\,/\,m,n):\mathrm{G})(t,\alpha),f_1\le f_2$$

(b)
$$\tilde{R}_w(((f_2 \,/\, (r,s)), k_2) \,/\, m,n) : \mathrm{G})(t,\alpha) \le \tilde{R}_w(((f_2 \,/\, (r,s)), k_1) \,/\, m,n) : \mathrm{G})(t,\alpha), k_1 \le k_2$$

Proof: From equation (14) fuzzy reliability of weighted-((*f* / (*r*, *s*)), *k*)/ (*m*, *n*): G system is given as

$$\tilde{R}_w(((f \,/\, (r,s)), k) \,/\, m,n) : \mathrm{G}) = \Pr\left\{A \ge k;\ B \ge f\right\}$$

(a) (a) For $f_1 \le f_2$, we have

$$\begin{aligned}\Pr\left\{A \ge k;\ B \ge f_1\right\} &= \Pr\left\{A \ge k; B \ge f_2\right\} + \Pr\left\{A \ge k; f_1 \le B < f_2\right\} \\ &\ge \Pr\left\{A \ge k; B \ge f_2\right\}\end{aligned}$$

Thus from equation (14),

$$\tilde{R}_w(((f_2 \,/\, (r,s)), k) \,/\, m,n) : \mathrm{G})(t,\alpha) \le \tilde{R}_w(((f_1 \,/\, (r,s)), k) \,/\, m,n) : \mathrm{G})(t,\alpha), f_1 \le f_2$$

(b) (b) For $k_1 \le k_2$, we get

$$\begin{aligned}\Pr\left\{A \ge k_1;\ B \ge f\right\} &= \Pr\left\{A \ge k_2; B \ge f\right\} + \Pr\left\{k_1 \le A < k_2; B < f\right\} \\ &\ge \Pr\left\{A < k_2; B < f\right\}\end{aligned}$$

Hence on using equation (14), we have

$$\tilde{R}_w(((f_2 \,/\, (r,s)), k_2) \,/\, m,n) : \mathrm{G})(t,\alpha) \le \tilde{R}_w(((f_2 \,/\, (r,s)), k_1) \,/\, m,n) : \mathrm{G})(t,\alpha), k_1 \le k_2 .$$

Preposition 7: If $\tilde{T}_m\left\{((f/(r,s)),k)/m,n):\mathrm{G}\right\}(\alpha)$ be the FMTTF of weighted-((*f* / (*r*, *s*)), *k*)/ (*m*, *n*): G system, then the variations on FMTTF of the system with changes in *f* and *k* is calculated as

(a)

$$\tilde{T}_m\left\{((f_2/(r,s)),k)/m,n):\mathrm{G}\right\}(\alpha_1)\leq\tilde{T}_m\left\{((f_1/(r,s)),k)/m,n):\mathrm{G}\right\}(\alpha_2), f_1\leq f_2$$

(b)

$$\tilde{T}_m\left\{((f/(r,s)),k_2)/m,n):\mathrm{G}\right\}(\alpha_1)\leq\tilde{T}_m\left\{((f/(r,s)),k_1)/m,n):\mathrm{G}\right\}(\alpha_2), k_1\leq k_2$$

Proof: (a) For $f_1\leq f_2$, we can write

$$\tilde{R}_w(((f_2/(r,s)),k)/m,n):\mathrm{G})(t,\alpha)\leq\tilde{R}_w(((f_1/(r,s)),k)/m,n):\mathrm{G})(t,\alpha), f_1\leq f_2$$

$$\int_0^\infty\tilde{R}_w(((f_2/(r,s)),k)/m,n):\mathrm{G})(t,\alpha)dt\leq\int_0^\infty\tilde{R}_w(((f_1/(r,s)),k)/m,n):\mathrm{G})(t,\alpha)dt$$

Thus, equation (15) yields

$$\tilde{T}_m\left\{((f_2/(r,s)),k)/m,n):\mathrm{G}\right\}(\alpha_1)\leq\tilde{T}_m\left\{((f_1/(r,s)),k)/m,n):\mathrm{G}\right\}(\alpha_2), f_1\leq f_2$$

(b) For $k_1\leq k_2$, we have

$$\tilde{R}_w(((f_2/(r,s)),k_2)/m,n):\mathrm{G})(t,\alpha)\leq\tilde{R}_w(((f_2/(r,s)),k_1)/m,n):\mathrm{G})(t,\alpha), k_1\leq k_2$$

$$\int_0^\infty\tilde{R}_w(((f_2/(r,s)),k_2)/m,n):\mathrm{G})(t,\alpha)dt\leq\int_0^\infty\tilde{R}_w(((f_2/(r,s)),k_1)/m,n):\mathrm{G})(t,\alpha)dt$$

Thus, from equation (15), we get

$$\tilde{T}_m\left\{((f/(r,s)),k_2)/m,n):\mathrm{G}\right\}(\alpha_1)\leq\tilde{T}_m\left\{((f/(r,s)),k_1)/m,n):\mathrm{G}\right\}(\alpha_2), k_1\leq k_2$$

Figure 1. A weighted-((f /2, 2), k)-out-of-(2, 3): G system

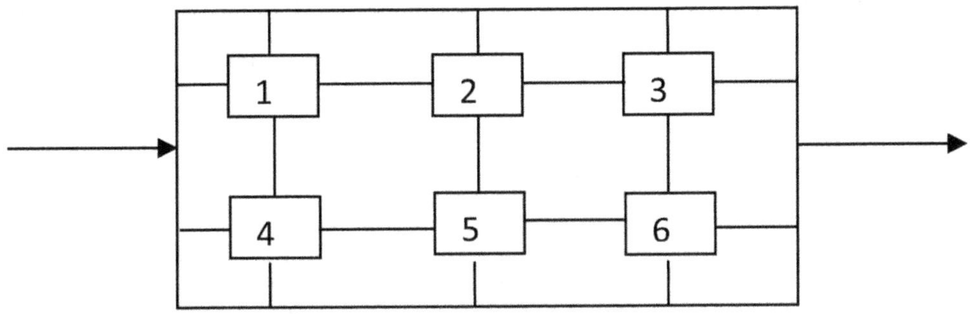

ILLUSTRATIVE EXAMPLE

A sensor is a transducer which is used to detect events or changes in its environments and provides a corresponding output, generally in the form of electrical or optical signals such as biosensors in airport, automatic mobile etc. Now suppose there are six identical sensors arranged in a matrix of type (2, 3) as shown in Figure 1. Let

Table 2. Interval of confidence of $\tilde{\lambda}_1, \tilde{\lambda}_2, \tilde{\lambda}_3, \tilde{\lambda}_4, \tilde{\lambda}_5$ and $\tilde{\lambda}_6$

A	$\tilde{\lambda}_1$	$\tilde{\lambda}_2$	$\tilde{\lambda}_3$	$\tilde{\lambda}_4$	$\tilde{\lambda}_5$	$\tilde{\lambda}_6$
0	[0.01, 0.09]	[0.35, 0.5]	[0.2, 0.35]	[0.15, 0.3]	[0.25, 0.55]	[0.01,0.28]
0.1	[0.02,0.085]	[0.3625, 0.4937]	[0.2071, 0.3428]	[0.1583, 0.2928]	[0.2833,0.5166]	[0.0357,0.2542]
0.2	[0.03, 0.08]	[0.375, 0.4875]	[0.2143, 0.3357]	[0.1666, 0.2857]	[0.3166,0.4833]	[0.0614,0.2285]
0.3	[0.04, 0.075]	[0.3875, 0.4812	[0.2214, 0.3285]	[0.175, 0.2785]	[0.35, 0.45]	[0.0871,0.2028]
0.4	[0.05, 0.07]	[0.4, 0.475]	[0.2285, 0.3214]	[0.1833, 0.2714]	[0.3833,0.4166]	[0.1128,0.1771]
0.5	[0.06, 0.065]	[0.4125, 0.4687]	[0.2357, 0.3142]	[0.1916, 0.2642]	[0.4166, 0.3833]	[0.1385,0.1514]
0.6	[0.07, 0.06]	[0.425, 0.4625]	[0.2428, 0.3071]	[0.2, 0.2571]	[0.45, 0.35]	[0.1642,0.1257]

us consider there are some uncertainties associated with each sensor and each having two possible states: working or failed. Let $\tilde{p}_i$ and w_i be the fuzzy probability of the success and weight of the sensor i (i = 1, 2, 3, 4, 5, 6) respectively. Let each sensor

Table 3. Fuzzy reliability of weighted-((1/ (2, 2), k)/-of-(2, 3): G system with respect of time

α	**k=1,2,3,4 (Lower bound)**	**k=1,2,3,4 (Upper bound)**	**k=5 (Lower bound)**	**k=5 (Upper bound)**	**k=6 (Lower bound)**	**k=6 (Upper bound)**	**k=7 (Lower bound)**	**k=7 (Upper bound)**
0.0	0.365592	0.999327	0.299865	0.999189	0.921804	0.129402	0.876473	0.120446
0.1	0.435023	0.992300	0.371856	0.990181	0.888491	0.168385	0.848272	0.156687
0.2	0.511192	0.974609	0.452553	0.967509	0.842630	0.215864	0.808021	0.200222
0.3	0.592239	0.944098	0.539993	0.928690	0.784170	0.272634	0.754554	0.251098
0.4	0.674921	0.900371	0.630770	0.873703	0.715110	0.338089	0.689248	0.307705
0.5	0.755461	0.844643	0.720532	0.804685	0.63839	0.411539	0.614914	0.368191
0.6	0.829218	0.779556	0.803916	0.725401	0.557966	0.490314	0.535861	0.428798

Table 4. Fuzzy reliability of weighted-((1 /(2, 2), k)/-of-(2, 3): G system with respect of time (t)

t	**k=1,2,3,4 (Lower bound)**	**k=1,2,3,4 (Upper bound)**	**k=5 (Lower bound)**	**k=5 (Upper bound)**	**k=6 (Lower bound)**	**k=6 (Upper bound)**	**k=7 (Lower bound)**	**k=7 (Upper bound)**
0	1	1	1	1	1	1	1	1
1	0.999451	1	0.999123	1	0.998408	0.999994	0.99684	0.999899
2	0.949092	0.999992	0.927679	0.999989	0.881004	0.998899	0.860989	0.995058
3	0.692341	0.999874	0.619838	0.999841	0.455327	0.983937	0.434002	0.962811
4	0.364557	0.999327	0.29883	0.999189	0.128367	0.921804	0.119411	0.876473
5	0.152842	0.998002	0.123556	0.997705	0.023756	0.793809	0.02104	0.739041
6	0.05374	0.995818	0.045535	0.995402	0.003181	0.623711	0.002596	0.578831
7	0.016153	0.992915	0.014546	0.992479	0.000315	0.452114	0.000229	0.424275
8	0.004184	0.988993	0.003952	0.988993	2.31E-05	0.249713	1.47E-05	0.234445
9	0.000932	0.985318	0.000907	0.985318	1.25E-06	0.165673	6.79E-07	0.159166
10	0.000177	0.981273	0.000175	0.981273	4.97E-08	0.106084	2.27E-08	0.103354

i (i = 1, 2, 3, 4, 5, 6) of the system follow fuzzy Rayleigh distribution with failure

Table 5. Variation on Fuzzy reliability of weighted-((f /2, 2), 8)/(2, 3): G system with respect of α

A	*f*=1,2 (Lower bound)	*f*=1,2 (upper bound)	*f*=1,2 (Lower bound)	*f*=1,2 (upper bound)	*f*=1,2 (Lower bound)	*f*=1,2 (upper bound)	*f*=1,2 (Lower bound)	*f*=1,2 (upper bound)	*f*=1,2 (Lower bound)	*f*=1,2 (upper bound)
0	0.0026	0.6223	0.0017	0.5065	0.00004	0.2090	1.49E-05	0.052476	1.46E-05	0.035157
0.1	0.0049	0.5362	0.0034	0.4208	0.00010	0.1406	3.14E-05	0.027595	3.06E-05	0.018252
0.2	0.0091	0.4404	0.0065	0.3332	0.00025	0.0917	6.36E-05	0.015165	6.15E-05	0.01057
0.3	0.0160	0.3442	0.0116	0.2508	0.00054	0.0585	0.000127	0.009473	0.000121	0.007383
0.4	0.0269	0.2560	0.0198	0.1794	0.00632	0.0367	0.000257	0.006755	0.000236	0.005858
0.5	0.0437	0.1808	0.0328	0.1216	0.00734	0.0226	0.000548	0.005069	0.00048	0.004697
0.6	0.0687	0.1215	0.0526	0.0778	0.00892	0.0134	0.00124	0.003721	0.001019	0.00357

rates $\tilde{\lambda}_1, \tilde{\lambda}_2, \tilde{\lambda}_3, \tilde{\lambda}_4, \tilde{\lambda}_5$ and $\tilde{\lambda}_6$ respectively. If failure rates are GrTFNs say $\tilde{\lambda}_1 = (0.01,$

Table 6. Changes on fuzzy reliability of weighted-((f /2, 2), 8)/(2, 3): G system with respect of time (t)

t	*f*=1,2 (Lower bound)	*f*=1,2 (upper bound)	*f*=1,2 (Lower bound)	*f*=1,2 (upper bound)	*f*=1,2 (Lower bound)	*f*=1,2 (upper bound)	*f*=1,2 (Lower bound)	*f*=1,2 (upper bound)	*f*=1,2 (Lower bound)	*f*=1,2 (upper bound)
0	1	1	1	1	1	1	1	1	1	1
1	0.9981	0.9999	0.9902	0.9998	0.9835	0.9995	0.8831	0.9886	0.8403	0.9842
2	0.8701	0.9988	0.7590	0.9925	0.6621	0.9810	0.3525	0.8705	0.2525	0.8299
3	0.4336	0.9838	0.3035	0.9469	0.1808	0.8763	0.0564	0.6063	0.0323	0.5221
4	0.1182	0.9213	0.0715	0.8347	0.0207	0.6564	0.0048	0.3275	0.0034	0.2477
5	0.0210	0.7928	0.0127	0.6737	0.0012	0.4041	0.0003	0.1427	0.0002	0.0978
6	0.0026	0.6223	0.0017	0.5065	4.3E-5	0.2090	1.4E-05	0.0524	1.4E-8	0.0351
7	0.0002	0.4505	0.0001	0.3617	9.4E-7	0.0933	4.2E-07	0.0167	4.2E07	0.0116
8	1.4E-5	0.2480	1.2E-5	0.2480	1.2E-8	0.0185	6.6E-09	0.0044	6.6E-9	0.0043
9	6.8E-7	0.1643	6.1E07	0.1643	8.8E-11	0.0062	5.6E-11	0.0010	5.6E-11	0.0010
10	2.2E-08	0.1050	2.1E-08	0.1050	3.6E-13	0.0019	2.5E-13	0.0002	2.5E-13	0.0002

$0.05, 0.07, 0.09; 0.4)$, $\tilde{\lambda}_2 = (0.35, 0.4, 0.45, 0.5; 0.5)$, $\tilde{\lambda}_3 = (0.2, 0.25, 0.3, 0.35; 0.7)$, $\tilde{\lambda}_4 = (0.5, 0.2, 0.25, 0.3; 0.3)$, $\tilde{\lambda}_5 = (0.25, 0.35, 0.45, 0.55; 0.3)$ and $\tilde{\lambda}_6 = (0.01, 0.1,$

Table 7. Changes on FMTTF of weighted-((1 /2, 2), k) /(2, 3): G system with respect of α

α	*k*=1,2,3,4 (Lower bound)	*k*=1,2,3,4 (Upper bound)	*k*=5 (Lower bound)	*k*=5 (Upper bound)	*k*=6 (Lower bound)	*k*=6 (Upper bound)	*k*=7 (Lower bound)	*k*=7 (Upper bound)
0	7.334674	58.25695	5.143124	53.26841	3.041745	8.715741	2.728678	7.519205
0.1	7.657891	25.34724	5.411277	21.58521	3.195341	7.402209	2.866151	6.458584
0.2	8.057048	17.61754	5.736083	14.63648	3.367049	6.444382	3.01952	5.665239
0.3	8.549393	13.94484	6.134276	11.50485	3.560977	5.717184	3.192353	5.051582
0.4	9.161341	11.7124	6.629984	9.67153	3.780544	5.148086	3.387558	4.565723
0.5	9.935206	10.17346	7.260699	8.440071	4.031836	4.689583	3.610354	4.170841
0.6	10.93995	9.028428	8.086863	7.540759	4.320058	4.311859	3.865099	3.842826

Table 8. Variation on FMTTF of weighted-((f/2, 2), 8)/(2, 3): G system with respect to α

α	*f*=1,2 (Lower bound)	*f*=1,2 (upper bound)	*f*=1,2 (Lower bound)	*f*=1,2 (upper bound)	*f*=1,2 (Lower bound)	*f*=1,2 (upper bound)	*f*=1,2 (Lower bound)	*f*=1,2 (upper bound)	*f*=1,2 (Lower bound)	*f*=1,2 (upper bound)
0.0	1.3394	3.2369	1.2175	2.7302	0.8077	1.6825	0.6033	1.3094	0.4830	1.0309
0.1	1.4010	2.8391	1.2686	2.4558	0.8385	1.5450	0.6313	1.1809	0.5037	0.9372
0.2	1.4692	2.5380	1.3243	2.2343	0.8718	1.4289	0.6620	1.0756	0.5260	0.8591
0.3	1.5455	2.3002	1.3857	2.0513	0.9080	1.3294	0.6959	0.9878	0.5506	0.7929
0.4	1.6314	2.1071	1.4534	1.8976	0.9475	1.2433	0.7334	0.9135	0.5774	0.7363
0.5	1.7290	1.9464	1.5288	1.7662	0.9909	1.1678	0.7752	0.8496	0.6071	0.6872
0.6	1.8413	1.8099	1.6131	1.6525	1.0387	1.1011	0.8221	0.7942	0.6400	0.6442

0.19, 0.28; 0.35), then probabilities of success of sensor i (i = 1, 2, 3, 4, 5, 6) are given by

$$\tilde{p}_1 = e^{-(\tilde{\lambda}_1 t)^2}, \tilde{p}_2 = e^{-(\tilde{\lambda}_2 t)^2}, \tilde{p}_3 = e^{-(\tilde{\lambda}_3 t)^2}, \tilde{p}_4 = e^{-(\tilde{\lambda}_4 t)^2}, \tilde{p}_5 = e^{-(\tilde{\lambda}_5 t)^2}, \tilde{p}_6 = e^{-(\tilde{\lambda}_6 t)^2} \text{ repectively.}$$

Now, from equation (6), interval of confidence of $\tilde{\lambda}_1, \tilde{\lambda}_2, \tilde{\lambda}_3, \tilde{\lambda}_4, \tilde{\lambda}_5$ and $\tilde{\lambda}_6$ is obtained and demonstrated in Table 2.

The FUGF and fuzzy reliability of the weighted-((1 /2, 2), *k*)-out-of-(2, 3): G system is calculated by equations (11) and (14) respectively. Now, effect of *k* on the fuzzy reliability of the illustrated system with respect to α and time (t) is provided in Tables 3 and 4 respectively.

Changes on the fuzzy reliability of the illustrated system for different values of f with respect to α and time is computed in Tables 5 and 6 respectively.

Lower and upper bounds of FMTTF are calculated by equations (17) and (18) and the variations of these bounds of FMTTF with respect to α for different values of f and k are given in Tables 7 and 8 respectively.

CONCLUSION

In this chapter we have studied weighted-((f / (r, s)), k)/ (m, n): G system with independent and non-identical components having uncertainties in it. Due to the uncertainties in the system it is difficult to find out exact reliability characteristics of the system. Therefore a combined approach of FUGF and fuzzy Rayleigh distribution is applied to the proposed system to obtain the fuzzy reliability characteristics such as fuzzy reliability and FMTTF. Here it is supposed that the components of the system follow fuzzy Rayleigh distribution having GTrFN failure rates. Also some conditions on the weight of the components have been imposed and the fuzzy reliabilities and FMTTF of the system have been examined by developing prepositions. Critical examination of prepositions 3-7 reveals that

1. Lower and upper bounds of fuzzy reliability decreases with the increment on time (for fixed alpha).
2. As alpha increases (fixed time), lower and upper bounds of fuzzy reliability and FMTTF increases and decreases respectively.
3. The values of fuzzy reliability and FMTTF of the proposed system are inversely proportional to the values of k (total weight of the working components in matrix $m \times n$) and f (total weight of the working components in sub-matrix $r \times s$).

Here it is noticeable that the results obtained in the illustrated example are in tune with the developed prepositions. In Future this work can be extended in the case of multi-state system.

REFERENCES

Baloui, J. E. (2011). *An evaluation of the systems reliability using fuzzy lifetime distribution*. Academic Press.

Bohra, K. S., & Singh, S. B. (2015). Evaluating fuzzy system reliability using intuitionistic fuzzy Rayleigh lifetime distribution. *Journal MESA, 6*(2), 245-254.

Bollinger, R. C. (1982). Direct computation for consecutive-k-out-of-n: F systems. *IEEE Transactions on Reliability*, *31*(5), 444–446. doi:10.1109/TR.1982.5221424

Cai, K. Y., Wen, C. Y., & Zhang, M. L. (1993). Fuzzy states as a basis for a theory of fuzzy reliability. *Microelectronics and Reliability*, *33*(15), 2253–2263. doi:10.1016/0026-2714(93)90065-7

Chen, S. H. (1985). Operations on fuzzy numbers with function principle. *Tamkang Journal of Management Sciences*, *6*(1), 13–26.

Chen, S. M. (1994). Fuzzy system reliability analysis using fuzzy number arithmetic operations. *Fuzzy Sets and Systems*, *64*(1), 31–38. doi:10.1016/0165-0114(94)90004-3

Chung, W. K. (1988). A k-out-of-N: G redundant system with dependent failure rates and common-cause failures. *Microelectronics and Reliability*, *28*(2), 201–203. doi:10.1016/0026-2714(88)90351-4

da Costa Bueno, V., & do Carmo, I. M. (2007). Active redundancy allocation for a k-out-of-n: F system of dependent components. *European Journal of Operational Research*, *176*(2), 1041–1051. doi:10.1016/j.ejor.2005.09.012

Ding, Y., & Lisnianski, A. (2008). Fuzzy universal generating functions for multi-state system reliability assessment. *Fuzzy Sets and Systems*, *159*(3), 307–324. doi:10.1016/j.fss.2007.06.004

Habib, A. S., Yuge, T., Al-Seedy, R. O., & Ammar, S. I. (2010). Reliability of a consecutive (r, s)-out-of-(m, n): F lattice system with conditions on the number of failed components in the system. *Applied Mathematical Modelling*, *34*(3), 531–538. doi:10.1016/j.apm.2009.06.022

Karpíšek, Z., Štěpánek, P., & Jurák, P. (2010). Weibull fuzzy probability distribution for reliability of concrete structures. *Engineering Mechanics, 17*(5-6), 363-372.

Kumar, A., Yadav, S. P., & Kumar, S. (2006). Fuzzy reliability of a marine power plant using interval valued vague sets. *International Journal of Applied Science and Engineering*, *4*(1), 71–82.

Lam, Y., & Ng, H. K. T. (2001). A general model for consecutive-k-out-of-n: F repairable system with exponential distribution and (k− 1)-step Markov dependence. *European Journal of Operational Research*, *129*(3), 663–682. doi:10.1016/S0377-2217(99)00474-9

Levitin, G. (2004). A universal generating function approach for the analysis of multi-state systems with dependent elements. *Reliability Engineering & System Safety*, *84*(3), 285–292. doi:10.1016/j.ress.2003.12.002

Levitin, G. (2005). *The universal generating function in reliability analysis and optimization* (Vol. 6). Springer.

Li, W., & Zuo, M. J. (2008). Reliability evaluation of multi-state weighted k-out-of-n systems. *Reliability Engineering & System Safety*, *93*(1), 160–167. doi:10.1016/j.ress.2006.11.009

Li, W., Zuo, M. J., & Ding, Y. (2008). Optimal design of binary weighted k-out-of-n systems. *International Journal of Reliability Quality and Safety Engineering*, *15*(05), 425–440. doi:10.1142/S0218539308003143

Liu, Y., & Huang, H. Z. (2010). Reliability assessment for fuzzy multi-state systems. *International Journal of Systems Science*, *41*(4), 365–379. doi:10.1080/00207720903042939

Makri, F. S., & Psillakis, Z. M. (1996). Bounds for reliability of k-within two-dimensional consecutive-r-out-of-n failure systems. *Microelectronics and Reliability*, *36*(3), 341–345. doi:10.1016/0026-2714(95)00102-6

Malinowski, J., & Preuss, W. (1996). Lower and upper bounds for the reliability of connected-(r, s)-out-of-(m, n): F lattice systems. *IEEE Transactions on Reliability*, *45*(1), 156–160. doi:10.1109/24.488935

Negi, S., & Singh, S. B. (2015). Reliability analysis of non-repairable complex system with weighted subsystems connected in series. *Applied Mathematics and Computation*, *262*, 79–89. doi:10.1016/j.amc.2015.03.119

Noguchi, K., Sasaki, M., Yanagi, S., & Yuge, T. (1996). Reliability of connected-(*rs*)-out-of-(*mn*): F lattice systems. *IEICE Transaction on Reliability*, *J79-A*, 954–962.

Pak, A., Parham, G. A., & Saraj, M. (2014). Reliability estimation in Rayleigh distribution based on fuzzy lifetime data. *International Journal of System Assurance Engineering and Management*, *5*(4), 487–494. doi:10.100713198-013-0190-5

Salvia, A. A., & Lasher, W. C. (1990). Two-dimensional consecutive-k-out-of-n: F models. *IEEE Transactions on Reliability*, *39*(3), 382–385. doi:10.1109/24.103023

Shaw, A. K., & Roy, T. K. (2011). Generalized trapezoidal fuzzy number with its arithmetic operations and its application in fuzzy system reliability analysis. *International Journal of Pure and Applied Sciences and Technology*, *5*(2), 60–76.

Ushakov, I. A. (1986). A universal generating function. *Soviet Journal of Computer and Systems Sciences*, *24*(5), 118–129.

Wang, Y., Li, L., Huang, S., & Chang, Q. (2012). Reliability and covariance estimation of weighted k-out-of-n multi-state systems. *European Journal of Operational Research*, *221*(1), 138–147. doi:10.1016/j.ejor.2012.02.037

Wu, J. S., & Chen, R. J. (1994). An algorithm for computing the reliability of weighted-k-out-of-n systems. *IEEE Transactions on Reliability*, *43*(2), 327–328. doi:10.1109/24.295016

Yamamoto, H., & Miyakawa, M. (1995). Reliability of a linear connected-(r, s)-out-of-(m, n): F lattice system. *IEEE Transactions on Reliability*, *44*(2), 333–336. doi:10.1109/24.387391

Yuge, T., Dehare, M., & Yanagi, S. (2000). Reliability and availability of a repairable lattice system. *IEICE TRANSACTIONS on Fundamentals of Electronics. Communications and Computer Sciences*, *83*(5), 782–787.

Zadeh, L. A. (1965). Fuzzy sets. *Information and Control*, *8*(3), 338–353. doi:10.1016/S0019-9958(65)90241-X

Zuo, M. J., Lin, D., & Wu, Y. (2000). Reliability evaluation of combined k-out-of-n: F, consecutive-k-out-of-n: F and linear connected-(r, s)-out-of-(m, n): F system structures. *IEEE Transactions on Reliability*, *49*(1), 99–104. doi:10.1109/24.855542

Chapter 3
Fuzzy Hybrid FMEA for Risk Assessment in Service Industry:
An Integrated Intuitionistic Fuzzy AHP and TOPSIS Approach

Nihan Kabadayi
https://orcid.org/0000-0003-1950-4877
Istanbul University, Turkey

ABSTRACT

Service products are mostly produced and consumed simultaneously through interaction between customer and service providers. To prevent external failures in service operations, it is important to identify potential risks and take relevant actions to eliminate or reduce the occurrence. Therefore, risk assessment is vital to customer satisfaction in any service organization. Failure mode and effects analysis (FMEA) is an effective and useful tool for risk assessment. Although FMEA has been extensively studied in the manufacturing literature, there are a limited number of studies considering the application of FMEA in the hospitality industry. In traditional FMEA, the risk priority of failure modes is determined by generating a crisp risk priority number (RPN). However, it has been claimed in the literature that crisp RPN doesn't have a good performance in reflecting real-life situations. To overcome this shortcoming, a fuzzy hybrid FMEA method is developed. The proposed method has been tested on a case study in a five-star hotel to assess its applicability and benefits.

DOI: 10.4018/978-1-7998-7564-2.ch003

INTRODUCTION

Recently, market conditions have become more competitive as a result of the technological developments and globalization. Recently, consumers have wished to acquire goods and services of a high quality, low prices, and a more customized way. Under these circumstances, it is important to design and execute successful business processes in order to gain a competitive advantage. Business process improvement is a strategically important activity for every organization as well as hotel organizations. There are a number of business improvement tools and techniques used in process improvement activities. Some of the most commonly used techniques are: process mapping, process flowcharting, cause and effect diagrams, define; review; identify; verify; and execute (DRIVE), Pareto analysis, brainstorming charting, Failure mode and effects analysis (FMEA), etc.

There is always a vast amount of potential failure risk and effect for business processes. The risk assessment is a useful method for appropriately identifying and managing potential risks in order to carry out flawless business processes. FMEA is one of the most popular risk assessment tools, which is studied extensively in the literature. The purpose of the risk assessment is to identify potential failure risks and present possible solutions and consequences for them.

The service operations management is a complex task due to the nature of the service product. The intangibility of the service product and the customer participation in the production process are some of the critical issues about the service product delivery process. In addition, the service product is produced and consumed simultaneously, and it may be impossible to prevent poor quality issues before the service is delivered. Therefore, it is possible to say that maintaining high-quality in-service operations is a challenging task (Rotondaro & De Oliveira, 2001).

To improve the performance of service operations, it is important to analyze potential risks related to that process. By doing so, the potential risks and consequences can be identified. Therefore, risk assessment tools are significantly important for service companies to deliver service products in a high quality. The FMEA method is a well-known and suitable tool for the risk assessment. FMEA has been a popular method for the risk assessment in the manufacturing industry to maintain a reliable system. Nevertheless, preventing and dealing with failures is also very important for service companies.

This book chapter is organized as follows: Firstly, a literature review related to the FMEA method is presented. Secondly, under the main topic of this chapter, a brief information about the traditional FMEA method is given. Thirdly, the solution methodology, which contains the intuitionistic fuzzy (IF)-AHP and intuitionistic fuzzy (IF)-TOPSIS, is described. Fourthly, a case study is presented in order to illustrate the proposed method in a real-life problem. Finally, conclusions and future research ideas were discussed.

BACKGROUND

It is possible to trace the roots of the FMEA studies back to the late 1940s. However, the traditional FMEA was developed for the first time as a formal design method by the aerospace industry in the 1960s. Since then, FMEA has been applied in a wide range of industries including electronics, automotive, aerospace, chemistry, construction, and healthcare (Liu et al., 2013). Although the application of the traditional FMEA method is very simple, it has some shortcomings in the measurement scale, the calculation of RPNs, risk factor weights, and so on (Song et al., 2013). Thus, the alternative approaches such as MCDM methods are developed by researchers in order to overcome some of these setbacks. One of the first studies that takes into account the combination of FMEA and MCDM was proposed by Bevilacqua and Braglia (2000). The proposed method was used to prioritize the causes of failures in determining the best maintenance strategy in a refrigerator manufacturing company in Italy. Seyed-Hosseini et al. (2006) applied the DEMATEL method to reprioritize the failure modes in an FMEA. Rau et al. (2007) used the TOPSIS method in order to determine the importance weights of the risks in the packaging unit of a computer company. Geum et al. (2011) introduced a new model to adapt the traditional FMEA method to service operations by developing an RPN that encompasses service features. They applied grey relational analysis (GRA) to determine the risk priority of each service failure. Zammori and Gabbrielli (2012) integrated the FMEA and analytic network process (ANP) methods in order to analyze the relationships between risk criteria in the decision-making process. Liu et al. (2012) presented a fuzzy FMEA based on the fuzzy set theory and VIKOR method in order to prioritize failure modes. Song et al. (2013) presented a new hybrid FMEA based on the fuzzy TOPSIS method. They showed that the new FMEA method is capable of ranking the failure modes more accurately than the conventional one and other hybrid methods. Maheswaran and Loganathan (2013) determined the weights of each risk factor using AHP and prioritized the failure modes by applying the Preference ranking organization method for enrichment evaluation (PROMETHEE) method. Liu et al. (2015) developed an integrated method based on a combination of the FMEA and fuzzy DEMATEL methods. The application results showed that a more accurate risk assessment could be performed by the application of the fuzzy OWA and the fuzzy DEMATEL methods to FMEA. Wang et al. (2016) studied a novel FMEA model that integrates the Complex proportional assessment (COPRAS) method and ANP method in order to assess and rank the risk of failure modes under the interval-valued intuitionistic fuzzy context. They tested the new FMEA model with a case study conducted in a hospital service process. The TOPSIS method was also used by Bian et al. (2018) to prioritize the failure modes in FMEA. Tooranloo et al. (2016) applied the FMEA model based on the IF-TOPSIS method to rank the failure modes for quality of internet banking services. Sisman (2017) used the fuzzy AHP method to determine the importance level of risk factors while ranking the failure modes by the fuzzy VIKOR. Fattahi and Khalilzadeh (2018) developed a novel fuzzy hybrid FMEA method in which the weights of the three factors and the weights of failure modes are calculated by the extended fuzzy AHP and fuzzy Multi-objective optimization on the basis of ratio Analysis (MULTIMOORA) methods. The applicability of the proposed method is tested on a case study. Lo and Liou (2018) investigated a new hybrid FMEA method that uses MCDM in combination with the grey theory. Additionally, they applied the best-worst method to calculate RPN. Dag et al. (2018) applied the FMEA method to improve the hotel management process. They classified the various hazards that could occur in a hotel in different groups according to their level of risk. And then, they implemented some preventive actions to reduce high-risk levels. Jahan (2019) implemented the fuzzy FMEA and fuzzy TOPSIS methods in order to investigate the risks occurring in the process of the post open-heart surgery.

Only few works in the literature applied the hybrid FMEA methods to service risk assessment problems. There are a few FMEA applications in hotel service, even though the hybrid FMEA methods are not used in this area. The hotel management success depends to a great extent on customers' satisfaction. Therefore, it is a critical task for the hotel management to identify the possible risks of failure and take action to prevent them from occurring. Therefore, this study aimed to propose a risk assessment model for the hotel management by using an intuitionistic fuzzy hybrid service-specific FMEA method and extend the existing limited service operations risk management literature.

MAIN FOCUS OF THE CHAPTER

This chapter aims to propose an efficient and practical model for hotel managers. Since maintaining high quality is a critical task for service operations, effective mathematical methods should be used to solve problems related to this issue. Additionally, Kilic and Ok (2012) conducted a research in a five-star hotel and, the results of their study showed that 43.8% of the hotel customer complaints are related to the front-desk operations. For this reason, in this study, the proposed method was applied to prioritize the failure modes that may occur during the front office operations of a five-star hotel.

Failure Mode and Effects Analysis (FMEA)

FMEA is a technique, in which the technology and experience of people are combined in order to identify potential failure modes of a product or service process and plan activities to eliminate them (Liu et al., 2012). The main objective of FMEA is to provide preventive actions in order to identify and eliminate potential failure modes of a product in the design and manufacture stages. Thus, it reduces the risk of delivering the defective product to customers (Song et al., 2013).

In the first step of the traditional FMEA, all the possible failure modes of the product/service or system are identified. Following this, these predetermined failures are analyzed by considering the risk factors called as occurrence (O), severity (S), and detection (D). In the literature, the RPN score is often used to identify and prioritize risks. In the traditional FMEA, RPN is calculated by multiplying three risk factors; namely, occurrence (O), severity (S), and detection (D) as shown below:

$RPN = O \times S \times D$.To calculate the RPN of a potential failure mode, the three risk factors are evaluated using the 10-point scale given in Tables 1-3. The value of the RPN of a failure mode determines the importance or the seriousness of the failure mode. Generally, the higher RPN of a failure mode indicates that the failure mode is highly risky. According to the scores of RPNs, the failure modes can be ranked, and then proper actions can be taken on the high-risk failure modes (Liu et al., 2012).

Table 1. The traditional FMEA scale for occurrence

Probability of failure	Possible failure rates	Rank
Extremely high: Failure almost inevitable	≥.in 2	10
Very high	1 in 3	9
Repeated failures	1 in 8	8
High	1 in 20	7
Moderately high	1 in 80	6
Moderate	1 in 400	5
Relatively Low	1 in 2000	4
Low	1 in 15000	3
Remote	1 in 150000	2
Nearly impossible	≤. in 1,500.000	1

The application of a traditional FMEA method contains nine steps as follows (Chiozza & Ponzetti, 2009):

Table 2. The traditional FMEA scale for severity

Effects	Criteria: Severity of effects	Rank
Hazardous	Failure is hazardous, and occurs without warning. It suspends operation of the system and/or involves noncompliance with government regulations or standards	10
Serious	The consequences of failure are hazardous and/or result in non-compliance with government regulations or standards.	9
Extreme	Failure has a significant negative effect on the primary function of the service. The system may not operate.	8
Major	Failure severely affects service product performance, but it can perform product functions.	7
Significant	The secondary function of the product is affected by the failure. The result is in loss of comfort and convenience.	6
Moderate	Product performance is affected moderately. There may be need for a repair for the product.	5
Low	Unimportant effect on product performance. There is no need for a repair.	4
Minor	Minor effect on product or system performance	3
Very Minor	Very minor effect on product or system performance	2
None	No effect	1

Table 3. The traditional FMEA scale for detection

Detection	Criteria: Probability of detection by design control	Rank
Absolute uncertainty	Potential failure cause or subsequent failure mode cannot be detected by design control.	10
Very remote	Potential cause of a failure or subsequent failure mode can be detected with a very remote chance.	9
Remote	Potential cause of a failure or subsequent failure mode can be detected with a remote chance.	8
Very low	Potential cause of a failure or subsequent failure mode can be detected with a very low chance.	7
Low	Potential cause of a failure or subsequent failure mode can be detected with a low chance.	6
Moderate	Potential cause of a failure or subsequent failure mode can be detected with a moderate chance.	5
Moderately high	Potential cause of a failure or subsequent failure mode can be detected with a moderately high chance.	4
High	Potential cause of a failure or subsequent failure mode can be detected with a high chance.	3
Very high	Potential cause of a failure or subsequent failure mode can be detected with a very high chance.	2
Almost certain	Potential cause of a failure or subsequent failure mode can be detected precisely.	1

Step 1: Review the process or product that is chosen to apply FMEA.
Step 2: Gather the decision makers and conduct a brainstorming in order to determine potential failure modes.
Step 3: Identify potential effects of each failure mode.
Step 4: Determine an occurrence raking for each failure mode using the scale presented in Table 1.
Step 5: Determine a severity raking for each failure mode using the scale presented in Table 2.
Step 6: Determine a detection raking for each failure mode using the scale presented in Table 3.
Step 7: Calculate the RPN of each failure mode.
Step 8: Prioritize the failure modes according to RPNs.
Step 9: Take appropriate actions to eliminate or reduce the risk of the failure mode.

There are four types of FMEA methods.

1. **System FMEA:** The main objective of this method is to maintain the reliability and quality of the system. In this method, the potential risk factors resulted from the structure of the system is determined.
2. **Design FMEA***:* This type of FMEA is used in the early design process of the product. It is useful for analyzing failure modes that affect product quality and prioritizing design improvements.
3. **Process FMEA:** The goal of this method is to analyze and minimize the risk factors that may appear in the process of production or assembly of a product.
4. **Service FMEA:** This type of FMEA aims to minimize the risk factors in the service processes. A service mode failure can be identified as any situation or event that negatively influences the customer satisfaction during the service process (Geum et al., 2011). Table 4 presents the type of service failure modes, the potential causes that may lead to the corresponding failure mode, and examples for these failure modes.

SOLUTION METHODOLOGY

Despite the common usage in the literature, several authors have stated that the RPN has some drawbacks. One of the major shortcomings of RPN is that it neglects some critical criteria for the risk assessment such as cost, quality, etc., and therefore, it has not been able to well-perform in real-life examples (Zammori & Gabbrielli, 2012). To overcome these shortcomings, the hybrid methods that are the combination of

Table 4. Type of service failure mode

Type of failure mode	Cause	Examples
Service delivery system failures	Incorrect service process design	Product defects
		Slow/unavailable services
		Facility problems
		Unclear policies
		Stockout
Inappropriate employee actions	Service employee mistakes	Rude behaviors
		Incorrect orders
		Lost orders
		Mischarge
Inappropriate customer actions	Customer errors	Mistake prior to use
		Wrong usage
		Irresponsive customer

Source: (Geum et al., 2011)

the classic FMEA with MCDM methods have been developed. MCDMs are simple and effective methods suitable for solving complicated decision-making problems. In the literature, different combinations of the AHP and TOPSIS methods are successfully applied to solve complicated decision-making problems. Furthermore, the risk evaluation problem may involve vagueness derived mainly from the nature of risk management, and the expert evaluations of potential risk alternatives can contain uncertainty and vagueness. Since the fuzzy logic approach is a powerful tool to handle vagueness, in this chapter, an integrated intuitionistic fuzzy hybrid FMEA approach is proposed to analyze the operational failure modes in the front office of a five-star hotel. In the proposed model, IF-AHP was used to determine the weights of risk factors, and then the IF-TOPSIS method was applied to rank the potential operational failure modes. Figure 1 presents the flow chart of the proposed method.

Intuitionistic Fuzzy Analytic Hierarchy Process (IF-AHP)

AHP is one of the most well-known MCDM methods. It is a very popular method to solve decision-making problems thanks to its efficiency and simplicity. This method allows decision-makers to compare a set of alternatives with respect to a predefined goal. The relative importance weight of a set of alternatives is identified by conducting pairwise comparisons along with expert judgments. Due to this feature of the method, expert comments can be incorporated on the solution, and complicated

Figure 1. Flow chart of the proposed method

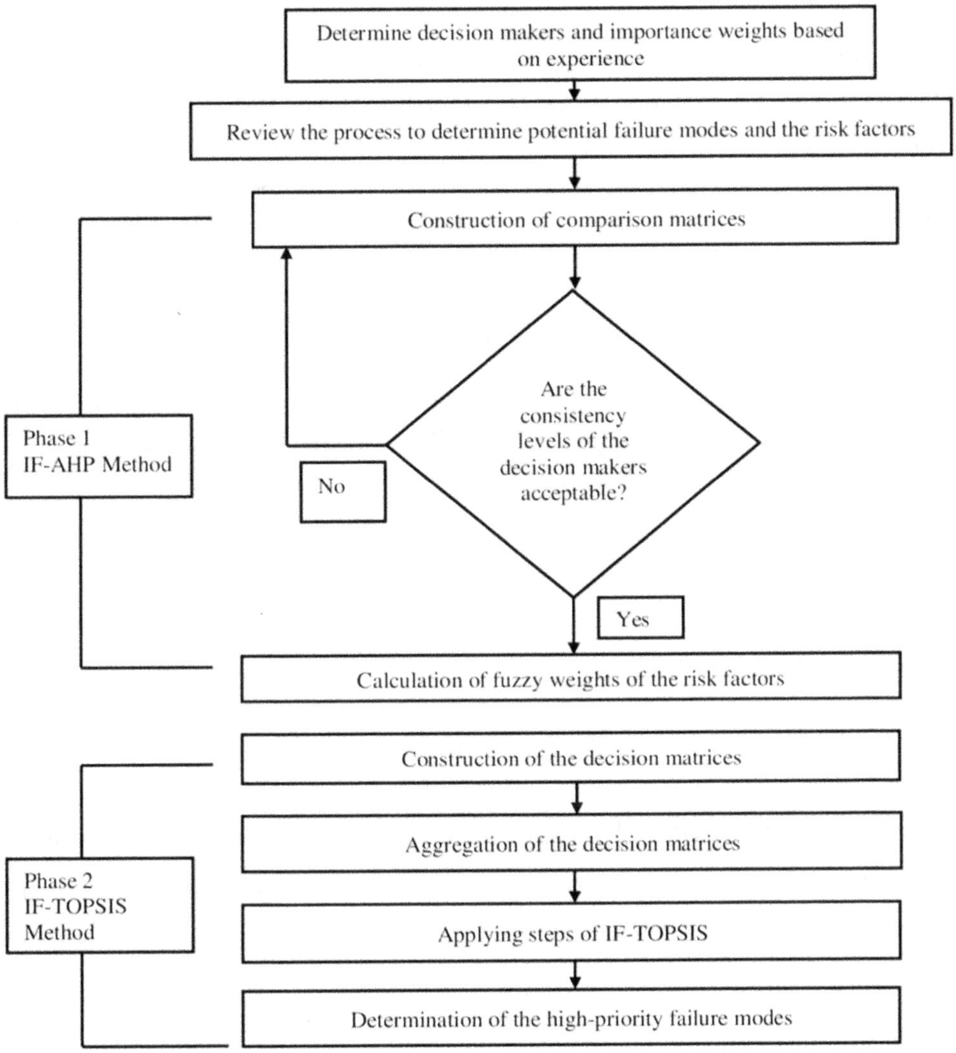

real-life problems can be efficiently solved. AHP has been extensively applied to solve various problems in different fields since it was developed by Saaty in 1980.

Since the pairwise comparisons require a qualitative assessment of human beings, the decision-making process is dominated by vagueness in AHP. One of the best ways to handle the vagueness is to use the fuzzy set theory. It was firstly proposed by Zadeh in 1965. The fuzzy theory claims that the key elements in human thought are linguistic terms, not numbers. Therefore, fuzzy set labels are necessary to measure non-numerical human judgments (Srichetta & Thurachon, 2012). The fuzzy analytic hierarchy process (FAHP) is basically the combination of AHP method and the fuzzy set theory and the extension of Saaty's theory. In FAHP, the pairwise comparisons are constructed using linguistic terms that can be represented by fuzzy numbers. There have been different fuzzy number sets developed in the literature. Among them, the triangular fuzzy numbers and trapezoidal fuzzy numbers are frequently used in the literature. However, using these fuzzy numbers may have some shortcomings such as not being able to express the support and objection evidences simultaneously, due to having a single-valued membership function. For this reason, Zadeh's (1965) fuzzy set has been extended to intuitionistic fuzzy set (IFS) by Atanassov (1983). IFSs provide a proper approach to handling vagueness. In additional to the membership degree, IFSs offer a solution to the problem of vagueness with a non-membership degree (Xu & Liao, 2013). There are four different types of IFS: single-valued, interval-valued, triangular, and trapezoidal IFS. In this chapter, triangular IFSs were used. Assume that A is an IFS in X:

$$A=\left\{x,\mu_A\left(x\right),\upsilon_A\left(x\right)\middle|x\in X\right\}. \tag{1}$$

where:

$\mu_A : X \rightarrow [0,1]$ .and $\upsilon_A : X \rightarrow [0,1]$. $0 \leq \mu_A(x)+\upsilon_A(x) \leq 1$ .for all $x \in X$.

In IFS A, $\mu_A(x)$ .and $\upsilon_A(x)$ .respectively indicate the membership and non-membership degrees. Additionally, triangular IFSs have also another value, $\pi_A(x)$. which is called as intuitionistic index or hesitancy degree. $\pi_A(x)$ .value indicates certainty of x.

$$\pi_A(x)=1-\mu_A(x)-\upsilon_A(x). \tag{2}$$

and $0 \leq .\pi_A(x) \leq 1$ .for all $x \in X$.

$\pi_A(x)$ .value indicates the certainty level of the information concerning x . When the value of $\pi_A(x)$ .is a small number, information is more confident.

Application steps of IF-AHP is given below.

Step 1: Determine the objective, evaluation criteria, sub-criteria, and alternatives of the decision-making problem, and then construct the hierarchy of the related problem.

Step 2: Construct the intuitionistic preference relations by pairwise comparisons demonstrated by the experts. Linguistic pairwise comparisons are transformed to corresponding IFNs.

Step 3: Determine the importance level of each decision maker (DM)s evaluation based on the experience and the department of the DM. Since some decision-makers have different education, knowledge, and experience levels, their opinions may have a different importance level accordingly. Assume that $D_k = [\mu_k, v_k, \pi_k]$.is the intuitionistic fuzzy number for ranking of kth .DM. Therefore, the weight of the kth .DM can be calculated as:

$$\lambda_k = \frac{\left(\mu_k + \pi_k\left(\frac{\mu_k}{\mu_k + v_k}\right)\right)}{\sum_{k=1}^{l}\left(\mu_k + \pi_k\left(\frac{\mu_k}{\mu_k + v_k}\right)\right)}., \sum_{k=1}^{l}\lambda_k = 1. \tag{3}$$

Step 4: Evaluate the consistency of each intuitionistic comparison matrix derived by the experts. In this step, the consistency ratio (CR) for each comparison matrix is calculated. The CR value must be less than 0.1 in order to validate the consistency of comparisons. Otherwise, the preference relations may lead to a misleading solution. Therefore, a perfect multiplicative consistent IPR $\bar{R} = (\bar{r}_{ik})_{nxn}$ .developed by Xu and Liao (2013) is computed as follows in Equations (4-8). If the consistency threshold value of a comparison matrix is greater than 0.10, go to Step 5; otherwise, go to Step 6.

For $k > i+1, let\ \bar{r}_{ik} = (\bar{\mu}_{ik}, \bar{\upsilon}_{ik})$. where

$$\bar{\mu}_{ik} = \frac{\sqrt[k-i-1]{\prod_{t=i+1}^{k-1}\mu_{it}\mu_{tk}}}{\sqrt[k-i-1]{\prod_{t=i+1}^{k-1}\mu_{it}\mu_{tk}} + \sqrt[k-i-1]{\prod_{t=i+1}^{k-1}(1-\mu_{it})(1-\mu_{tk})}}. \tag{4}$$

$$\bar{\upsilon}_{ik} = \frac{\sqrt[k-i-1]{\prod_{t=i+1}^{k-1}\upsilon_{it}\upsilon_{tk}}}{\sqrt[k-i-1]{\prod_{t=i+1}^{k-1}\upsilon_{it}\upsilon_{tk}} + \sqrt[k-i-1]{\prod_{t=i+1}^{k-1}(1-\upsilon_{it})(1-\upsilon_{tk})}}. \tag{5}$$

For $k = i+1, let\ \bar{r}_{ik} = r_{ik}$. (6)

For $k < i+1, let\ \bar{r}_{ik} = (\bar{\upsilon}_{ik}, \bar{\mu}_{ik})$. (7)

R is an acceptable multiplicative consistent IPR, where $d\left(R^{(p)}, \bar{R}\right) < r$ .and,

$$d\left(R^{(p)}, \bar{R}\right) = \frac{1}{2(n-1)(n-2)} \sum_{i=1}^{n} \sum_{k=1}^{n} \left(\left| \bar{\mu}_{ik} - \mu_{ik}^{(p)} \right| + \left| \bar{\upsilon}_{ik} - \upsilon_{ik}^{(p)} \right| + \left| \bar{\pi}_{ik} - \pi_{ik}^{(p)} \right| \right). \tag{8}$$

where, d$\left(R^{(p)}, \bar{R}\right)$.is the distance measure between the given intuitionistic preference relation R, its corresponding perfect multiplicative consistent intuitionistic preference relation $\bar{R}, r$ .is the consistency threshold, and p .is the number of iterations.

Step 5: Repair the inconsistent intuitionistic comparisons, a new IRP ought to be calculated using Equation (9) and Equation (10) as follows:

$$\tilde{\mu}_{ik} = \frac{(\mu_{ik}^{(p)})^{1-\sigma} (\bar{\mu}_{ik})^{\sigma}}{(\mu_{ik}^{(p)})^{1-\sigma} (\bar{\mu}_{ik})^{\sigma} + (1-\mu_{ik}^{(p)})^{1-\sigma} (1-\bar{\mu}_{ik})^{\sigma}}, i,k = 1,2,\ldots,n. \tag{9}$$

$$\tilde{\upsilon}_{ik} = \frac{(\upsilon_{ik}^{(p)})^{1-\sigma} (\bar{\upsilon}_{ik})^{\sigma}}{(\upsilon_{ik}^{(p)})^{1-\sigma} (\bar{\upsilon}_{ik})^{\sigma} + (1-\upsilon_{ik}^{(p)})^{1-\sigma} (1-\bar{\upsilon}_{ik})^{\sigma}}., \text{ i, k= 1, 2, …, n} \tag{10}$$

where σ .is a controlling parameter determined by the decision-making team.

Step 6: Construct the aggregated intuitionistic fuzzy decision matrix according to the DM opinions. Assume, $A = \{A_1, A_2, A_3, \ldots A_m\}$ .is a set of alternatives.

An Intuitionistic Fuzzy Weighted Averaging (IFWA) operator can be applied to aggregate the group decisions as follows:

$R = \left(r_{ij}\right)_{mxn}$., where

$$r_{ij} = IFWA_{\lambda}\left(r_{ij}^{(1)}, r_{ij}^{(2)}, \ldots .. r_{ij}^{(l)}\right). = \lambda_1 r_j^{(1)} \oplus \lambda_2 r_j^{2} \oplus, .., \oplus \lambda_k r_j^{(l)}.$$

$$=\left[1-\prod_{k=1}^{l}\left(1-\mu_{ij}^{(k)}\right)^{\lambda k},\prod_{k=1}^{l}\left(v_{ij}^{(k)}\right)^{\lambda k},\prod_{k=1}^{l}\left(1-\mu_{ij}^{(k)}\right)^{\lambda k}-\prod_{k=1}^{l}\left(v_{ij}^{(k)}\right)^{\lambda k}\right]. \quad (11)$$

Where $r_{ij}=\left(\mu_{Ai}\left(x_j\right),\upsilon_{Ai}\left(x_j\right),\pi_{Ai}\left(x_j\right)\right)\left(i=1,2,\ldots,m;j=1,2,\ldots,n\right)$. .According to the Equation (9), the aggregated intuitionistic fuzzy decision matrix can be presented as follows:

R=

$$\begin{matrix}\mu_{A1}(x_1),\upsilon_{A1}(x_1),\pi_{A1}(x_1) & \mu_{A1}(x_2),\upsilon_{A1}(x_2),\pi_{A1}(x_2)\ldots\ldots\mu_{A1}(x_n),\upsilon_{A1}(x_n),\pi_{A1}(x_n)\\ \mu_{A2}(x_1),\upsilon_{A2}(x_1),\pi_{A2}(x_1) & \mu_{A2}(x_2),\upsilon_{A2}(x_2),\pi_{A2}(x_2)\ldots\ldots\mu_{A2}(x_n),\upsilon_{A2}(x_n),\pi_{A2}(x_n)\\ \vdots & \vdots \quad \ldots\ddots \\ \mu_{Am}(x_1),\upsilon_{Am}(x_1),\pi_{Am}(x_1) & \mu_{Am}(x_2),\upsilon_{Am}(x_2),\pi_{Am}(x_2)\ldots\ldots\mu_{Am}(x_n),\upsilon_{Am}(x_n),\pi_{Am}(x_n)\end{matrix}$$

$$R=\begin{bmatrix} r_{11} & \cdots & r_{1m} \\ \vdots & \ddots & \vdots \\ r_{n1} & \cdots & r_{nm} \end{bmatrix}.$$

Step 7: Determine the weights of the criteria. A new method developed by Xu and Liao (2013) is adapted in this step to calculate the intuitionistic fuzzy weights of the criteria.

$$\omega_i=\left(\frac{\sum_{k=1}^{n}\mu_{ik}}{\sum_{i=1}^{n}\sum_{k=1}^{n}\left(1-v_{ik}\right)},1-\frac{\sum_{k=1}^{n}\left(1-v_{ik}\right)}{\sum_{i=1}^{n}\sum_{k=1}^{n}\mu_{ik}}\right)., \text{ i=1, 2, …, n} \quad (12)$$

Step 8: Rank overall weights using the formula proposed by Szmidt and Kacprzyk (2009), which is as follows:

$$\rho(\alpha)=0.5(1+\pi\alpha)(1-\mu\alpha). \quad (13)$$

According to this formula, the smaller the value of $\rho(\alpha)$, the greater the IFV α in terms of the amount of positive information included and the reliability of the information.

Intuitionistic Fuzzy TOPSIS Method (IF-TOPSIS)

The TOPSIS method is one of the most popular MCDMs. It was firstly developed by Hwang and Yoon in 1981. According to this method, the best alternative should have the shortest distance from the positive-ideal solution and the farthest distance from the negative-ideal solution. The positive ideal solution is a hypothetical alternative that maximizes the benefit criteria and minimizes the cost criteria, while the negative ideal solution minimizes the benefit criteria and maximizes the cost criteria (Behzadian et al., 2012; Liu et al., 2015; Nyaoga et al., 2016).

IF-TOPSIS has been applied to solve many decision-making problems such as location analysis, supplier selection, project evaluation, investment selection, 3PL provider selection, portfolio selection, machine selection, and risk evaluation (Boran et al., 2009; Büyükozan & Güleryüz, 2016; Güler et al., 2019; Liu et al., 2015; Memari et al., 2019; Senel et al., 2018; Tooranloo et al., 2016; Uyanık et al., 2020).

The application steps of IF-TOPSIS consists of the following:

Step 1: Calculate the importance level of each DMs evaluation based on the experience and the departmental position of the DM via Equation (3).

Step 2: Construct the aggregated intuitionistic fuzzy decision matrix according to the DMs opinions. Assume that, $A=\{A_1, A_2, A_3, \dots A_m\}$ .is a set of alternatives.

An IFWA operator can be applied to aggregate the group decisions as follows:

$R=\left(r_{ij}\right)_{mxn}$., where

$$r_{ij}=IFWA_\lambda\left(r_{ij}^{(1)}, r_{ij}^{(2)}, \dots r_{ij}^{(l)}\right).=\lambda_1 r_j^{(1)} \oplus \lambda_2 r_j^2 \oplus, .., \oplus \lambda_k r_j^{(l)}.$$

$$=\left[1-\prod_{k=1}^{l}\left(1-\mu_{ij}^{(k)}\right)^{\lambda k}, \prod_{k=1}^{l}\left(v_{ij}^{(k)}\right)^{\lambda k}, \prod_{k=1}^{l}\left(1-\mu_{ij}^{(k)}\right)^{\lambda k}-\prod_{k=1}^{l}\left(v_{ij}^{(k)}\right)^{\lambda k}\right]. \quad (14)$$

Where

According to the Eq. (12), the aggregated intuitionistic fuzzy decision matrix can be presented as below:

R=[

$$\begin{matrix} \mu_{A1}(x_1),\upsilon_{A1}(x_1),\pi_{A1}(x_1) & \mu_{A1}(x_2),\upsilon_{A1}(x_2),\pi_{A1}(x_2)\ldots\mu_{A1}(x_n),\upsilon_{A1}(x_n),\pi_{A1}(x_n) \\ \mu_{A2}(x_1),\upsilon_{A2}(x_1),\pi_{A2}(x_1) & \mu_{A2}(x_2),\upsilon_{A2}(x_2),\pi_{A2}(x_2)\ldots\mu_{A2}(x_n),\upsilon_{A2}(x_n),\pi_{A2}(x_n) \\ \vdots & \vdots \quad \ldots \ddots \\ \mu_{Am}(x_1),\upsilon_{Am}(x_1),\pi_{Am}(x_1) & \mu_{Am}(x_2),\upsilon_{Am}(x_2),\pi_{Am}(x_2)\ldots\ldots\mu_{Am}(x_n),\upsilon_{Am}(x_n),\pi_{Am}(x_n) \end{matrix}$$

$$R=\begin{bmatrix} r_{11} & \cdots & r_{1m} \\ \vdots & \ddots & \vdots \\ r_{n1} & \cdots & r_{nm} \end{bmatrix}.$$

Step 3: Calculate the aggregated weighted intuitionistic fuzzy decision matrix using the calculated criteria weights (W) and the aggregated intuitionistic fuzzy decision matrix as shown below:

$$R\otimes W=\{x,\mu_{Ai}(x).\mu_w(x),\upsilon_{Ai}(x)+\upsilon_w(x)-\upsilon_{Ai}(x).\upsilon_w(x)Ix\in X\}. \tag{15}$$

Next,

$$\pi_{Ai}w(x)=1-\upsilon_{Ai}(x)-\upsilon_w(x)-\mu_{Ai}(x).\mu_w(x)+\upsilon_{Ai}(x).\upsilon_w(x). \tag{16}$$

And then, the aggregated weighted intuitionistic fuzzy decision matrix is determined as:

$$R'=\begin{bmatrix} r'_{11} & r'_{12} & r'_{13} & \cdots & r'_{1m} \\ r'_{21} & r'_{22} & r'_{23} & \cdots & r'_{2m} \\ \vdots & \vdots & \vdots & \ddots & \vdots \\ r'_{n1} & r'_{n2} & r'_{n3} & \cdots & r'_{nm} \end{bmatrix}.$$

An element of the aggregated weighted intuitionistic fuzzy decision matrix can be symbolized as

$$r'_{ij} = \left(\mu'_{ij}, \upsilon'_{ij}, \pi'_{ij}\right) = \left(\mu_{A_iw}\left(x_j\right), \upsilon_{A_iw}\left(x_j\right), \pi_{A_iw}\left(x_j\right)\right).$$

Step 4: Calculate the distances from the positive and negative ideal points. Let J_1 .and J_2 .represent benefit criteria and cost criteria, respectively. Here, A^+ .is intuitionistic fuzzy positive-ideal solution and A^- .is intuitionistic fuzzy negative-ideal solution. So, A^+ .and A^- .can be calculated as:

$$A^+ = ((\mu_{A^+w}\left(x_j\right), \left(\upsilon_{A^+w}\left(x_j\right)\right) \text{.nd, } A^- = ((\mu_{A^-w}\left(x_j\right), \left(\upsilon_{A^-w}\left(x_j\right)\right). \quad (17)$$

where,

$$\mu_{A^+w}\left(x_j\right) = ((max_i \mu_{A_i.w}\left(x_j\right) | j \in J_1), (min_i \mu_{A_{i.w}} | j \in J_2). \quad (18)$$

$$\upsilon_{A^+w}\left(x_j\right) = ((max_i \upsilon_{A_i.w}\left(x_j\right) | j \in J_1), (min_i \upsilon_{A_{i.w}} | j \in J_2). \quad (19)$$

$$\mu_{A^-w}\left(x_j\right) = ((min_i \mu_{A_i.w}\left(x_j\right) | j \in J_1), (max_i \mu_{A_{i.w}} | j \in J_2). \quad (20)$$

$$\upsilon_{A^-w}\left(x_j\right) = ((min_i \upsilon_{A_i.w}\left(x_j\right) | j \in J_1), (max_i \upsilon_{A_{i.w}} | j \in J_2). \quad (21)$$

Step 5: Determine the separation measures of IFSs of the available alternatives. There are different proposed methods to calculate the separation measures. In this chapter the normalized Euclidean distance was used. The distance of each alternative from the positive and negative ideal points are computed as follows:

$$S^+ = \sqrt{\frac{1}{2n}\sum_{j=1}^{n}\left[(\mu_{Ai.w}\left(x_j\right) - \mu_{A^+w}\left(x_j\right))^2 + (\left(\upsilon_{A_i.w}\left(x_j\right) - \upsilon_{A^+w}\left(x_j\right)\right)^2 + \left(\pi_{A_iw}\left(x_j\right) - \pi_{A^+w}\left(x_j\right)\right)^2\right]}. \quad (22)$$

$$S^{-} = \sqrt{\frac{1}{2n}\sum_{j=1}^{n}\left[(\mu_{Ai.w}(x_j)-\mu_{A^{-}w}(x_j))^2+((\upsilon_{A_i.w}(x_j)-\upsilon_{A^{-}w}(x_j))^2+(\pi_{A_iw}(x_j)-\pi_{A^{-}w}(x_j))^2\right]}. \tag{23}$$

Step 6: Calculate the relative closeness coefficient (CC_i).for the intuitionistic ideal solution. The relative CC_i .f an alternative A_i .with respect to A^{+} .is identified as follows:

$$CC_i = \frac{S_i^{-}}{S_i^{-}+S_i^{+}}. \text{ where } 0 \leq CC_i \leq 1. \tag{24}$$

Step 7: Rank the alternatives according to descending order of CC_i .values.

CASE STUDY

In this section, a real-life application was employed to analyze the applicability and advantages of the proposed method. The FMEA method was considered in a service industry, and the proposed model was applied in a five-star business hotel in Istanbul, Turkey. The hotel has 210 guest rooms and suites, and 14 meeting rooms. This chapter presents the outcome of applying a hybrid fuzzy FMEA method for the risk assessment in a hotel. The front office is chosen because its interaction with customers is very high. Additionally, the service quality in the front office has a significant effect on customer satisfaction. In this context, identifying potential failure risks in hotel front office operations and taking actions to prevent them have a positive impact on the competitiveness of the hotel. A two-phased method is proposed, in which the IF-AHP model is constructed to prioritize the failure modes and the IF-TOPSIS method is used to rank the identified hotel service failure modes. Decision-making team consists of three front-office workers (one front-office manager with 15-year experience, two assistants of front-office managers with 10-year experience). As shown in Table 5, 14 potential major risk modes were identified through the literature review and interviews conducted with decision makers.

Table 5. Potential failure modes

Code	Failure Mode
FM1	Inaccurate check-in (providing an already occupied room key to the customer)
FM2	No availability of room (wrong allocation of customers, unregistered room to the system)
FM3	Inaccurate check-out
FM4	Inconvenient check-in/out waiting experience
FM5	Mischarge (charging customers the wrong amount)
FM6	Rude response to the customer
FM7	Billing or invoice error
FM8	Long waiting time
FM9	Uninspected room (checking-in customers into uncleaned room)
FM10	Check-in without pre-authorization payment
FM11	Using inaccurate/duplicated profile in check-in
FM12	Unregistered check-in
FM13	Not being able to provide comprehensive information to the customer
FM14	Unregistered room changings

Calculating the Weights of Risk Factors

In the first phase of the integrated proposed method, weights of risk factors (Occurrence, Severity, and Detection) are designated by the IF-AHP method. The first risk factor 'occurrence' expresses the frequency or likelihood of a service failure. The 'severity' factor shows how much a failure affects customer satisfaction when a product or process failure occurs. The 'detection' factor is the noticeability degree of a failure before it is experienced by the customer. To determine the weights of these three risk factors, the IF-AHP method is applied as follows:

Initially, the objective of the proposed problem (identifying the most important risk factor) and the risk factors are determined, and then the hierarchy of the proposed problem is constructed.

The intuitionistic fuzzy judgment matrix of the risk factors is constructed according to the questionnaires responded by the decision-making team. Firstly, the pairwise comparisons are conducted among the risk factors. And then, linguistic pairwise comparisons are transformed to the corresponding IFNs using the scale presented in Table 6.

Table 6. Linguistic scale for IFNs

Linguistic variables	IFNs
Very High (VH)	(0.95, 0.05, 0.00)
High (H)	(0.75, 0.15, 0.10)
Medium (M)	(0.50, 0.40, 0.10)
Equal (E)	(0.50, 0.50, 0.00)
Low (L)	(0.25, 0.65, 0.10)
Very Low (VL)	(0.05, 0.95, 0.00)

Table 7. Intuitionistic fuzzy comparison matrix derived by decision maker 1

	Occurrence	Severity	Detection
Occurrence	(0.5, 0.5, 0)	(0.15, 0.75, 0.1)	(0.15, 0.75,0 .10)
Severity	(0.75, 0.15, 0.1)	(0.5, 0.5, 0)	(0.4, 0.5, 0.10)
Detection	(0.75, 0.15, 0.1)	(0.5, 0.4, 0.1)	(0.5, 0.5, 0)

The three decision makers construct the intuitionistic fuzzy comparison matrix in order to determine the importance level of each risk factor as shown in Tables 7-9.

Table 8. Intuitionistic fuzzy comparison matrix derived by decision maker 2

	Occurrence	Severity	Detection
Occurrence	(0.5, 0.5, 0)	(0.25, 0.65, 0.10)	(0.15, 0.75, 0.10)
Severity	(0.65, 0.25, 0.1)	(0.5, 0.5, 0)	(0.5, 0.5, 0)
Detection	(0.75, 0.15, 0.1)	(0.5, 0.5, 0)	(0.5, 0.5, 0)

Table 9. Intuitionistic fuzzy comparison matrix derived by decision maker 3

	Occurrence	Severity	Detection
Occurrence	(0.5, 0.5, 0)	(0.15, 0.75, 0.10)	(0.15, 0.75, 0.10)
Severity	(0.75, 0.15, 0.1)	(0.5, 0.5, 0)	(0.4, 0.5, 0.1)
Detection	(0.75, 0.15, 0.1)	(0.5, 0.4, 0.1)	(0.5, 0.5, 0)

Table 10. Transformed intuitionistic fuzzy comparison matrix for decision maker 1

	Occurrence	Severity	Detection
Occurrence	(0.5, 0.5, 0)	(0.15, 0.75, 0.10)	(0.105, 0.750, 0.145)
Severity	(0.75, 0.15, 0.1)	(0.5, 0.5, 0)	(0.4,0 .5, 0.10)
Detection	(0.750, 0.105, 0.145)	(0.5, 0.4, 0.1)	(0.5, 0.5, 0)

The consistency of each intuitionistic comparison matrix derived by the experts is evaluated. In this step, the judgments of each expert were checked in terms of consistency of the IPR, and the inconsistent IPRs of the experts are repaired automatically via Equations (9-10).

Therefore, a perfect multiplicative consistent IPR $\bar{R} = \left(\bar{r}_{ik}\right)_{nxn}$.was computed for decision maker 1 as shown below. Firstly, the modified IPR $\bar{R} = \left(\bar{r}_{ik}\right)_{nxn}$.was calculated using Equation (4) – Equation (8) as shown in Table 10.

Here, $\bar{r}_{13}$.was calculated as an example to illustrate the process of transformation of intuitionistic fuzzy comparison matrix.

$$\bar{\mu}_{13} = \frac{\sqrt[3-1-1]{\prod_{t=1+1}^{k-1} \mu_{12}\mu_{23}}}{\sqrt[3-1-1]{\prod_{t=1+1}^{3-1} \mu_{12}\mu_{23}} + \sqrt[3-1-1]{\prod_{t=1+1}^{3-1} (1-\mu_{13})(1-\mu_{23})}}.$$

$$\bar{\upsilon}_{13} = \frac{\sqrt[3-1-1]{\prod_{t=i+1}^{3-1} \upsilon_{12}\upsilon_{23}}}{\sqrt[3-1-1]{\prod_{t=1+1}^{3-1} \upsilon_{12}\upsilon_{23}} + \sqrt[3-1-1]{\prod_{t=1+1}^{3-1} (1-\upsilon_{12})(1-\upsilon_{23})}}.$$

$$\bar{\mu}_{13} = \frac{0.15 \times 0.4}{(0.15 \times 0.4) + (0.85 \times 0.60)} = 0.105.$$

$$\bar{\upsilon}_{13} = \frac{0.75 \times 0.5}{(0.75 \times 0.5) + (0.25 \times 0.50)} = 0.75.$$

And then, the distance between the given intuitionistic preference relation R and its corresponding perfect multiplicative consistent intuitionistic preference relation $\bar{R}$.was calculated using Equation (8). The calculation result shows that d $\left(R^1, \bar{R}\right) = 0.033 < 0.1$. which means the intuitionistic fuzzy comparison is of

Table 11. Linguistic terms for ranking the importance of the DMs

Linguistic Term	IFS
Very Important (VI)	(0.9,0.10)
Important (I)	(0.75,0.20)
Medium (M)	(0.5,0.45)
Unimportant (U)	(0.35, 0.60)
Very Unimportant (VU)	(0.1, 0.90)

acceptable consistency. The consistency checking was conducted using the same process for the other IPR of risk factors according to the other decision makers. The calculations revealed that the consistency of other IPR of risk factors are acceptable, and there is no need to repair them.

The IFWA operator was used to aggregate the judgement of decision makers based on the importance degree of experts. Decision makers' importance level was determined based on the experience, and the position of the decision maker using linguistic terms given in Table 11 is calculated as shown below.

The weights of three decision makers were respectively determined using Equation (3) as follows:

$$\lambda_1 = \frac{\left(0.9+0\left(\frac{0.9}{0.9+0.1}\right)\right)}{\left(0.9+0\left(\frac{0.9}{0.9+0.1}\right)\right)+\left(0.75+0.05\left(\frac{0.75}{0.75+0.20}\right)\right)+\left(0.75+0.05\left(\frac{0.75}{0.75+0.20}\right)\right)}=0.357^{\cdot}$$

$$\lambda_2 = \frac{\left(0.75+0.05\left(\frac{0.75}{0.75+0.20}\right)\right)}{\left(0.9+0\left(\frac{0.9}{0.9+0.1}\right)\right)+\left(0.75+0,05\left(\frac{0.75}{0.75+0.20}\right)\right)+\left(0.75+0.05\left(\frac{0.75}{0.75+0.20}\right)\right)}=0.321^{\cdot}$$

$$\lambda_3 = \frac{\left(0.75+0.05\left(\frac{0.75}{0.75+0.20}\right)\right)}{\left(0.9+0\left(\frac{0.9}{0.9+0.1}\right)\right)+\left(0.75+0,05\left(\frac{0.75}{0.75+0.20}\right)\right)+\left(0.75+0.05\left(\frac{0.75}{0.75+0.20}\right)\right)}=0.321^{\cdot}$$

Table 12. The aggregated intuitionistic fuzzy comparison matrix

	Occurrence	Severity	Detection
Occurrence	(0.5, 0.5, 0)	(0.18, 0.72, 0.10)	(0.15, 0.75, 0.10)
Severity	(0.72, 0.18, 0.10)	(0.5, 0.5, 0)	(0.43, 0.50, 0.067)
Detection	(0.75, 0.15, 0.1)	(0.5, 0.43, 0.067)	(0.5, 0.5, 0)

Therefore, the aggregated intuitionistic fuzzy comparison matrix was generated by Equation (11) by combining the judgments of all experts as shown in Table 12.

The consistency of the aggregated intuitionistic fuzzy comparison matrix was calculated as 0.036 < 0.10, which means the consistency is acceptable.

And then, the weights of the criteria were computed using Equation (12). The intuitionistic fuzzy weights of the criteria were calculated as follows:

$$\omega_1 = (0.175, 0.756, 0.069), \omega_2 = (0.3462, 0.5709, 0.083), \omega_3 = (0.367, 0.547, 0.086).$$

Finally, to rank overall weights, Equation (13) was used, and the weights of each of the risk factors were calculated as follows:

$$\rho(\omega_1) = 0.5(1 + 0.069)(1 - 0.175) = 0.441.$$

$$\rho(\omega_2) = 0.5(1 + 0.083)(1 - 0.3462) = 0.354.$$

$$\rho(\omega_3) = 0.5(1 + 0.086)(1 - 0.367) = 0.344.$$

Since the lower the value of $\rho(\alpha)$, the higher the IFV α in terms of the amount of positive information included and the reliability of the information, the weights of each risk factor can be ranked as $\omega_3 > \omega_2 > \omega_1$. Thus, according to IF-AHP application results, 'detection' is the most important risk factor for the hotel service.

Calculating Ranking of Failure Modes for the Front Office Operations

In this phase of the proposed method, an IF-TOPSIS model was developed to rank the potential service failure modes. Firstly, decision makers were asked to evaluate the failure modes using the scale shown in Table 13. And then, these linguistic

Table 13. Intuitionistic fuzzy linguistic variables

Failure occurrence probability		Failure severity probability		Failure detectability probability	
Linguistic Terms	**IFNs**	**Linguistic Terms**	**IFNs**	**Linguistic Terms**	**IFNs**
Very High	(0.9, 0.1)	Risky without Warning	(1, 0)	Absolutely Impossible	(1, 0)
High	(0.75, 0.1)	Risky with Warning	(0.9, 0.1)	Highly Unlikely	(0.9, 0.1)
Medium	(0.5, 0.45)	Very High	(0.8, 0.1)	Unlikely	(0.8, 0.1)
Low	(0.35, 0.6)	High	(0.7, 0.2)	Very Low	(0.7, 0.2)
Very Low	(0.1, 0.9)	Medium	(0.6, 0.3)	Low	(0.6, 0.3)
		Low	(0.5, 0.4)	Medium	(0.5, 0.4)
		Very Low	(0.4, 0.5)	Relatively High	(0.4, 0.5)
		Insignificant	(0.25, 0.6)	High	(0.25, 0.6)
		Very Insignificant	(0.1, 0.75)	Very High	(0.1, 0.75)
		None	(0.1, 0.9)	Absolutely Possible	(0.1, 0.9)

Source: (Senel et al., 2018)

variables were transformed into intuitionistic fuzzy numbers as presented in Table 14. The weights of the decision-makers calculated in the previous stage were used to calculate the aggregated intuitionistic fuzzy decision matrix at this stage as well.

And then, the aggregated intuitionistic fuzzy decision matrix given in Table 15 was constructed according to the importance weight of each decision-maker. In this step, the IFWA operator was applied to aggregate the group decisions.

The aggregated weighted intuitionistic fuzzy decision matrix is constructed using the risk factor weights (W) calculated by IF-AHP and the aggregated intuitionistic fuzzy decision matrix via Equation (13) and Equation (14). The aggregated weighted intuitionistic fuzzy decision matrix is shown in Table 16.

Intuitionistic fuzzy positive-ideal solution (A^+ .) and intuitionistic fuzzy negative-ideal solution A^- .are calculated as in Table 17.

The separation measures of the IFS of the failure modes were calculated using Equation (20) and Equation (21), and the relative closeness coefficient $\left(CC_i\right)$ .alue of each failure mode was computed using Equation (22). The separation measures

Table 14. Evaluation of failure modes by intuitionistic fuzzy numbers

	Decision Maker 1			Decision Maker 2			Decision Maker 3		
Failure Mode	**O**	**S**	**D**	**O**	**S**	**D**	**O**	**S**	**D**
FM 1	(0.35, 0.6, 0.05)	(1, 0, 0)	(0.8, 0.1, 0.1)	(0.35, 0.6, 0.05)	(1, 0, 0)	(0.8, 0.1, 0.1)	(0.35, 0.6, 0.05)	(1, 0, 0)	(0.8, 0.1, 0.1)
FM 2	(0.5, 0.45, 0.05)	(0.9, 0.1, 0)	(0.8, 0.1, 0.1)	(0.35, 0.6, 0.05)	(0.9, 0.1, 0)	(0.7, 0.2, 0.1)	(0.35, 0.6, 0.05)	(0.9, 0.1, 0)	(0.7, 0.2, 0.1)
FM 3	(0.35, 0.6, 0.05)	(0.5, 0.4, 0.1)	(0.25, 0.6, 0.15)	(0.75, 0.1, 0.15)	(0.6, 0.3, 0.1)	(0.7, 0.2, 0.1)	(0.35, 0.6, 0.05)	(0.5, 0.4, 0.1)	(0.25, 0.6, 0.15)
FM 4	(0.75, 0.1, 0.15)	(0.9, 0.1, 0)	(0.25, 0.6, 0.15)	(0.35, 0.6, 0.05)	(0.5, 0.4, 0.1)	(0.1, 0.75, 0.15)	(0.5, 0.45, 0.05)	(0.9, 0.1, 0)	(0.1, 0.75, 0.15)
FM 5	(0.5, 0.45, 0.05)	(0.6, 0.3, 0.1)	(0.8, 0.1, 0.1)	(0.75, 0.1, 0.15)	(0.6, 0.3, 0.1)	(0.7, 0.2, 0.1)	(0.35, 0.6, 0.05)	(0.9, 0.1, 0)	(0.8, 0.1, 0.1)
FM 6	(0.1, 0.9, 0)	(1, 0, 0)	(0.9, 0.1, 0)	(0.35, 0.6, 0.05)	(0.6, 0.3, 0.1)	(0.7, 0.2, 0.1)	(0.35, 0.6, 0.05)	(1, 0, 0)	(0.9, 0.1, 0)
FM 7	(0.5, 0.45, 0.05)	(0.6, 0.3, 0.1)	(0.8, 0.1, 0.1)	(0.75, 0.1, 0.15)	(0.5, 0.4, 0.1)	(0.25, 0.6, 0.15)	(0.5, 0.45, 0.05)	(0.7, 0.2, 0.1)	(0.5, 0.4, 0.1)
FM 8	(0.75, 0.1, 0.15)	(0.6, 0.3, 0.1)	(0.25, 0.6, 0.15)	(0.5, 0.45, 0.05)	(0.5, 0.4, 0.1)	(0.25, 0.6, 0.15)	(0.35, 0.6, 0.05)	(0.7, 0.2, 0.1)	(0.25, 0.6, 0.15)
FM 9	(0.35, 0.6, 0.05)	(1, 0, 0)	(0.9, 0.1, 0)	(0.1, 0.9, 0)	(1, 0, 0)	(0.25, 0.6, 0.15)	(0.35, 0.6, 0.05)	(1, 0, 0)	(0.9, 0.1, 0)
FM 10	(0.5, 0.45, 0.05)	(0.5, 0.4, 0.1)	(0.6, 0.3, 0.1)	(0.75, 0.1, 0.15)	(0.1, 0.9, 0)	(0.6, 0.3, 0.1)	(0.5, 0.45, 0.05)	(0.5, 0.4, 0.1)	(0.7, 0.2, 0.1)
FM 11	(0.75, 0.1, 0.15)	(0.8, 0.1, 0.1)	(0.7, 0.2, 0.1)	(0.75, 0.1, 0.15)	(0.8, 0.1, 0.1)	(0.25, 0.6, 0.15)	(0.75, 0.1, 0.15)	(0.7, 0.2, 0.1)	(0.7, 0.2, 0.1)
FM 12	(0.75, 0.1, 0.15)	(0.9, 0.1, 0)	(0.25, 0.6, 0.15)	(0.5, 0.45, 0.05)	(0.9, 0.1, 0)	(0.6, 0.3, 0.1)	(0.75, 0.1, 0.15)	(0.9, 0.1, 0)	(0.25, 0.6, 0.15)
FM 13	(0.75, 0.1, 0.15)	(0.7, 0.2, 0.1)	(0.9, 0.1, 0)	(0.5, 0.45, 0.05)	(0.6, 0.3, 0.1)	(0.6, 0.3, 0.1)	(0.35, 0.6, 0.05)	(0.7, 0.2, 0.1)	(0.9, 0.1, 0)
FM 14	(0.5, 0.45, 0.05)	(1, 0, 0)	(0.25, 0.6, 0.15)	(0.35, 0.6, 0.05)	(1, 0, 0)	(0.6, 0.3, 0.1)	(0.5, 0.45, 0.05)	(1, 0, 0)	(0.6, 0.3, 0.1)

Table 15. The aggregated intuitionistic fuzzy decision matrix

Failure Mode	Occurrence	Severity	Detection
FM 1	(0.35, 0.60, 0.05)	(1.00, 0.00, 0.00)	(0.80, 0.10, 0.10)
FM 2	(0.41, 0.54, 0.05)	(0.90, 0.10, 0.00)	(0.74, 0.16, 0.10)
FM 3	(0.52, 0.34, 0.14)	(0.53, 0.36, 0.10)	(0.44, 0.42, 0.14)
FM 4	(0.58, 0.29, 0.14)	(0.83, 0.16, 0.01)	(0.16, 0.69, 0.15)
FM 5	(0.56, 0.30, 0.13)	(0.74, 0.21, 0.05)	(0.77, 0.12, 0.10)
FM 6	(0.27, 0.69, 0.04)	(1.00, 0.00, 0.00)	(0.86, 0.12, 0.02)
FM 7	(0.60, 0.28, 0.12)	(0.61, 0.29, 0.10)	(0.59, 0.28, 0.13)
FM 8	(0.58, 0.29, 0.14)	(0.61, 0.29, 0.10)	(0.25, 0.60, 0.15)
FM 9	(0.28, 0.68, 0.04)	(1.00, 0.00, 0.00)	(0.81, 0.18, 0.01)
FM 10	(0.60, 0.28, 0.12)	(0.40, 0.52, 0.08)	(0.64, 0.26, 0.10)
FM 11	(0.75, 0.10, 0.15)	(0.77, 0.12, 0.10)	(0.60, 0.28, 0.12)
FM 12	(0.69, 0.16, 0.15)	(0.90, 0.10, 0.00)	(0.39, 0.48, 0.13)
FM 13	(0.40, 0.55, 0.05)	(0.67, 0.23, 0.10)	(0.84, 0.14, 0.01)
FM 14	(0.46, 0.49, 0.05)	(1.00, 0.00, 0.00)	(0.50, 0.38, 0.12)

Table 16. The aggregated weighted intuitionistic fuzzy decision matrix

Failure Mode	Occurrence	Severity	Detection
FM 1	(0.06, 0.90, 0.04)	(0.35, 0.57, 0.08)	(0.29, 0.59, 0.11)
FM 2	(0.07, 0.89, 0.04)	(0.31, 0.61, 0.07)	(0.27, 0.62, 0.11)
FM 3	(0.09, 0.84, 0.07)	(0.19, 0.73, 0.09)	(0.16, 0.74, 0.10)
FM 4	(0.10, 0.83, 0.07)	(0.29, 0.64, 0.07)	(0.06, 0.86, 0.08)
FM 5	(0.10, 0.83, 0.07)	(0.26, 0.66, 0.08)	(0.28, 0.60, 0.11)
FM 6	(0.05, 0.93, 0.03)	(0.35, 0.57, 0.08)	(0.31, 0.60, 0.08)
FM 7	(0.10, 0.82, 0.07)	(0.21, 0.69, 0.09)	(0.22, 0.67, 0.11)
FM 8	(0.10, 0.83, 0.07)	(0.21, 0.69, 0.09)	(0.09, 0.82, 0.09)
FM 9	(0.05, 0.92, 0.03)	(0.35, 0.57, 0.08)	(0.30, 0.63, 0.08)
FM 10	(0.10, 0.82, 0.07)	(0.14, 0.79, 0.07)	(0.23, 0.67, 0.10)
FM 11	(0.13, 0.78, 0.09)	(0.27, 0.62, 0.11)	(0.22, 0.68, 0.10)
FM 12	(0.12, 0.80, 0.08)	(0.31, 0.61, 0.07)	(0.14, 0.76, 0.09)
FM 13	(0.07, 0.89, 0.04)	(0.23, 0.67, 0.10)	(0.31, 0.61, 0.08)
FM 14	(0.08, 0.88, 0.04)	(0.35, 0.57, 0.08)	(0.18, 0.72, 0.10)

Table 17. The Intuitionistic fuzzy positive-ideal and negative-ideal solutions

Risk Factors	A^{+}	A^{-}
Occurrence	(0.13, 0.78, 0.09)	(0.05, 0.93, 0.03)
Severity	(0.35, 0.57, 0.08)	(0.14, 0.79, 0.07)
Detection	(0.31, 0.59, 0.09)	(0.06, 0.86, 0.08)

Table 18. The relative closeness coefficient and separation measures of each failure mode

Code	Failure Mode	S^{+}	S^{-}	CC_i	Rank
FM1	Inaccurate check-in (providing an already occupied room key to the customer)	0.062	0.192	0.755	1
FM2	No availability of room (wrong allocation of customers, unregistered room to the system)	0.062	0.168	0.729	3
FM3	Inaccurate check-out	0.129	0.086	0.400	12
FM4	Inconvenient check-in/out waiting experience	0.157	0.101	0.391	13
FM5	Mischarge (charging customers the wrong amount)	0.059	0.165	0.734	2
FM6	Rude response to the customer	0.073	0.193	0.726	4
FM7	Billing or invoice error	0.094	0.124	0.568	9
FM8	Long waiting time	0.151	0.074	0.329	14
FM9	Uninspected room (checking in customers into uncleaned room)	0.073	0.184	0.715	5
FM10	Check-in without pre-authorization payment	0.134	0.118	0.468	11
FM11	Using inaccurate/duplicated profile in check-in	0.065	0.152	0.698	6
FM12	Unregistered check-in	0.102	0.132	0.563	10
FM13	Not being able to provide comprehensive information to the customer	0.082	0.159	0.658	7
FM14	Unregistered room changings	0.089	0.148	0.625	8

and the relative closeness coefficient (CC_i).values of each failure mode are shown in Table 18.

According to the CC_i .values, the ranking of failure mode was determined as given in Table 18. Therefore, FM1 'inaccurate check-in' is the most critical failure for the hotel front office service quality. It is needless to say that allocating a guest into an already occupied room is a fatal error for the hotel. This failure mode can cause unacceptable results for both guest parties. Therefore, the front office manager should take some serious actions to prevent this failure. Following this, FM5 'mischarge' and FM2 'no availability of room' have a high priority in the failure of the hotel front office service quality.

SOLUTIONS AND RECOMMENDATIONS

In this study, an integrated hybrid FMEA method is proposed to solve the risk evaluation problem. Since the traditional FMEA method has some drawbacks, in this study a hybrid fuzzy FMEA method is proposed. In the literature, it has been proved that IFS can provide a proper approach to handle vagueness since it offers a solution to the problem of vagueness with a non-membership degree. Therefore, the proposed model is developed based on IFS, as the expert evaluations of potential risk alternatives contain uncertainty and vagueness. On the other hand, the integration of the AHP and TOPSIS methods produce an excellent evaluation methodology under fuzziness. For this reason, in the proposed integrated model in order to determine the importance weights of the risk factors, the IF-AHP method was applied, and then the IF-TOPSIS method was used to rank the failure modes. According to the IF-AHP application, 'detection' is the most important risk factor for the hotel service risk assessment. 'Detection' is of great importance in terms of service quality as it can be defined as the degree of recognition of a failure before it is experienced by the customer. And the implementation results of the IF-TOPSIS method revealed that inaccurate check-in, that is, allocating customers into an occupied room or a wrong room is the most important failure mode for the hotel service operations. Needless to say, the consequences of that failure are severe, and correcting this failure is almost impossible.

FUTURE RESEARCH DIRECTIONS

As a future study, different MCDMs such as MOORA, ELECTRE, VIKOR and GRA may be combined with the FMEA method to solve the risk classification problem. Besides, it can also be suggested to use extensions of fuzzy sets such as type-2 fuzzy sets, hesitant fuzzy sets for the same problem.

CONCLUSION

The risk evaluation is of vital importance for the service company's success. In general, due to the nature of the service product, it is not possible to detect a service failure without reaching the customer. On the other hand, risk management techniques such as FMEA can help service producers improve their service quality by identifying high-risk failure modes before they occur and then taking the right action to prevent them. However, only few works in the literature applied the hybrid FMEA methods to the service risk assessment problems. Additionally, a few FMEA applications have been conducted to analyze hotel service risks, even though the hybrid FMEA methods are not used in this area.

The success of hotel management depends heavily on customer satisfaction. Therefore, it is a critical task for the hotel management to identify the possible risks of failure and act to prevent them from occurring. The aim of the study was to propose a risk assessment model for hotel management by using a fuzzy hybrid FMEA method. For this reason, in this study, a two-phase method was developed in which, IF-AHP is used to calculate the weights of risk factors, and IF-TOPSIS was applied to rank the failure modes. To analyze the performance of the proposed method, a case study was conducted in a five-star business hotel. Since the front-office operations are critical in terms of the hotel customer satisfaction, the proposed method is applied to prioritize the failure modes that may occur during the front office operations of a five-star hotel. The application results were discussed with experts, and findings revealed that the proposed method is an effective alternative method to evaluate the failure modes in service operations. Therefore, the proposed integrated fuzzy FMEA method provides a simple and practical solution to service companies dealing with complex risk evaluation problems. The application of the proposed method is relatively easy to implement since it does not require any complex calculation knowledge as it can be easily programmed into a spreadsheet.

REFERENCES

Atanassov, K. (1983). *Intuitionistic fuzzy sets. VII.* ITKR's Session. Sofia, June 1,1983. Deposed in Central Science—Technology Library of Bulgaria Academy of Science—1697/84.

Behzadian, M., Otaghsara, S. K., Yazdani, M., & Ignatius, J. (2012). A state-of the-art survey of TOPSIS applications. *Expert Systems with Applications*, *39*(17), 13051–13069. doi:10.1016/j.eswa.2012.05.056

Bevilacqua, M., & Braglia, M. (2000). The analytic hierarchy process applied to maintenance strategy selection. *Reliability Engineering & System Safety*, *70*(1), 71–83. doi:10.1016/S0951-8320(00)00047-8

Bian, T., Zheng, H., Yin, L., & Deng, Y. (2018). Failure mode and effects analysis based on D numbers and TOPSIS. *Quality and Reliability Engineering International*, *34*(4), 501–515. doi:10.1002/qre.2268

Boran, F. E., Genç, S., Kurt, M., & Akay, D. (2009). A multi-criteria intuitionistic fuzzy group decision making for supplier selection with TOPSIS method. *Expert Systems with Applications*, *36*(8), 11363–11368. doi:10.1016/j.eswa.2009.03.039

Büyüközkan, G., & Güleryüz, S. (2016). Multi criteria group decision making approach for smart phone selection using intuitionistic fuzzy TOPSIS. *International Journal of Computational Intelligence Systems*, *9*(4), 709–725. doi:10.1080/18756891.2016.1204119

Chiozza, M. L., & Ponzetti, C. (2009). FMEA: A model for reducing medical errors. *Clinica Chimica Acta*, *404*(1), 75–78. doi:10.1016/j.cca.2009.03.015 PMID:19298799

Dag, M., Yalçın, Z. G., Oz, A., & Aydoğmuş, E. (2018). Risk Analysis by FMEA Method in Hotel. *Journal of Physical Chemistry and Functional Materials*, *1*(1), 63–71.

Fattahi, R., & Khalilzadeh, M. (2018). Risk evaluation using a novel hybrid method based on FMEA, extended MULTIMOORA, and AHP methods under fuzzy environment. *Safety Science*, *102*, 290–300. doi:10.1016/j.ssci.2017.10.018

Geum, Y., Cho, Y., & Park, Y. (2011). A systematic approach for diagnosing service failure: Service-specific FMEA and grey relational analysis approach. *Mathematical and Computer Modelling*, *54*(11-12), 3126–3142. doi:10.1016/j.mcm.2011.07.042

Güler, E., Avci, S., & Aladağ, Z. (2019). Multi-Criteria Decision Approach With AHP And IF-TOPSIS Methods For R&D Project Selectıon Process. *Journal of Engineering Studies and Research*, *25*(3), 22–32. doi:10.29081/jesr.v25i3.328

Jahan, A. (2019). Analysis of failure modes and solutions in the process of open-heart surgery using fuzzy FMEA and fuzzy TOPSIS techniques. *Japan-Hospitals*, *18*(1), 31–42.

Kilic, B., & Ok, S. (2012). Otel işletmelerinde müşteri şikayetleri ve şikayetlerin değerlendirilmesi. *Journal of Yasar University*, *25*(7), 4189–4202.

Liu, H. C., Liu, L., & Liu, N. (2013). Risk evaluation approaches in failure mode and effects analysis: A literature review. *Expert Systems with Applications*, *40*(2), 828–838. doi:10.1016/j.eswa.2012.08.010

Liu, H. C., Liu, L., Liu, N., & Mao, L. X. (2012). Risk evaluation in failure mode and effects analysis with extended VIKOR method under fuzzy environment. *Expert Systems with Applications*, *39*(17), 12926–12934. doi:10.1016/j.eswa.2012.05.031

Liu, H. C., You, J. X., Lin, Q. L., & Li, H. (2015). Risk assessment in system FMEA combining fuzzy weighted average with fuzzy decision-making trial and evaluation laboratory. *International Journal of Computer Integrated Manufacturing*, *28*(7), 701–714. doi:10.1080/0951192X.2014.900865

Liu, H. C., You, J. X., Shan, M. M., & Shao, L. N. (2015). Failure mode and effects analysis using intuitionistic fuzzy hybrid TOPSIS approach. *Soft Computing*, *19*(4), 1085–1098. doi:10.100700500-014-1321-x

Lo, H. W., & Liou, J. J. (2018). A novel multiple-criteria decision-making-based FMEA model for risk assessment. *Applied Soft Computing*, *73*, 684–696. doi:10.1016/j.asoc.2018.09.020

Maheswaran, K., & Loganathan, T. (2013). A novel approach for prioritization of failure modes in FMEA using MCDM. *International Journal of Engineering Research and Applications*, *3*(4), 733–739.

Memari, A., Dargi, A., Jokar, M. R. A., Ahmad, R., & Rahim, A. R. A. (2019). Sustainable supplier selection: A multi-criteria intuitionistic fuzzy TOPSIS method. *Journal of Manufacturing Systems*, *50*, 9–24. doi:10.1016/j.jmsy.2018.11.002

Nyaoga, R., Magutu, P., & Wang, M. (2016). Application of Grey-TOPSIS approach to evaluate value chain performance of tea processing chains. *Decision Science Letters*, *5*(3), 431–446. doi:10.5267/j.dsl.2016.1.002

Rau, H., Liao, C. P., Shiang, W. J., & Lin, C. H. (2007, June). Using fuzzy theory for packaging attribute deployment for new notebook computer introduction. In *International Conference on Industrial, Engineering and Other Applications of Applied Intelligent Systems* (pp. 250-259). Springer. 10.1007/978-3-540-73325-6_25

Rotondaro, R. G., & De Oliveira, C. L. (2001). *Using failure mode effect analysis (FMEA) to improve service quality service operations management*. Academic Press.

Saaty, T. L. (1980). *The analytic Hierarchy Process*. McGraw-Hill.

Senel, M., Senel, B., & Havle, C. A. (2018). Risk analysis of ports in Maritime Industry in Turkey using FMEA based intuitionistic Fuzzy TOPSIS Approach. In *ITM Web of Conferences* (Vol. 22, p. 01018). EDP Sciences.

Seyed-Hosseini, S. M., Safaei, N., & Asgharpour, M. J. (2006). Reprioritization of failures in a system failure mode and effects analysis by decision making trial and evaluation laboratory technique. *Reliability Engineering & System Safety*, *91*(8), 872–881. doi:10.1016/j.ress.2005.09.005

Sisman, B. (2019). Risk Evaluating by Fuzzy AHP and Fuzzy VIKOR methods in failure mode and effects analysis for automotive sector. *Mehmet Akif Ersoy Üniversitesi Sosyal Bilimler Enstitüsü Dergisi*, *9*(18), 234–250.

Song, W., Ming, X., Wu, Z., & Zhu, B. (2013). Failure modes and effects analysis using integrated weight-based fuzzy TOPSIS. *International Journal of Computer Integrated Manufacturing*, *26*(12), 1172–1186. doi:10.1080/0951192X.2013.785027

Srichetta, P., & Thurachon, W. (2012). Applying fuzzy analytic hierarchy process to evaluate and select product of notebook computers. *International Journal of Modeling and Optimization*, *2*(2), 168–173. doi:10.7763/IJMO.2012.V2.105

Szmidt, E., & Kacprzyk, J. (2009). Amount of information and its reliability in the ranking of Atanassov's intuitionistic fuzzy alternatives. In *Recent advances in decision making* (pp. 7–19). Springer. doi:10.1007/978-3-642-02187-9_2

Tooranloo, H. S., & Sadat Ayatollah, A. (2016). A model for failure mode and effects analysis based on intuitionistic fuzzy approach. *Applied Soft Computing*, *49*, 238-247.

Uyanik, C., Tuzkaya, G., Kalender, Z. T., & Oguztimur, S. (2020). An integrated DEMATEL–IF-TOPSIS methodology for logistics centers' location selection problem: An application for Istanbul Metropolitan area. *Transport*, *0*(0), 1–9. doi:10.3846/transport.2020.12210

Wang, L. E., Liu, H. C., & Quan, M. Y. (2016). Evaluating the risk of failure modes with a hybrid MCDM model under interval-valued intuitionistic fuzzy environments. *Computers & Industrial Engineering*, *102*, 175–185. doi:10.1016/j.cie.2016.11.003

Xu, Z., & Liao, H. (2013). Intuitionistic fuzzy analytic hierarchy process. *IEEE Transactions on Fuzzy Systems*, *22*(4), 749–761. doi:10.1109/TFUZZ.2013.2272585

Zadeh, L. A. (1965). Fuzzy sets. *Information and Control*, *8*(3), 338–353. doi:10.1016/S0019-9958(65)90241-X

Zammori, F., & Gabbrielli, R. (2012). ANP/RPN: A multi criteria evaluation of the risk priority number. *Quality and Reliability Engineering International*, *28*(1), 85–104. doi:10.1002/qre.1217

ADDITIONAL READING

Carlson, C. S. (2014, January). Understanding and applying the fundamentals of FMEAs. In *Annual Reliability and Maintainability Symposium* (Vol. 10, pp. 1-35).

Gilchrist, W. (1993). Modelling failure modes and effects analysis. *International Journal of Quality & Reliability Management*, *10*(5), 02656719310040105. doi:10.1108/02656719310040105

Kahraman, C. (Ed.). (2008). *Fuzzy multi-criteria decision making: theory and applications with recent developments* (Vol. 16). Springer Science & Business Media. doi:10.1007/978-0-387-76813-7

Mandal, S., & Maiti, J. (2014). Risk analysis using FMEA: Fuzzy similarity value and possibility theory-based approach. *Expert Systems with Applications*, *41*(7), 3527–3537. doi:10.1016/j.eswa.2013.10.058

McCain, C. (2006). Using an FMEA in a service setting. *Quality Progress*, *39*(9), 24.

Saaty, T. L. (2000). *Fundamentals of decision making and priority theory with the analytic hierarchy process* (Vol. 6). RWS publications.

Stevenson, W. J., Hojati, M., & Cao, J. (2007). *Operations management* (Vol. 8). McGraw-Hill/Irwin.

Sutrisno, A., & Lee, T. J. (2011). Service reliability assessment using failure mode and effect analysis (FMEA): Survey and opportunity roadmap. *International Journal of Engineering Science and Technology*, *3*(7), 25–38.

KEY TERMS AND DEFINITIONS

Failure Mode and Effect Analysis: A structured approach to identify potential product or process failures.

Fuzzy Logic: A form of multi-valued logic obtained from the fuzzy set theory to deal with reasoning that is approximate rather than precise.

Multi-Criteria Decision Making: Decision-making methods that use a set of criteria to evaluate alternatives based on expert opinions.

Quality: Fitness for purpose or degree to meet customer requirements of a product or service.

Risk: The unexpected situations that can cause undesirable or negative consequences.

Risk Assessment: Evaluation of potential failure modes.

Service Operations: All the activities that are organized and performed to provide a service product.

Chapter 4
Fuzzy Reliability Theory in the Decision-Making Process

Halid Kaplan
University of Texas at San Antonio, USA

Muhammed Can
https://orcid.org/0000-0002-6717-2611
University of Minho, Portugal

ABSTRACT

In the process of dealing with various problems that are related to the control and management of complex systems, uncertainty is always the issue. Most of the decisions to be made by engineers and governors are subject to a lack of data that causes uncertainty. Some information is not always accessible and insufficient at the time of decision-making (DM). It is important to use the concept of a human being's expert knowledge, so developing models to estimate and imitate human decision-making systems has been becoming necessary. This chapter will critically assess the fuzzy reliability theory and systems to demonstrate how fuzzy logic represents human behavior and non-linearity. The authors created a fuzzy inference system to model two different complex systems. Fuzzy approaches to real problems in DM are effective alternatives to traditional approaches. Fuzzy integrations improve DM models in five principal features: (1) expert knowledge, (2) uncertainty handling, (3) human and government behavior modeling, (4) flexible modeling, and (5) simpler representations.

DOI: 10.4018/978-1-7998-7564-2.ch004

INTRODUCTION

In the process of dealing with various problems that are related to the control and management of complex systems, uncertainty is always the issue. The management of uncertainties has been one of the main issues for decision makers, including engineers and governor for years. Most of the decisions to be made by engineers and governors are subject to lack of data that causes the uncertainty. Some information is not always accessible and insufficient at the time of decision making (Derrida, 2001). Expert judgment is essential in many decisions-making models, where the parameters and their relationships are complex. There are numerous decision-making tools developed for dealing with the uncertain parameters. Type of decision-making methods can be divided into two - probabilistic approach and possibilists approach. Decision-making methods and approaches use the assumption. It is important to use the concept of a human being's expert knowledge, so developing models to estimate and imitate human decision-making systems has been becoming necessary. As a result, numerous studies have been achieved to make decision models to be used as human-thinking. Uncertainties of human behaviour and having an insufficient information to make a decision are the biggest issue in decision-making models.

Uncertainty can be demonstrated in many forms;

- Fuzzy (not crisp, unclear, imprecise, estimated),
- Vague (not specific, unstructured),
- Ambiguous (too many choices),
- Natural variability (contradictory, random, chaotic, unpredictable).

Decision-making model uses fuzzy logic can handle the uncertainties of human behavior in decision-making. Fuzzy logic provides to employ vague decision-making problems (Zadeh,1965). The first progresses of alternatives to probability theory and traditional logic to handle uncertainty started in the twentieth century. The discrete logic was developed by Jan Lukasiewicz in 1930 (Ross, 2016). In the 1967, theory of evidence that process under lack of information was developed by Arthur Dempster (Ross, 2016). Researches of recent years illustrate the usefulness of fuzzy theory for considering varied kinds of uncertainty (Zimmermann, 2010). Fuzzy logic is introduced as a mathematical discipline by L. A. Zadeh in 1965 in his article (Zadeh,1965) which contradicts the traditional logic based on True or False. His concept handles with impreciseness and uncertainty. The fuzzy logic is derivative of the probability theory this makes fuzzy theory more suitable to be used in fields such as engineering, management, social and governance. In recent years, abundant research papers have been published on fuzzy theory, and the concept of the fuzzy theory has advanced in diverse directions. Fuzzy theory has high potential to understand the systems that are complex systems. The systems have human related parameters such as biological and cognitive can be counted as a complex system. Social and political systems can be an example of complex systems since the relationships between inputs and outputs are imprecise. In this complex system, the relationships between inputs and outputs is hard to determine but can be observed. One of the most characteristic concepts of fuzzy reliability theory is decision-making models. In practical, most of the decision-making models don't require exact outputs. In this case, we could use a fuzzy theory, which can handle uncertainties and nonlinearities Fuzzy reliability theory carries a new understanding of how AI and decision-making models can be implemented with higher efficiency and reliability.

FUZZY LOGIC AND RELIABILITY THEORY

The applications of fuzzy logic started after Zadeh's publications in 1965 (Zadeh,1965). These applications are in control systems, artificial intelligence, home electronics, automobile industry etc. (Zadeh,1978). Artificial intelligence (AI) researchers use fuzzy logic as a human behaviour model to simulate human behaviour and perception. Fuzzy logic models and fuzzy control methods are used in numerous engineering systems such as automobile sectors, robotics, and navigation systems. The strength of the fuzzy reliability theory in dealing with complex nonlinear systems comes forward due to its expert knowledge-based structures (Huang, 1997). Reliability of the systems have precise measurements can be calculated when all probabilities are determinable. In the most practical applications, it is difficult to have determinable possibilities because of imprecise parameters. An improvement to the imprecise reliability applications of imprecise probabilities in terms of reliability is provided by (Utkin & Coolen, 2007). System variables and their interrelationships are fuzzy. Reliability is one of the key factors to get better performance of a system which is designed to deal with the uncertainty. Fuzzy sets provide a way to represent impreciseness mathematically (Zadeh, 1978). Determining the optimal type and method of membership function, fuzzification and defuzzification effects the performance of the fuzzy analysis. In this chapter, the theoretical background of fuzzy sets and their basic processes will be described.

Fuzzy processing is consisted of three components; fuzzification, fuzzy inference (Rules, membership functions) and defuzzification (Zadeh,1965). In the fuzzification process, input variables are transformed into fuzzy sets that the elements have a varying degree of membership (between 0 and 1). When imprecision is occurred in systems, the variable will be fuzzy which can be represented by membership function. Scalar inputs should be converted into membership function by fuzzification process, and these inputs can be used in a fuzzy system. The next step is fuzzy inference which have sets of fuzzy if-then-else rules used to process inputs. In this process fuzzy inference generates the outputs by processing inputs depending on creating rules. The last step is defuzzification, where fuzzy outputs are converting to crisp results. A fuzzy set is any set having an interval between zero to one, allows its member to have a different degree of membership. Information in the fuzzy set is described by its membership function. It is significant to determine the feature of membership function.

DECISION MAKING (DM) MODELS

Making decision has a big significance in the most scientific and social environments. The goal of decision making is to be able to make reliable and correct selections in uncertain cases. The issues in decision making under uncertainty are that the information about the possible outputs, changeable conditions and preferences for each process is fuzzy. The growth of methodologies to improve the ability in decision-making models concerning non-probabilistic uncertainty is inevitable. Fuzzy analysis has applications in decision-making problems. In addition to that, Fuzzy analysis has proved to be applicable in expert systems and other AI applications to model and capture human decision procedures. Fuzzy is modeled after the human cognitive process like using perceptions, knowledge and experiences. Fuzzy logic uses the if-then types of production rules such as human problem-solving ability (Simon & Newell, 1971). The rule-based systems require many rules which should be appropriate with the sets of inputs of fuzzy sets (Cai et al., 1993). After a problem statement, the next step of Fuzzy reliability in DM is identifying the variables, possibilities and principles. Rules define the behavior of the DM by the conditions. Creating rules using human experts is the most effective and time-consuming part of the fuzzy models. We have considered reliability problem of systems and planned to implement fuzzy models to find optimal system reliability in decision making (DM) to decrease the system cost. Effective implementation should certainly benefit to improve the process of DM with respect to the defined goal. Fuzzy logic allows to find a solution for the certain decision-making cases on the basis created rules being set for similar cases.

EXPERIMENTAL DESIGN OF FUZZY RELIABILITY IN DECISION MAKING

In this chapter we would like to illustrate the process of fuzzy reliability in decision-making models and survey the possible fields of DM with fuzzy logic. Additionally, we would like to illustrate how fuzzy reliability theory is useful in DM by modelling the fuzzy inferences for both engineering control systems and foreign policy decision. The capability of creating fuzzy rules depending on datasets and human experts directly affects the performance of the DM models. To get better performance in DM models, considering all possible variables and other conditions to improve the fuzzy model is crucial. In this chapter we would like to demonstrate that fuzzy reliability theory and associated decision making have found widespread application in many disciplines since the influential contributions of Zadeh and his followers.

Making decision in political and social systems method was introduced by using fuzzy cognitive mapping in (Axelrod, 1977). FCMs was implemented for potential impact of Brexit (Dincer et al.,2018). Fuzzy cognitive maps (FCMs) model interrelationships and provide knowledge to make predictions and strategic plans. FCMs are useful to manage the uncertainty and complexity by visualizing relationships. This helps researchers to validate their rational models compare to reality. FCMs were used to create an estimate model about electric power demands (Lincoln and Ross, 2012). Like estimations of power demands, solar and wind forecasting play important role decision making process for renewable energy plants (Kaplan, 2014). Fuzzy cognitive maps offer an empirical approach for evaluating complex systems with clustering techniques and improve the DM models.

Agent-based modeling (ABM) was introduced to mimic the flocking behavior of birds by Craig Reynolds in 1987 (Ross, 2016). Another key modelling method for making decision is Agent-based modeling (ABM) provides a methodology to represent the complex systems (Kabiri Chimeh et al., 2019). ABM is a unique method uses both empirical and inductive reasoning to determining the rules governing complex systems. Unreasonable decision making is often appeared because of the lack of knowledge on concerning the costs of the decisions. Using fuzzy logic in ABM can be decrease the rate of unreasonable decisions. Fuzzy logic provides a hypothetical concept to handle the presence of the uncertainty which is the main reason of unreasonable decisions.

The model was created in (Ross, 2016) illustrates how fuzzy rule-based approach improves agent's ability to get desired outputs. Fuzzy approach is more like human thinking and reasoning that increase performance of decision-making model in complex environments.

Decision Making Model with Fuzzy in Engineering Areas

The uncertainty issue has been one of the main problems of the decision makers such as engineers and scientists for years (Attoh-Okine & Ayyub, 2005). Fuzzy logic control theory has been used widely in engineering applications such as robotics, information and communication technology, image processing, energy-based forecasting and control systems. We can classify the principal fuzzy engineering applications under five; fuzzy control systems, fuzzy optimization, fuzzy classification, fuzzy cognitive mapping and system analyzing. The electrical fault classifier uses fuzzy reinforcement learning was implemented for transformer faults that can be helpful in decision making process of the electrical transmission lines (Malik et al., 2020). Wind turbines are complex systems have sensors, and complex components require the faults detection to protect the systems by making healthy decisions. (EL Bakri & Boumhidi 2018). The optimized IoT based decision-making model for autonomous vehicles illustrates the how AI tools provide decision-making models (Sahba, 2019). Fuzzy provides advantages to expert knowledge without needing to have mathematical structures of the systems. In the energy sector, decision maker meets with a high level of uncertainty in datasets. By combining fuzzy tools with neural network, performance of the decision-making model can be improved. An adaptive fuzzy-neuro approach is created to design the maintenance service network (Ren, 2020).

In this chapter we will illustrate how fuzzy theory can be used in energy and control systems with by designing a basic fuzzy model for decision making. Current and voltage data were obtained from experimental electric vehicle (EV) to use as fuzzy sets. We used MATLAB for implementation of the fuzzy system. Dataset is created with 12,000 current information.

Table 1. Description of the fuzzy system

Linguistic Variable	Conditions	Membership Function
Voltage	Low	Trapezoidal
	Optimum	Trapezoidal
	High	Trapezoidal
Currents	Low	Trapezoidal
	Optimum	Trapezoidal
	High	Trapezoidal

Proposed method can be used as power management and fault control systems by designating the limitation for current and voltage. This can be used to protect system for electrical fault by detecting the abnormal current/voltage changes. In addition to that, this approach allows us to power saving.

Figure 1. Input variable a) Voltage, b) Current

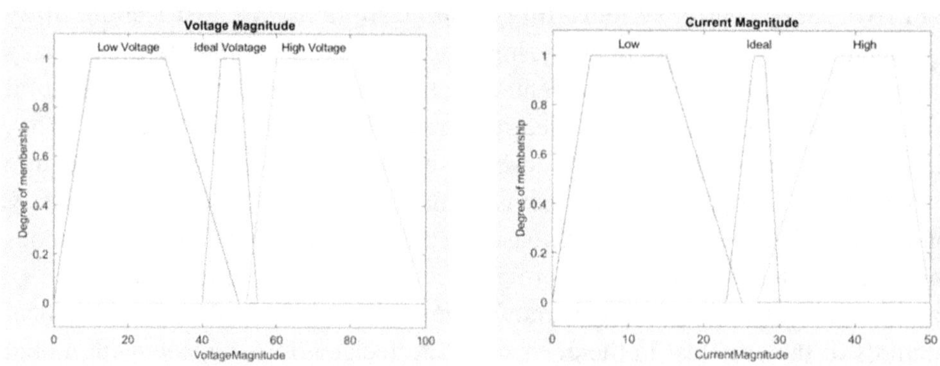

Figure 2. Fuzzy inference system

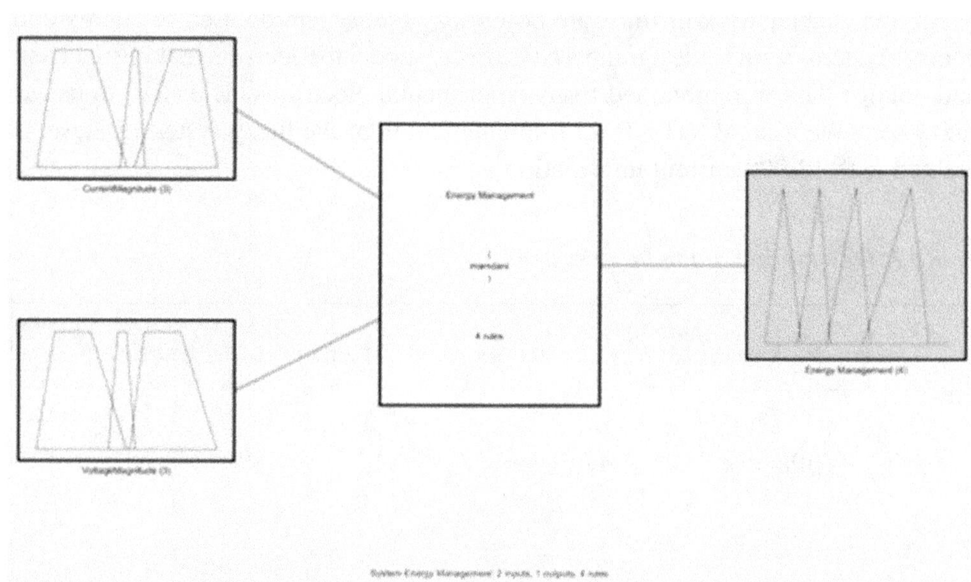

To validate our fuzzy model ANFIS (Artificial Neural Fuzzy Inference System) is simulated same current data set. We got training root-mean-square error (RMSE) around 2.35 and checking RMSE as 5.35. In figure 5, errors are mostly around 5 and reach 15. This rate of error is acceptable for current around 200. Results illustrate that performance of this approach is well with an acceptable error rate.

Figure 3. Outputs of fuzzy model

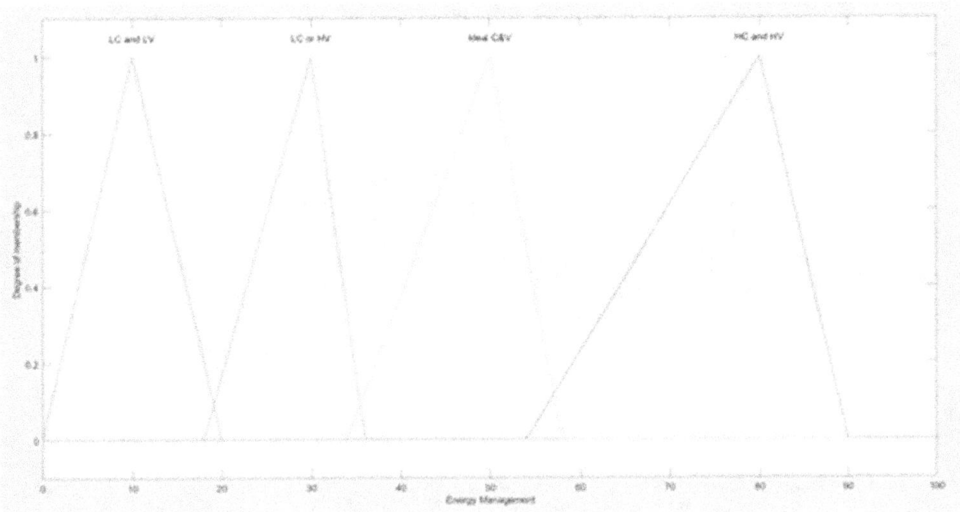

Figure 4. Current estimation with ANFIS and actual current

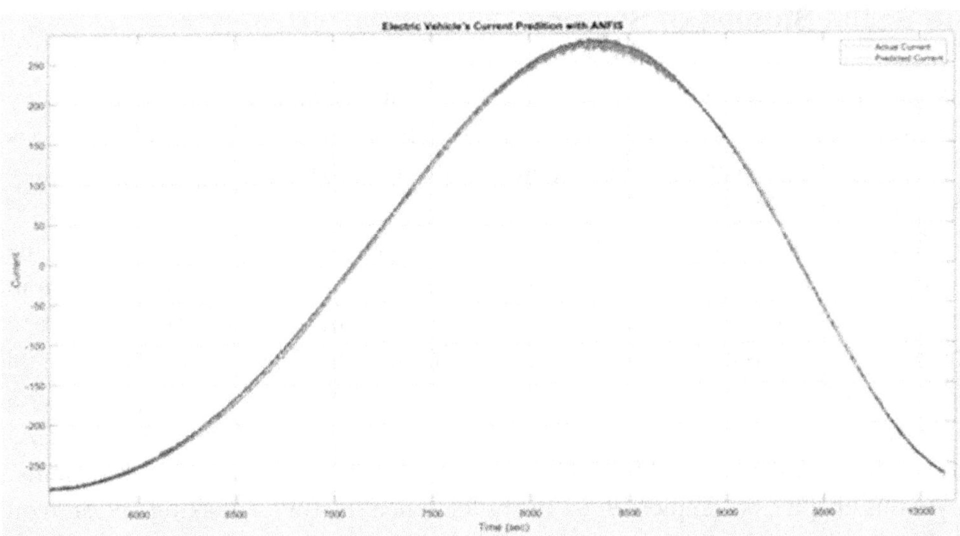

Figure 5. Prediction error

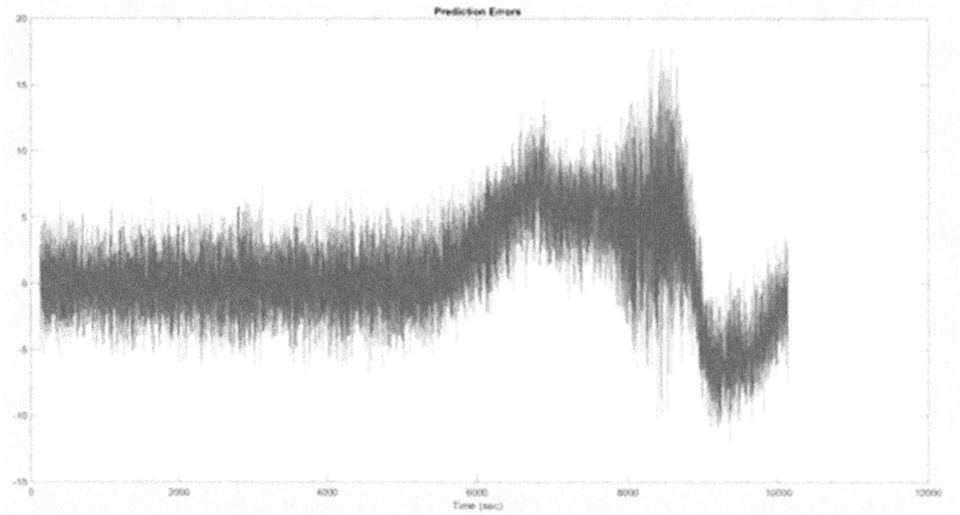

Decision Making Model with Fuzzy in the Social Studies of Science

The literatures on models for decision making in management; in economic policy (Dincer et al., 2016), climate policy (Shackley, 2001); in human reliability analysis (HRA) (Baziuk, 2016); financial markets (MacKenzie, 2006), and health-care policy (Mansnerus, 2012).

Decision making in politics and management differs from in engineering one. The main reason is the amount of uncertainty because of human behaviors. As seen in figure 6, fuzzy theory has been used mostly in engineering and industries compare to social science that is only% 1(Baziuk, 2016). We point to the next step of research on decision-making models pointing to contribute to all possible developing literature. Knowledge about the future is incomplete but this can be estimated by using AI such as fuzzy logic, inevitability of decision for future is not same for all systems.

In this chapter, we implemented fuzzy inference to provide making decision on international politics. We created a pragmatic model about the global position of China and US in politics.

We consider five main power of countries; economic power, technological power, military power, global political power and soft power. Consequently, there are explicit ways to example operative models and validate their performances.

Proposed fuzzy model was updated for decision making for future global position of countries. Fuzzy rules need the human expert. This can be historical data, past

Figure 6. Comparison between fuzzy in social science and principal fuzzy applications

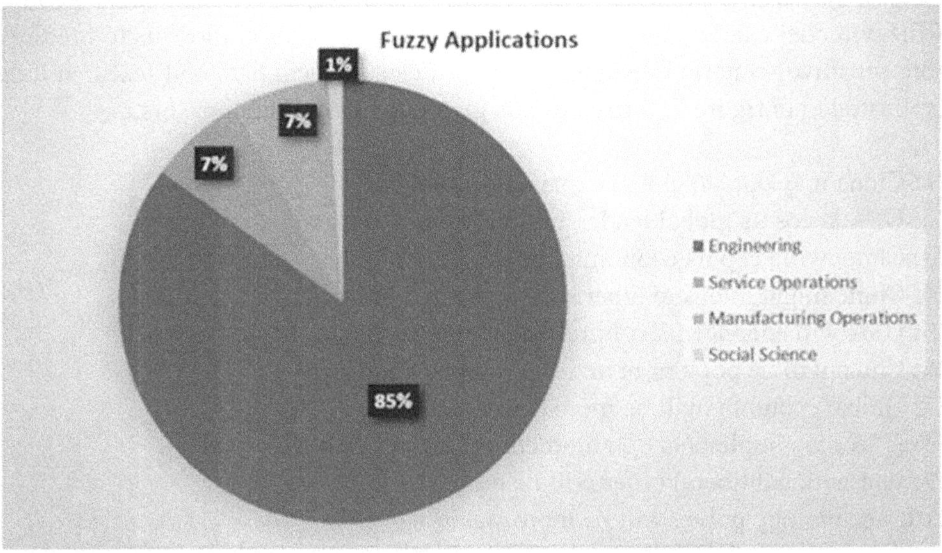

events, government behavior and statistical information etc. Weight of the fuzzy inference system assigned depending on the past government reactions and politics. As outputs we determined ten empirical cases.

We created ten cases which are represented as C1 to C 10 by using countries' economic power (EP), military power (MP), technological power (TP), global

Table 2. Fuzzy Rules

		China									
		Low EP	High EP	Low TP	High TP	Low MP	High MP	Low GP	High GP	Low SP	High SP
USA	Low EP		C1								
	High EP	C 7	C2								
	Low TP										
	High TP			C 2, C 10							
	Low MP						C 4				
	High MP					C 8					
	Low GP								C 3		
	High GP							C 9			
	Low SP										C6
	High SP									C 5	

position power (GP) and soft power (SP) in table 2. For each rule have weights in fuzzy systems. By adding more scenarios and proper weights, the capability of fuzzy model can be increased. Rule operators like and/or allow us to predict a more sensitive scenario depending on previous statistical data and research data. As illustrated in figure 7, fuzzy outputs give ideas to decision makers.

C1: China may pursue global expansionist policy
C2: USA keeps its global leadership position in Tech
C3: China will keep its economic prowess
C4: China might interfere other states' internal politics
C5: USA will enhance its cultural hegemony
C6: China will be popular in underdeveloped countries
C7: Global economy will be more stable
C8: USA may implement containment policy in south china sea
C9: International liberal order will be more stable
C10: Technology policy will be more stable

Figure 7. The output surface for the fuzzy inference system

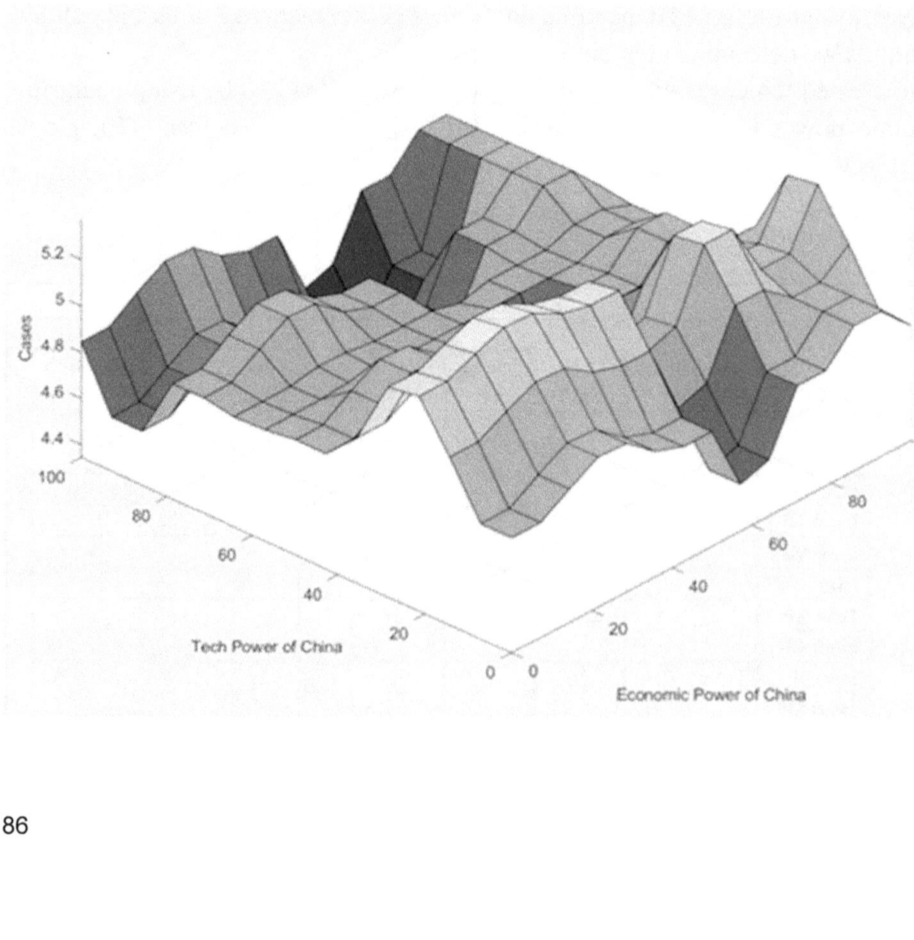

CONCLUSION

Due to the amount of the reachable data is growing exponentially, the structure of data is becoming much more complex. However, this causes the new challenge for DM processes to deal with this complexity, applying more useful data in fuzzy systems improves the performance of DM is advanced with Fuzzy theory. Fuzzy models are not that different from more other AI models. In some cases, fuzzy gives better performance, and in some case it doesn't. It is significant to get vital variables for each system. The performance of the model may not be same for all cases and all inputs. Depending on inputs, rules need to be updated. After the model is built validation and testing are also important steps. Validation of model is difficult because of collecting improper empirical data. However, fuzzy provides ideas on possible conclusion to make a decision. It is important to examine the decisions that can be suitable with models. There are two main factors to improve fuzzy integrated DM models; (1) an efficient data collection and (2) proper fuzzy rules. In addition to that, when uncertainties prevent decision makers to build an efficient model, fuzzy allows them to create a model that gives exact or approximate results in decision making.

As a consequence, we illustrated how fuzz theory can be integrated into the decision-making models and demonstrate the strengths of the fuzzy theory in DM in both social and scientific fields with the experimental results. Fuzzy approaches to real problems in DM are an effective alternative to traditional approaches. Fuzzy integrations improve DM models in five principal features: (1) expert knowledge, (2) uncertainty handling, (3) human and government behavior modelling, (4) flexible modelling (5) simpler representations.

The main advantage of using fuzzy sets in DM modelling for social science is to process the uncertainty associated with subjective variables and empirical variables. DM with fuzzy reliability, probability and binary logic are replaced by possibility and fuzzy sets. In the existing literature, currency, stock and credit score estimation is translated into a decision to take action. Definitely, although the integration of fuzzy in DM models may suffer from numerous boundaries, there is a significant involvement in DM for political and management, not just engineering. An FT enhances the DM models by allowing concepts to be represented vague with a fuzzy set, instead of needing them to be exact.

REFERENCES

Attoh-Okine, N., & Ayyub, B. (2005). Applied Research in Uncertainty Modeling and Analysis. In *Applied Research in Uncertainty Modeling and Analysis* (1. Aufl., Vol. 20). Springer Netherlands. doi:10.1007/b101807

Axelrod, R. (1977). *Structure of decision: the cognitive maps of political elite*. Princeton University Press.

Bakri, E. L. A., & Boumhidi, I. (2018). Fuzzy model-based faults diagnosis of the wind turbine benchmark. *Procedia Computer Science, 127*, 464–470. doi:10.1016/j.procs.2018.01.144

Baziuk, P., Rivera, S., & Núñez McLeod, J. (2016). Fuzzy Human Reliability Analysis: Applications and Contributions Review. *Advances in Fuzzy Systems, 2016*, 1–9. https://doi.org/10.1155/2016/4612086

Cai, K., Wen, C., & Zhang, M. (1993). Fuzzy states as a basis for a theory of fuzzy reliability. *Microelectronics and Reliability, 33*(15), 2253–2263. doi:10.1016/0026-2714(93)90065-7

Derrida, J. (2001). *Limited Inc*. Passagen.

Dincer, H., Yüksel, S., & Senel, S. (2018). Analyzing the Global Risks for the Financial Crisis after the Great Depression Using Comparative Hybrid Hesitant Fuzzy Decision-Making Models: Policy Recommendations for Sustainable Economic Growth. *Sustainability (Basel, Switzerland), 10*(9), 3126– doi:10.3390u10093126

Huang, H. (1997). Fuzzy multi-objective optimization decision-making of reliability of series system. *Microelectronics and Reliability, 37*(3), 447–449. doi:10.1016/S0026-2714(96)00040-6

Kabiri Chimeh, M., Heywood, P., Pennisi, M., Pappalardo, F., & Richmond, P. (2019). Parallelisation strategies for agent based simulation of immune systems. *BMC Bioinformatics, 20*(S6), 579–14. https://doi.org/10.1186/s12859-019-3181-y

Kaplan, H. (2014). *Data analytic studies for Turkey's energy forecast*. ProQuest LLC.

Lincoln, D., & Ross, T. (2012). Evaluation of commercial facility demand response using genetically evolved fuzzy cognitive mapping. *Int. J. Distributed Energy Resources, 8*(4), 315–342.

MacKenzie, D. (2006). *An engine, not a camera: how financial models shape markets*. MIT Press.

Malik, H., Sharma, R., & Mishra, S. (2020). Fuzzy reinforcement learning based intelligent classifier for power transformer faults. *ISA Transactions, 101*, 390–398. https://doi.org/10.1016/j.isatra.2020.01.016

Mansnerus, E. (2012). Understanding and Governing Public Health Risks by Modeling. In Handbook of Risk Theory (pp. 213–237). Springer Netherlands; https://doi.org/10.1007/978-94-007-1433-5_9.

Ren, W., Wu, K., Gu, Q., & Hu, Y. (2020). Intelligent decision making for service providers selection in maintenance service network: An adaptive fuzzy-neuro approach. *Knowledge-Based Systems*, *190*, 105263. https://doi.org/10.1016/j.knosys.2019.105263

Ross, T. (2016). Fuzzy Logic with Engineering Applications. In *Fuzzy Logic with Engineering Applications*. John Wiley & Sons, Incorporated.

Sahba, A., Sahba, R., Rad, P., & Jamshidi, M. (2019). *Optimized IoT Based Decision Making for Autonomous Vehicles in Intersections*. doi:10.1109/UEMCON47517.2019.8992978

Shackley, S. (2001). Epistemic Lifestyles in Climate Change Modeling. In *Changing the Atmosphere: Expert Knowledge and Environmental Governance* (pp. 107–133). Cambridge: MIT Press.

Simon, H., & Newell, A. (1971). Human problem solving: The state of the theory in 1970. *The American Psychologist*, *26*(2), 145–159. doi:10.1037/h0030806

Utkin, L., & Coolen, F. (2007). Imprecise reliability: An introductory overview. In *Intelligence in Reliability Engineering* (Vol. 40, pp. 261–306). Springer Berlin Heidelberg. doi:10.1007/978-3-540-37372-8_10

Zadeh, L. (1965). Fuzzy sets. *Information and Computation*, *8*, 338–353.

Zadeh, L. (1978). Fuzzy sets as a basis for a theory of possibility. *Fuzzy Sets and Systems*, *1*(1), 3–28. doi:10.1016/0165-0114(78)90029-5

Zimmermann, H. (2010). Fuzzy set theory. *Wiley Interdisciplinary Reviews: Computational Statistics*, *2*(3), 317–332. doi:10.1002/wics.82

Ziv, G., Watson, E., Young, D., Howard, D., Larcom, S., & Tanentzap, A. (2018). The potential impact of Brexit on the energy, water and food nexus in the UK: A fuzzy cognitive mapping approach. *Applied Energy*, *210*, 487–498. https://doi.org/10.1016/j.apenergy.2017.08.033

Chapter 5
Risk Evaluation of a Construction Building Using a Hybrid Approach

Burak Efe
Necmettin Erbakan University, Turkey

Mustafa Kurt
Gazi University, Turkey

ABSTRACT

This chapter aims basically to ensure two benefits, which focus on method and application. Firstly, this chapter proposes a new approach, which is TOPSIS (technique for order preference by similarity to ideal solution) based on interval type-2 trapezoidal fuzzy (IT2TrF) numbers. Secondly, the proposed method is applied to failure mode and effects analysis (FMEA) problem to overcome the drawbacks of traditional FMEA, which determines possible failure modes. In traditional FMEA, the risk priority number method, which is obtained as the product of occurrence, severity, and detection risk factors, has been debated to have several drawbacks. The implementation of risk assessment is realized to indicate the validity of the suggested approach in a construction building. The results of the proposed approach and some other methods are compared. The aim and resources are optimized by using linear programming.

DOI: 10.4018/978-1-7998-7564-2.ch005

INTRODUCTION

An occupational accident is an event that concerns not only the employees, but also the family, the firm and the state. Those who are exposed to work accidents and occupational diseases are employees but also have costs for the company and society. The material and moral difficulties experienced by the person and his / her family who suffered a work accident, the companies that suffered serious financial losses due to work accidents, the state which lost both the healthy and productive work force and the ones who lost the material were the parties who suffered as a result of these accidents. Failure mode and effects analysis (FMEA) is a widely utilized risk assessment method to identify and eliminate potential defects in manufacturing and service systems. The failure mode is defined as the situations in which a potential failure may occur in the part, subsystem, system, or process when performing the design purpose (Wang et al., 2009).

The difference between FMEA and other risk assessment tools is that, instead of finding a solution after a failure occurs, problems are eliminated in connection with the system's predetermination. This helps decision makers to correct existing programs, to take action to reduce the likelihood of failure, and to avoid dangerous accidents. Today, FMEA is widely used in many fields of industry (Efe et al., 2017).

A system, design, process or service usually has many failure modes and effects. In this case, it is necessary to prioritize and evaluate each failure mode and effect according to the risk values. The high-risk (or most dangerous) failure mode should be corrected first. The traditional FMEA uses risk priority number (RPN) method to determine the risk priorities of failure modes. The conventional FMEA determines the risk priorities of the failure modes with RPN, which is the product of the probability (O), severity (S), and detectability (D) of a failure mode.

$$RPN = O \times S \times D \tag{1}$$

This chapter consists of five parts. The second part handles literature review about FMEA. The used methods are presented in Part three. Part four shows an application to test the validity of the presented approach. The last part introduces the conclusion of the study.

LITERATURE REVIEW

Different methods have been presented to cope with the drawbacks of the traditional FMEA. The related papers are introduced in Table 1.

Table 1. Papers about FMEA

Methods	Authors
Sezgisel bulanık küme- DEMATEL	Chang and Cheng (2010)
Bulanık AHP- Bulanık TOPSIS	Kutlu and Ekmekçioğlu (2012)
Bulanık AHP- Entropi-Bulanık VIKOR	Liu et al. (2015)
Sezgisel bulanık küme- Öklid mesafe operatörü	Liu et al. (2014)
Fuzzy evidential reasoning and belief rule-based approach	Liu et al. (2013)
Fuzzy inference system	Jee et al. (2015)
Fuzzy and grey theories	Zhou and Thai (2016)
Hesitant 2-tuple linguistic term sets and an extended QUALIFLEX approach	Liu et al. (2016)
Z numbers based AHP, entropy and VIKOR methods	Mohsen and Fereshteh (2017)
Fuzzy PROMETHEE	Efe et al. (2016)
Intuitionistic fuzzy AHP - VIKOR methods	Efe et al. (2017)
Fuzzy best-worst, relative entropy, VIKOR	Tian et al. (2018)
MULTIMOORA (Multi-objective Optimization By Ratio Analysis) and AHP	Fattahi and Khalilzadeh (2018)
Quality function deployment and intuitionistic fuzzy VIKOR	Efe (2019b)
Fuzzy ANP	Yazdani et al. (2019)
Intuitionistic fuzzy best-worst method	Yazdi et al. (2020)
Interval type-2 fuzzy evidential reasoning method	Qin et al. (2020)
Regret theory and PROMETHEE under linguistic neutrosophic context	Zhu et al. (2020)

Fuzzy logic has been handled for different areas in the literature. Shtovba et al. (2002) presented fuzzy rule based system for diagnosis of stone construction cracks of buildings. Efe (2020) handled fuzzy AHP and fuzzy TOPSIS for joint health and safety unit selection. Efe and Efe (2018) introduced intuitionistic fuzzy AHP and intuitionistic fuzzy VIKOR approach for personnel selection. Maity et al. (2020) focused on a backorder economic order quantity model for cloud-type intuitionistic dense fuzzy demand rate.

The study has the following contributions:

1. In traditional FMEA, RPN, which is the product of O, S and D risk factors, has several limitations. In this study, IT2F number was used to determine the weights of O, S and S risk factors and the value of each failure mode. Therefore,

loss of information due to the utilization of fuzzy numbers or type-1 fuzzy number is prevented.

2. Although several multi-criteria decision-making approaches have been used, the TOPSIS approach with IT2F number has not previously been used in FMEA based risk assessment. The TOPSIS method uses crisp numbers while the fuzzy TOPSIS method uses fuzzy numbers to evaluate alternatives. In this study, TOPSIS method with IT2F number was used to model further uncertainty of subjective judgments.
3. The results of the multi-criteria decision-making approach based FMEA are inadequate because it does not take into account the firm capacity. This may cause some problems in the long term. This study proposes a linear programming approach that takes into account the firm capacities. Studies in the literature have only sought to improve FMEA and have ignored the firm capacities. This study both improved the FMEA approach and took into account the firm capacities. This study is a leading study for risk assessment studies based on FMEA.
4. This study presents 12 different situations in terms of an occupational health and safety policy. The construction firm can choose the most appropriate policy for itself. Since the firm consists of many processes, the flexibility to choose the most appropriate policy is very important.

Type-1 fuzzy numbers examine the exact degree of membership to express fuzzy numbers. However, IT2F number addresses more uncertainty than type-1 fuzzy numbers. IT2F number provides us with additional information to describe ambiguity and uncertainty in real life applications. This study aims to extend the TOPSIS method using IT2F number to model further uncertainty of subjective ideas. The goal of this work is to present a new TOPSIS approach based on IT2F number. To evaluate alternatives, IT2F number uses approximately positive and negative ideal solutions with trapezoidal fuzzy numbers. It uses possibility degrees instead of distance measurements in the current TOPSIS to determine a comparison index. The last stage of IT2F number based TOPSIS method achieves the possibility degree based closeness coefficient.

Another benefit of this study is that it overcomes the shortcomings of traditional FMEA. The RPN method, obtained by multiplying the values of risk factors such as O, S and D in traditional HTEA, has been criticized in the literature for its many deficiencies. In this study, the subjective opinions of decision makers are combined to determine the weights of these risk factors. These weights were used in the TOPSIS method to rank risks. The values obtained as a result of TOPSIS method are used in a linear programming which deals with firm capacities and an occupational health and safety policy is proposed for the firm. The proposed model is applied to assess the risks associated with ironwork in a construction.

THE PROPOSED METHOD

Interval Type-2 Fuzzy Sets

The addition, multiplication and exponent operations among $\tilde{\tilde{A}}_1$ and $\tilde{\tilde{A}}_2$ are defined below when k is a crisp number (Efe, 2019b):

$$\tilde{\tilde{A}}_1 \oplus \tilde{\tilde{A}}_2 = (\tilde{A}_1^U, \tilde{A}_1^L) \oplus (\tilde{A}_2^U, \tilde{A}_2^L) = ((a_{11}^U + a_{21}^U, a_{12}^U + a_{22}^U, a_{13}^U + a_{23}^U, a_{14}^U + a_{24}^U; \min\{H_1(\tilde{A}_1^U), H_2(\tilde{A}_1^U)\}), (a_{11}^L + a_{21}^L, a_{12}^L + a_{22}^L, a_{13}^L + a_{23}^L, a_{14}^L + a_{24}^L; \min\{H_1(\tilde{A}_1^L), H_2(\tilde{A}_1^L)\})) \quad (2)$$

$$\tilde{\tilde{A}}_1 \otimes \tilde{\tilde{A}}_2 = (\tilde{A}_1^U, \tilde{A}_1^L) \otimes (\tilde{A}_2^U, \tilde{A}_2^L) = ((a_{11}^U \times a_{21}^U, a_{12}^U \times a_{22}^U, a_{13}^U \times a_{23}^U, a_{14}^U \times a_{24}^U; \min\{H_1(\tilde{A}_1^U), H_2(\tilde{A}_1^U)\}), (a_{11}^L \times a_{21}^L, a_{12}^L \times a_{22}^L, a_{13}^L \times a_{23}^L, a_{14}^L \times a_{24}^L; \min\{H_1(\tilde{A}_1^L), H_2(\tilde{A}_1^L)\})) \quad (3)$$

$$k \times \tilde{\tilde{A}}_1 = k \times (\tilde{A}_1^U, \tilde{A}_1^L) = ((k \times a_{11}^U, k \times a_{12}^U, k \times a_{13}^U, k \times a_{14}^U; H_1(\tilde{A}_1^U)), (k \times a_{11}^L, k \times a_{12}^L, k \times a_{13}^L, k \times a_{14}^L; H_1(\tilde{A}_1^L))) \quad (4)$$

where $k>0$.

$$(\tilde{\tilde{A}}_1)^k = (\tilde{A}_1^U, \tilde{A}_1^L)^k = (((a_{11}^U)^k, (a_{12}^U)^k, (a_{13}^U)^k, (a_{14}^U)^k; H_1(\tilde{A}_1^U)), ((a_{11}^L)^k, (a_{12}^L)^k, (a_{13}^L)^k, (a_{14}^L)^k; H_1(\tilde{A}_1^L))) \quad (5)$$

If $\tilde{\tilde{A}} = (\tilde{A}^U, \tilde{A}^L) = ((a_1^U, a_2^U, a_3^U, a_4^U; h_U), (a_1^L, a_2^L, a_3^L, a_4^L; h_L))$ is a trapezoidal IT2F number, the lower possibility mean value is calculated below (Efe, 2019b):

$$M_*(\tilde{A}) = \frac{1}{6}(a_1^U + 2a_2^U)h_U^2 + \frac{1}{6}(a_1^L + 2a_2^L)h_L^2 \quad (6)$$

The upper possibility mean value is calculated below:

$$M^{*}(\tilde{A})=\frac{1}{6}(a_{4}^{U}+2a_{3}^{U})h_{U}^{2}+\frac{1}{6}(a_{4}^{L}+2a_{3}^{L})h_{L}^{2} \tag{7}$$

The possibility degree is calculated below:

$$p(\tilde{A}_1 \succ \tilde{A}_2)=\min\left\{\max\left(\frac{M^{*}(\tilde{A}_1)-M_{*}(\tilde{A}_2)}{M^{*}(\tilde{A}_1)-M_{*}(\tilde{A}_1)+M^{*}(\tilde{A}_2)-M_{*}(\tilde{A}_2)},0\right),1\right\} \tag{8}$$

Interval Type-2 Fuzzy TOPSIS Method

TOPSIS approach, which was proposed by Hwang and Yoon (1981), is used in this chapter. The presented approach in this study is shown below (Efe, 2019b):

Step 1: Determine the weight (W_l) of the criteria.

$$W_l=\left(\tilde{\tilde{w}}_i^l\right)_{1\times m}=\begin{matrix} f_1 & f_2 & \cdots & f_m \\ \end{matrix}\left[\begin{matrix}\tilde{\tilde{w}}_1^l & \tilde{\tilde{w}}_2^l & \cdots & \tilde{\tilde{w}}_m^l\end{matrix}\right] \tag{9}$$

$$\tilde{\tilde{w}}_i=\sum_{l=1}^{k} w_l\tilde{\tilde{w}}_i^l \tag{10}$$

w_l is the weight of the decision maker D_l. $\tilde{\tilde{w}}_i^l$ is the weight of criteria f_i according to the expert D_l.

Step 2: Determine the decision matrix Y_l of the expert D_l and establish the average decision matrix $\overline{Y}$.

$$Y_l=(\tilde{\tilde{f}}_{ij}^l)_{m\times n}=\begin{matrix} \\ f_1 \\ f_2 \\ \vdots \\ f_m \end{matrix}\begin{matrix} x_1 & x_2 & \cdots & x_n \\ \left[\begin{matrix} \tilde{\tilde{f}}_{11}^l & \tilde{\tilde{f}}_{12}^l & \cdots & \tilde{\tilde{f}}_{1n}^l \\ \tilde{\tilde{f}}_{21}^l & \tilde{\tilde{f}}_{22}^l & \cdots & \tilde{\tilde{f}}_{2n}^l \\ \vdots & \vdots & \vdots & \vdots \\ \tilde{\tilde{f}}_{m1}^l & \tilde{\tilde{f}}_{m2}^l & \cdots & \tilde{\tilde{f}}_{mn}^l \end{matrix}\right]\end{matrix} \tag{11}$$

$$\overline{Y}=(\overline{\tilde{\tilde{f}}}_{ij})_{m\times n} \tag{12}$$

$$\tilde{\tilde{f}}_{ij} = \sum_{l=1}^{k} w_l . \tilde{\tilde{f}}_{ij}^{l} \tag{13}$$

w_l is the weight of the decision maker D_l. $\tilde{\tilde{f}}_{ij}^{l}$ is the judgment of the decision maker D_l for $\tilde{\tilde{f}}_{ij}$.

Step 3: Determine the average weighted matrix $\bar{W}$.

$$\bar{W} = (\tilde{\tilde{v}}_{ij})_{m\times n} = \begin{matrix} & x_1 & x_2 & \cdots & x_n \\ f_1 & \tilde{\tilde{v}}_{11} & \tilde{\tilde{v}}_{12} & \cdots & \tilde{\tilde{v}}_{1n} \\ f_2 & \tilde{\tilde{v}}_{21} & \tilde{\tilde{v}}_{22} & \cdots & \tilde{\tilde{v}}_{2n} \\ \vdots & \vdots & \vdots & \vdots & \vdots \\ f_m & \tilde{\tilde{v}}_{m1} & \tilde{\tilde{v}}_{m2} & \cdots & \tilde{\tilde{v}}_{mn} \end{matrix} \tag{14}$$

where $\tilde{\tilde{v}}_{ij} = \tilde{\tilde{w}}_i \otimes \tilde{\tilde{f}}_{ij}, 1 \le i \le m,$ and $1 \le j \le n.$

Step 4: Determine the approximate positive and negative ideal solution $x^+ = (v_1^+, v_2^+, ..., v_m^+)$ and $x^- = (v_1^-, v_2^-, ..., v_m^-)$ according to benefit (F_1) and cost criteria (F_2).

Let be $\tilde{\tilde{v}} = (\tilde{v}^U, \tilde{v}^L) = ((v_1^U, v_2^U, v_3^U, v_4^U; h_U), (v_1^L, v_2^L, v_3^L, v_4^L; h_L)).$

$$\tilde{\tilde{v}}_i^+ = \begin{cases} \bigvee_{j=1}^{n} \tilde{\tilde{v}}_{1ij}^L, \bigvee_{j=1}^{n} \tilde{\tilde{v}}_{2ij}^L, \bigvee_{j=1}^{n} \tilde{\tilde{v}}_{3ij}^L, \bigvee_{j=1}^{n} \tilde{\tilde{v}}_{4ij}^L, \bigwedge_{j=1}^{n} \left\{H(\tilde{A}_{ij}^L)\right\}, \bigvee_{j=1}^{n} \tilde{\tilde{v}}_{1ij}^U, \bigvee_{j=1}^{n} \tilde{\tilde{v}}_{2ij}^U, \bigvee_{j=1}^{n} \tilde{\tilde{v}}_{3ij}^U, \bigvee_{j=1}^{n} \tilde{\tilde{v}}_{4ij}^U, \bigwedge_{j=1}^{n} \left\{H(\tilde{A}_{ij}^U)\right\}, \text{if } f_i \in F_1 \\ \bigwedge_{j=1}^{n} \tilde{\tilde{v}}_{1ij}^L, \bigwedge_{j=1}^{n} \tilde{\tilde{v}}_{2ij}^L, \bigwedge_{j=1}^{n} \tilde{\tilde{v}}_{3ij}^L, \bigwedge_{j=1}^{n} \tilde{\tilde{v}}_{4ij}^L, \bigwedge_{j=1}^{n} \left\{H(\tilde{A}_{ij}^L)\right\}, \bigwedge_{j=1}^{n} \tilde{\tilde{v}}_{1ij}^U, \bigwedge_{j=1}^{n} \tilde{\tilde{v}}_{2ij}^U, \bigwedge_{j=1}^{n} \tilde{\tilde{v}}_{3ij}^U, \bigwedge_{j=1}^{n} \tilde{\tilde{v}}_{4ij}^U, \bigwedge_{j=1}^{n} \left\{H(\tilde{A}_{ij}^U)\right\}, \text{if } f_i \in F_2 \end{cases} \tag{15}$$

$$\tilde{\tilde{v}}_i^- = \begin{cases} \bigwedge_{j=1}^{n} \tilde{\tilde{v}}_{1ij}^L, \bigwedge_{j=1}^{n} \tilde{\tilde{v}}_{2ij}^L, \bigwedge_{j=1}^{n} \tilde{\tilde{v}}_{3ij}^L, \bigwedge_{j=1}^{n} \tilde{\tilde{v}}_{4ij}^L, \bigwedge_{j=1}^{n} \left\{H(\tilde{A}_{ij}^L)\right\}, \bigwedge_{j=1}^{n} \tilde{\tilde{v}}_{1ij}^U, \bigwedge_{j=1}^{n} \tilde{\tilde{v}}_{2ij}^U, \bigwedge_{j=1}^{n} \tilde{\tilde{v}}_{3ij}^U, \bigwedge_{j=1}^{n} \tilde{\tilde{v}}_{4ij}^U, \bigwedge_{j=1}^{n} \left\{H(\tilde{A}_{ij}^U)\right\}, \text{if } f_i \in F_1 \\ \bigvee_{j=1}^{n} \tilde{\tilde{v}}_{1ij}^L, \bigvee_{j=1}^{n} \tilde{\tilde{v}}_{2ij}^L, \bigvee_{j=1}^{n} \tilde{\tilde{v}}_{3ij}^L, \bigvee_{j=1}^{n} \tilde{\tilde{v}}_{4ij}^L, \bigwedge_{j=1}^{n} \left\{H(\tilde{A}_{ij}^L)\right\}, \bigvee_{j=1}^{n} \tilde{\tilde{v}}_{1ij}^U, \bigvee_{j=1}^{n} \tilde{\tilde{v}}_{2ij}^U, \bigvee_{j=1}^{n} \tilde{\tilde{v}}_{3ij}^U, \bigvee_{j=1}^{n} \tilde{\tilde{v}}_{4ij}^U, \bigwedge_{j=1}^{n} \left\{H(\tilde{A}_{ij}^U)\right\}, \text{if } f_i \in F_2 \end{cases} \tag{16}$$

$\vee$ and $\wedge$ indicate max and min, respectively, and $1 \le i \le m.$

Step 5: Determine $p(\tilde{\tilde{v}}_i \succ \tilde{\tilde{v}}_i^+)$ between $\tilde{\tilde{v}}_i$ and $\tilde{\tilde{v}}_i^+$, $p(\tilde{\tilde{v}}_i \succ \tilde{\tilde{v}}_i^-)$ between $\tilde{\tilde{v}}_i$ and $\tilde{\tilde{v}}_i^-$:

$$p(\tilde{\tilde{v}}_i \succ \tilde{\tilde{v}}_i^+) = \min\left\{\max\left(\frac{M^*(\tilde{\tilde{v}}_i) - M_*(\tilde{\tilde{v}}_i^+)}{M^*(\tilde{\tilde{v}}_i) - M_*(\tilde{\tilde{v}}_i) + M^*(\tilde{\tilde{v}}_i^+) - M_*(\tilde{\tilde{v}}_i^+)}, 0\right), 1\right\} \tag{17}$$

$$p(\tilde{\tilde{v}}_i \succ \tilde{\tilde{v}}_i^-) = \min\left\{ \max\left(\frac{M^*(\tilde{\tilde{v}}_i) - M_*(\tilde{\tilde{v}}_i^-)}{M^*(\tilde{\tilde{v}}_i) - M_*(\tilde{\tilde{v}}_i) + M^*(\tilde{\tilde{v}}_i^-) - M_*(\tilde{\tilde{v}}_i^-)}, 0 \right), 1 \right\} \quad (18)$$

where

$$M_*(\tilde{\tilde{v}}_i) = \frac{1}{6}(v_{i1}^U + 2v_{i2}^U) H(\tilde{A}_i^U)^2 + \frac{1}{6}(v_{i1}^L + 2v_{i2}^L) H(\tilde{A}_i^L)^2 \quad (19)$$

and

$$M^*(\tilde{\tilde{v}}_i) = \frac{1}{6}(v_{i4}^U + 2v_{i3}^U) H(\tilde{A}_i^U)^2 + \frac{1}{6}(v_{i4}^L + 2v_{i3}^L) H(\tilde{A}_i^L)^2 \quad (20)$$

Step 6: Determine PC_i and rank the values of PC_i:

$$PC_i = \frac{\sum_{j=1}^{m} p(\tilde{\tilde{v}}_i \succ \tilde{\tilde{v}}_i^-)}{\sum_{j=1}^{m} \left(p(\tilde{\tilde{v}}_i \succ \tilde{\tilde{v}}_i^-) + p(\tilde{\tilde{v}}_i \succ \tilde{\tilde{v}}_i^+) \right)} \quad (21)$$

where $1 \leq i \leq n$.

PRACTICAL APPLICATION FOR RISK ASSESSMENT

Risk assessment process includes the determination of decision makers, determination of failure modes, determination of O, S and D risk factors examined during the evaluation stage, weighting risk factors and assessment of failure modes. A real application in building construction is presented to demonstrate the effectiveness of the proposed approach. The firm managers want to eliminate the failure modes related to ironwork in building construction in terms of occupational health and safety. Therefore, they appointed an expert committee of five people who are proficients in occupational safety area. After the first elimination, eleven failure modes were analyzed for further evaluation. A team of five experts consists of three occupational health and safety experts (E1, E2 and E3), a site chief (E4) and an academician (E5). They aim to identify the ironwork related failure mode with the highest risk in a building construction. These failure modes are determined by taking into account the expert opinions. They are the utilization of the bond wire (A1), stacking the irons (A2), setting down to the unique point of iron load set down to the groundwork (A3), the groundwork iron manufacturing in the same direction continuously (A4), setting down the load by crane operator (A5), disobeying to the instructions by worker (A6), connecting the tables to the iron in groundwork iron knitting (A7), the openness of extreme portion of the iron parts in concrete (A8), manual transport for heavy loads (A9), transporting as connecting with only bond wire without iron bond node (A10), transporting with incomplete chains for steel wicker and table transportation (A11). Table 2 shows the iron works related failure mode and effect analysis. The three risk criteria that affect risk assessment are the occurrence of the failure mode (O), the severity (S) of the failure mode, and the detectability (D) of the failure mode. The O and S risk criteria represent the cost criteria and the D risk criteria indicate the benefit criteria. The weights of E1, E2, E3, E4 and E5 experts were determined as (0.30, 0.20, 0.25, 0.15, 0.10), respectively. Experts express their subjective opinions with the help of linguistic variables while determining the weights of the criteria. The subjective opinions of experts are combined to determine the weights of these risk criteria. The IT2F number based TOPSIS method is used to rank criteria based failure modes and the risk assessment process is completed.

The proposed approach is applied to assess the risks associated with ironwork in building construction using FMEA. Using the linguistic scale shown in Table 3, the subjective opinions of the experts are determined to achieve the priorities of the criteria and to assess the failure modes. Five experts identified the weights of the criteria as shown in Table 4.

Based on the weights of experts, subjective opinions of decision makers are combined. The results of the combined opinions of the five experts for the severity of the criteria after the consolidation process are shown in Table 5.

Here, S risk criterion is examined as an example. The minimum values of the lower membership functions for H and VH were 0.77 and 0.95, respectively.

$$a_1^L = \left(0.30\times0.77+0.20\times0.95+0.25\times0.95+0.15\times0.77+0.10\times0.95\right)=0.869$$

The aggregation process was also performed for other criteria. TOPSIS approach has been handled to rank the failure modes in IT2F set. The five experts use the linguistic variables shown in Table 3 to define the values of criteria based alternatives. Subjective views of five decision makers for eleven failure modes based on three criteria are shown in Table 6. The linguistic assessments shown in Table 6 are converted to IT2F numbers using Table 3. Group opinion is obtained and the combined IT2F numbers value for the three criteria based failure modes are introduced in Table 8.

Table 2. Iron works related failure mode and effect analysis

System: Building construction Sub-system: Iron works Stage: Operation								
Part	**Failure mode**	**Direct effects**	**Causal factors**	**Failure ratio**	**System effects**	**Detection method**	**RPN**	**Suggestion Actions**
Material	A1	Injury, outpatient, first aid requirement	No use of gloves	Nonavailable	Loss of labor	Workplace physician reports	300	Training, ensuring the use of gloves
	A8	Serious injury, long term treatment	Do not bend the U shaped ends, no cut the extreme points	Nonavailable	Loss of labor	Control	150	The points where iron pieces come out should be closed or warning signs should be written.
	A10	Serious injury, long term treatment	Manual lifting of heavy iron loads by workers	Nonavailable	Loss of labor	Control	360	Lifting equipment must be used when lifting loads exceeding 25 kg, personnel must be trained.
Worker	A2	Serious injury, long term treatment	Lack of stacking order	Nonavailable	Loss of labor	Control	160	Stacks of iron to be at most three levels and wedges under the iron bundles
	A5	Mortal accident	Unloading load at a time	Nonavailable	Loss of labor, compensation	Control	240	Loads should not be rigid when lowering, but should be lowered slowly.
	A6	Serious injury, long term treatment	Continuation of training activities	Nonavailable	Loss of labor	Control	250	Training of workers
Construction structure	A3	Mortal accident	No set down evenly distribution of iron bonds	Nonavailable	Loss of labor, compensation	Control	180	Loads must be evenly distributed and lowered and the foundation bars must not be overloaded..
	A4	Serious injury, long term treatment	No reverse assembly in iron assembly	Nonavailable	Loss of labor	Control	135	In the case of iron mounting, it should be installed in the opposite direction at regular intervals.
	A7	Serious injury, long term treatment	No make the bonding wires in every connection	Nonavailable	Loss of labor	Control	225	Each connection must be made with a bond wire.
Management	A9	Serious injury, long term treatment	Manual lifting of heavy iron loads by workers	Nonavailable	Loss of labor	Control	210	Lifting equipment must be used when lifting loads exceeding 25 kg, personnel must be trained.
	A11	Serious injury, long term treatment	When transporting steel mesh, in some cases the mesh is not fully secured	Nonavailable	Loss of labor	Control	270	When transporting steel mats, they must be tightly tied at four corners and connected at least four points.

Table 3. The related linguistic terms.

Linguistic term	IT2TrF numbers
Very low (VL)	(0.08,0.11,0.15,0.18;0.8),(0.04,0.09,0.17,0.22;1)
Low (L)	(0.20,0.25,0.33,0.36;0.8),(0.17,0.22,0.38,0.43;1)
Medium (M)	(0.40,0.45,0.54,0.57;0.8),(0.30,0.40,0.60,0.66;1)
High (H)	(0.77,0.80,0.86,0.90;0.8),(0.72,0.75,0.90,0.95;1)
Very high (VH)	(0.95,0.97,0.98,0.99;0.8),(0.92,0.96,0.99,1.00;1)

Table 4. The weights of O, S and D criteria

Criteria	Decision makers				
	E1	E2	E3	E4	E5
O	M	H	H	M	H
S	H	VH	VH	H	VH
D	M	M	M	H	M

Table 5. The combined weights of O, S and D criteria.

Criteria	IT2TrF numbers
O	(0.604,0.643,0.716,0.752;0.8), (0.531,0.593,0.765,0.820;1)
S	(0.869,0.894,0.926,0.950;0.8), (0.830,0.866,0.950,0.978;1)
D	(0.456,0.503,0.588,0.620;0.8), (0.363,0.453,0.645,0.704;1)

Table 6. Criteria based failure modes.

	E1			E2			E3			E4			E5		
	C_1	C_2	C_3	C_1	C_2	C_3	C_1	C_2	C_3	C_1	C_2	C_3	C_1	C_2	C_3
FM1	VH	M	L	VH	L	L	M	L	M	M	M	M	VL	H	VH
FM2	H	VL	VH	M	H	VH	M	L	H	L	L	M	M	H	L
FM3	L	M	M	L	M	H	M	M	H	M	M	M	VL	M	VH
FM4	L	M	M	L	M	H	L	M	H	VL	VL	VL	VL	M	VH
FM5	H	VH	VL	L	VH	H	M	VH	VH	M	L	VH	M	H	VH
FM6	VH	H	H	VH	H	H	M	H	H	M	M	H	L	M	VH
FM7	M	M	M	VH	M	M	M	VH	H	M	L	VH	VL	H	VH
FM8	M	VL	M	M	H	H	L	H	H	H	L	H	VL	M	VH
FM9	H	L	VH	H	H	M	M	M	H	VH	M	VH	H	M	VH
FM10	H	H	H	M	H	M	H	H	VH	H	M	VH	H	H	VH
FM11	H	H	H	M	H	M	M	H	H	H	H	VH	M	H	VH

Table 7. The combined evaluations of failure modes.

FM1	O	((0.699,0.729,0.769,0.791;0.8),(0.647,0.702,0.797,0.830;1.0))
	S	((0.403,0.448,0.526,0.558;0.8),(0.347,0.407,0.576,0.629;1.0))
	D	((0.411,0.455,0.527,0.557;0.8),(0.360,0.419,0.574,0.623;1.0))
FM2	O	((0.481,0.525,0.605,0.638;0.8),(0.407,0.478,0.657,0.713;1.0))
	S	((0.335,0.373,0.435,0.468;0.8),(0.296,0.340,0.473,0.523;1.0))
	D	((0.748,0.778,0.819,0.842;0.8),(0.702,0.750,0.848,0.880;1.0))
FM3	O	((0.268,0.316,0.396,0.426;0.8),(0.209,0.279,0.447,0.501;1.0))
	S	((0.400,0.450,0.540,0.570;0.8),(0.300,0.400,0.600,0.660;1.0))
	D	((0.622,0.660,0.728,0.761;0.8),(0.551,0.614,0.774,0.825;1.0))
FM4	O	((0.170,0.215,0.285,0.315;0.8),(0.138,0.188,0.328,0.378;1.0))
	S	((0.352,0.399,0.482,0.512;0.8),(0.261,0.354,0.536,0.594;1.0))
	D	((0.574,0.609,0.670,0.702;0.8),(0.512,0.567,0.710,0.759;1.0))
FM5	O	((0.471,0.515,0.594,0.627;0.8),(0.400,0.469,0.646,0.701;1.0))
	S	((0.820,0.845,0.871,0.887;0.8),(0.788,0.828,0.890,0.910;1.0))
	D	((0.653,0.678,0.707,0.729;0.8),(0.616,0.657,0.726,0.756;1.0))
FM6	O	((0.655,0.690,0.739,0.759;0.8),(0.597,0.662,0.773,0.807;1.0))
	S	((0.678,0.713,0.780,0.818;0.8),(0.615,0.663,0.825,0.878;1.0))
	D	((0.788,0.817,0.872,0.909;0.8),(0.740,0.771,0.909,0.955;1.0))
FM7	O	((0.478,0.520,0.589,0.615;0.8),(0.398,0.481,0.635,0.684;1.0))
	S	((0.545,0.585,0.651,0.677;0.8),(0.478,0.548,0.695,0.740;1.0))
	D	((0.630,0.668,0.730,0.758;0.8),(0.560,0.628,0.773,0.818;1.0))
FM8	O	((0.374,0.419,0.497,0.528;0.8),(0.305,0.377,0.547,0.602;1.0))
	S	((0.441,0.476,0.536,0.570;0.8),(0.392,0.438,0.573,0.624;1.0))
	D	((0.677,0.712,0.776,0.810;0.8),(0.614,0.666,0.819,0.868;1.0))
FM9	O	((0.705,0.738,0.798,0.831;0.8),(0.645,0.694,0.839,0.885;1.0))
	S	((0.414,0.460,0.541,0.573;0.8),(0.345,0.416,0.594,0.649;1.0))
	D	((0.795,0.824,0.862,0.884;0.8),(0.746,0.796,0.890,0.920;1.0))
FM10	O	((0.696,0.730,0.796,0.834;0.8),(0.636,0.680,0.840,0.892;1.0))
	S	((0.715,0.748,0.812,0.851;0.8),(0.657,0.698,0.855,0.907;1.0))
	D	((0.786,0.815,0.856,0.879;0.8),(0.736,0.785,0.885,0.917;1.0))
FM11	O	((0.567,0.608,0.684,0.719;0.8),(0.489,0.558,0.735,0.791;1.0))
	S	((0.770,0.800,0.860,0.900;0.8),(0.720,0.750,0.900,0.950;1.0))
	D	((0.741,0.773,0.826,0.857;0.8),(0.686,0.733,0.863,0.905;1.0))

Table 8. The weighted combined evaluations of failure modes.

FM1	O	((0.422,0.468,0.551,0.594;0.8),(0.344,0.416,0.610,0.680;1.0))
	S	((0.350,0.400,0.487,0.530;0.8),(0.288,0.352,0.547,0.615;1.0))
	D	((0.187,0.228,0.310,0.345;0.8),(0.131,0.189,0.370,0.438;1.0))
FM2	O	((0.290,0.337,0.433,0.479;0.8),(0.216,0.283,0.503,0.584;1.0))
	S	((0.291,0.333,0.403,0.444;0.8),(0.246,0.294,0.449,0.511;1.0))
	D	((0.340,0.391,0.482,0.521;0.8),(0.255,0.339,0.547,0.619;1.0))
FM3	O	((0.162,0.203,0.284,0.320;0.8),(0.111,0.165,0.342,0.411;1.0))
	S	((0.348,0.402,0.500,0.541;0.8),(0.249,0.346,0.570,0.645;1.0))
	D	((0.283,0.331,0.428,0.471;0.8),(0.200,0.278,0.499,0.580;1.0))
FM4	O	((0.103,0.138,0.204,0.237;0.8),(0.073,0.111,0.251,0.309;1.0))
	S	((0.306,0.357,0.446,0.486;0.8),(0.217,0.306,0.508,0.581;1.0))
	D	((0.261,0.306,0.394,0.435;0.8),(0.186,0.257,0.458,0.534;1.0))
FM5	O	((0.284,0.331,0.425,0.471;0.8),(0.212,0.278,0.494,0.574;1.0))
	S	((0.712,0.755,0.806,0.842;0.8),(0.654,0.717,0.845,0.889;1.0))
	D	((0.297,0.341,0.416,0.452;0.8),(0.224,0.297,0.468,0.532;1.0))
FM6	O	((0.395,0.443,0.529,0.570;0.8),(0.317,0.392,0.591,0.661;1.0))
	S	((0.589,0.637,0.722,0.776;0.8),(0.510,0.573,0.783,0.858;1.0))
	D	((0.359,0.411,0.513,0.563;0.8),(0.269,0.349,0.586,0.672;1.0))
FM7	O	((0.288,0.334,0.422,0.462;0.8),(0.211,0.285,0.486,0.561;1.0))
	S	((0.473,0.523,0.602,0.642;0.8),(0.396,0.474,0.659,0.723;1.0))
	D	((0.287,0.335,0.429,0.469;0.8),(0.203,0.284,0.498,0.575;1.0))
FM8	O	((0.225,0.269,0.355,0.397;0.8),(0.162,0.223,0.418,0.493;1.0))
	S	((0.383,0.425,0.496,0.541;0.8),(0.325,0.379,0.544,0.610;1.0))
	D	((0.308,0.358,0.456,0.502;0.8),(0.223,0.301,0.528,0.611;1.0))
FM9	O	((0.425,0.474,0.571,0.624;0.8),(0.342,0.411,0.641,0.725;1.0))
	S	((0.360,0.411,0.501,0.544;0.8),(0.286,0.360,0.564,0.634;1.0))
	D	((0.362,0.414,0.507,0.547;0.8),(0.271,0.360,0.574, 0.647;1.0))
FM10	O	((0.420,0.469,0.570,0.627;0.8),(0.338,0.403,0.643, 0.731;1.0))
	S	((0.621,0.668,0.752,0.808;0.8),(0.545,0.604,0.812,0.886;1.0))
	D	((0.358,0.410,0.503,0.545;0.8),(0.267,0.355,0.571,0.645;1.0))
FM11	O	((0.342,0.390,0.490,0.540;0.8),(0.260,0.330,0.562,0.648;1.0))
	S	((0.669,0.715,0.796,0.855;0.8),(0.598,0.649,0.855,0.929;1.0))
	D	((0.338,0.388,0.486,0.531;0.8),(0.249,0.331,0.556,0.636;1.0))

Here, for example, failure mode 2 based on combined O criteria is presented. The subjective opinions of the five decision-makers were H,M, M, L and M, respectively. The minimum values of the lower membership functions $\left(a_1^L\right)$ for L, M and H are 0.2, 0.4 and 0.77, respectively. The weighted combined trapezoidal IT2F number values of the three criteria based failure modes are shown in Table 8. The weighted combined trapezoidal IT2F number value for failure mode 2 based on O criterion $\left(\bar{A}_{12}\right)$ is presented here as an example.

$$a_1^L = 0.77\times0.3+0.4\times0.2+0.4\times0.25+0.2\times0.15+0.4\times0.1=0.481$$

$$\bar{A}_{12} = w_1 \times A_{12} = \begin{pmatrix} \left(0.481\times0.604, 0.525\times0.643, 0.605\times0.716, 0.638\times0.752; \min\{0.8,0.8\}\right), \\ \left(0.407\times0.531, 0.478\times0.593, 0.657\times0.765, 0.713\times0.820; \min\{1.0,1.0\}\right) \end{pmatrix}$$
$$= \left(\left(0.290,0.337,0.433,0.479;0.8\right),\left(0.216,0.283,0.503,0.584;1.0\right)\right)$$

$$\tilde{\tilde{v}}_1^{+L} = \begin{pmatrix} \min\{0.422,0.290,0.162,0.103,0.284,0.395,0.288,0.225,0.425,0.420,0.342\}, \\ \min\{0.468,0.337,0.203,0.138,0.331,0.443,0.334,0.269,0.474,0.469,0.390\}, \\ \min\{0.551,0.433,0.284,0.204,0.425,0.529,0.422,0.355,0.571,0.570,0.490\}, \\ \min\{0.594,0.479,0.320,0.237,0.471,0.570,0.462,0.397,0.624,0.627,0.540\}, \\ \min\{0.8,0.8,0.8,0.8,0.8,0.8,0.8,0.8,0.8,0.8,0.8\} \end{pmatrix}$$
$$= \left(0.103,0.138,0.204,0.237,0.8\right)$$

Table 9 and Table 10 illustrates positive and negative ideal solutions, and lower and upper possibility mean values for each criterion, respectively.

Table 9. The related ideal solutions.

	Approximate positive ideal solution	Approximate negative ideal solution
O	((0.103,0.138,0.204,0.237;0.8), (0.073,0.111,0.251,0.309;1.0))	((0.425,0.474,0.571,0.627;0.8), (0.344,0.416,0.643,0.731;1.0))
S	((0.291,0.333,0.403,0.444;0.8), (0.217,0.294,0.449,0.511;1.0))	((0.712,0.755,0.806,0.855;0.8), (0.654,0.717,0.855,0.929;1.0))
D	((0.362,0.414,0.513,0.563;0.8), (0.271,0.360,0.586,0.672;1.0))	((0.187,0.228,0.310,0.345;0.8), (0.131,0.189,0.370,0.438;1.0))

Table 10. Possibility mean values.

	O (i=1)	S (i=2)	D (i=3)
$M_*(A_{1i})$	0,341	0,288	0,154
$M^*(A_{1i})$	0,497	0,445	0,299
$M_*(A_{2i})$	0,233	0,241	0,275
$M^*(A_{2i})$	0,408	0,368	0,444
$M_*(A_{3i})$	0,134	0,280	0,227
$M^*(A_{3i})$	0,277	0,462	0,405
$M_*(A_{4i})$	0,090	0,247	0,210
$M^*(A_{4i})$	0,204	0,413	0,372
$M_*(A_{5i})$	0,229	0,585	0,241
$M^*(A_{5i})$	0,401	0,691	0,382
$M_*(A_{6i})$	0,320	0,475	0,287
$M^*(A_{6i})$	0,481	0,641	0,477
$M_*(A_{7i})$	0,232	0,386	0,231
$M^*(A_{7i})$	0,395	0,537	0,404
$M_*(A_{8i})$	0,183	0,312	0,247
$M^*(A_{8i})$	0,340	0,447	0,429
$M_*(A_{9i})$	0,341	0,294	0,292
$M^*(A_{9i})$	0,523	0,459	0,466
$M_*(A_{10i})$	0,335	0,501	0,288
$M^*(A_{10i})$	0,524	0,665	0,463
$M_*(A_{11i})$	0,273	0,540	0,271
$M^*(A_{11i})$	0,457	0,701	0,452

The lower possibility value $M_*(A_{11})$ of failure mode 1 alternative based on O criterion is indicated as an example.

Table 11. The possibility degrees.

	O (i=1)	S (i=2)	D (i=3)
$p(\tilde{\tilde{v}}_{1i} \succ \tilde{\tilde{v}}_i^+)$	1,000	0,722	0,022
$p(\tilde{\tilde{v}}_{1i} \succ \tilde{\tilde{v}}_i^-)$	0,457	0,000	0,500
$p(\tilde{\tilde{v}}_{2i} \succ \tilde{\tilde{v}}_i^+)$	1,000	0,509	0,429
$p(\tilde{\tilde{v}}_{2i} \succ \tilde{\tilde{v}}_i^-)$	0,185	0,000	0,923
$p(\tilde{\tilde{v}}_{3i} \succ \tilde{\tilde{v}}_i^+)$	0,729	0,718	0,311
$p(\tilde{\tilde{v}}_{3i} \succ \tilde{\tilde{v}}_i^-)$	0,000	0,000	0,776
$p(\tilde{\tilde{v}}_{4i} \succ \tilde{\tilde{v}}_i^+)$	0,500	0,593	0,230
$p(\tilde{\tilde{v}}_{4i} \succ \tilde{\tilde{v}}_i^-)$	0,000	0,000	0,709
$p(\tilde{\tilde{v}}_{5i} \succ \tilde{\tilde{v}}_i^+)$	1,000	1,000	0,275
$p(\tilde{\tilde{v}}_{5i} \succ \tilde{\tilde{v}}_i^-)$	0,167	0,475	0,796
$p(\tilde{\tilde{v}}_{6i} \succ \tilde{\tilde{v}}_i^+)$	1,000	1,000	0,493
$p(\tilde{\tilde{v}}_{6i} \succ \tilde{\tilde{v}}_i^-)$	0,404	0,198	0,963
$p(\tilde{\tilde{v}}_{7i} \succ \tilde{\tilde{v}}_i^+)$	1,000	1,000	0,312
$p(\tilde{\tilde{v}}_{7i} \succ \tilde{\tilde{v}}_i^-)$	0,152	0,000	0,785
$p(\tilde{\tilde{v}}_{8i} \succ \tilde{\tilde{v}}_i^+)$	0,922	0,788	0,373
$p(\tilde{\tilde{v}}_{8i} \succ \tilde{\tilde{v}}_i^-)$	0,000	0,000	0,840
$p(\tilde{\tilde{v}}_{9i} \succ \tilde{\tilde{v}}_i^+)$	1,000	0,749	0,484
$p(\tilde{\tilde{v}}_{9i} \succ \tilde{\tilde{v}}_i^-)$	0,496	0,000	0,977
$p(\tilde{\tilde{v}}_{10i} \succ \tilde{\tilde{v}}_i^+)$	1,000	1,000	0,476
$p(\tilde{\tilde{v}}_{10i} \succ \tilde{\tilde{v}}_i^-)$	0,490	0,284	0,966
$p(\tilde{\tilde{v}}_{11i} \succ \tilde{\tilde{v}}_i^+)$	1,000	1,000	0,437
$p(\tilde{\tilde{v}}_{11i} \succ \tilde{\tilde{v}}_i^-)$	0,314	0,416	0,913

$p(\tilde{\tilde{v}}_i \succ \tilde{\tilde{v}}_i^+)$ between $\tilde{\tilde{v}}_i$ and $\tilde{\tilde{v}}_i^+$ is computed and indicated in Table 11. $p(\tilde{\tilde{v}}_i \succ \tilde{\tilde{v}}_i^-)$ between $\tilde{\tilde{v}}_i$ and $\tilde{\tilde{v}}_i^-$ is also computed and indicated in Table 12. The computation of $p(\tilde{\tilde{v}}_{11} \succ \tilde{\tilde{v}}_1^-)$ between failure mode 1 and the approximate positive ideal solution for criterion S is presented as a sample.

$$p(\tilde{\tilde{v}}_{21} \succ \tilde{\tilde{v}}_2^+) = \min\left\{\max\left(\frac{0.445-0.236}{0.445-0.288+0.368-0.236},0\right),1\right\} = 0.722$$

The closeness coefficient PC_i for each alternative is computed and indicated in Table 12. The computation of the possibility degree based closeness coefficient for failure mode 1 is presented as a sample.

$$PC_1 = \frac{0.457+0.000+0.500}{0.457+0.000+0.500+1.000+0.722+0.022} = 0.3544$$

The possibility based closeness coefficients of eleven failure modes are listed in ascending order and are shown in Table 12. Failure mode 8 (the openness of extreme portion of the iron parts in concrete) was determined according to their values as the highest risky failure mode related to iron works in building construction and it was determined that the risk that should be eliminated first. The order ranking of the other failure modes is failure modes 7, 3, 4, 1, 2, 6, 5, 9, 11 and 10.

Table 12. The closeness coefficients of failure modes .

	A1	**A2**	**A3**	**A4**	**A5**	**A6**	**A7**	**A8**	**A9**	**A10**	**A11**
PC_i	0,3544	0,3637	0,3063	0,3489	0,3872	0,3856	0,2883	0,2873	0,3974	0,4128	0,4026
Rank	5	6	3	4	8	7	2	1	9	11	10

Comparisons and Discussion

Traditional FMEA, fuzzy VIKOR, intuitionistic fuzzy grey relational analysis (IFGRA) methods have been investigated to demonstrate the effectiveness of the developed model. Table 13 indicates the obtained results

- The orders of other failure modes except failure modes 2, 3, 6, and 10 are different. In conventional FMEA, criteria weights are examined equally.

The proposed approach determines the risk factor weights using linguistic variables.

- The order of the other failure modes except failure modes 1, 2, and 10 are different. The VIKOR method examines minimum individual regrets and maximum group benefit. The parameter v in VIKOR plays an important role in sorting the failure modes. If these parameters change, the order may change.
- The results of the proposed approach and IFGRA approach are quite different. The orders of other failure modes except failure modes 2, 6, 8 and 10 are different. The grey relational analysis method allows measuring the grey grip between the reference point and the alternative. The best alternative is chosen according to the grey grip. The parameter ρ in the grey relational analysis plays an important role in the sequence of failure modes. Incorrect selection of this parameter may cause erroneous results.

Although the order of 11 failure modes for different methods is different, failure modes 1, 3, 4, 7 and 8 in building construction have always been included in the first 5 failure modes according to the results in Table 13. This result shows that the proposed approach is significant for the FMEA approach in building construction. Unlike other methods, the proposed approach uses IT2F number to evaluate expert opinions. IT2F number addresses the more ambiguity of subjective views in complex problems. Thus, more uncertainty of expert opinions was examined. This has enabled the proposed approach to achieve stronger and more accurate results than other methods. The results obtained from the proposed approach are supported by the results of linear programming.

Table 13. The comparison results.

	A1	A2	A3	A4	A5	A6	A7	A8	A9	A10	A11
Proposed approach	5	6	3	4	8	7	2	1	9	11	10
Traditional FMEA	4	6	3	1	10	7	5	2	8	11	9
FVIKOR	5	6	1	2	9	8	3	4	10	11	7
IFGRA	2	6	5	3	10	7	4	1	8	11	9

Linear Programming

Firm managers have established an occupational health and safety policy that includes certain requirements. The managers stated that the FMEA approach neglects the capacity of the firm. Moreover, the results of the presented method and the methods compared are different. This may cause some problems in the long term. In this chapter, linear programming approach is proposed to eliminate these limitations. According to the occupational health and safety policy, the firm presented the budget constraint, time constraint and the number of risks that it could correct. The firm has set a budget of 20000 Turkish Liras (TL), a period between 6-12 weeks and a limit of 5-8 correctable risks. 3 different times were determined as 6, 9 and 12 weeks. 4 different correctable risks, 5, 6, 7 and 8, were determined. A total of 12 different cases have been identified and presented to the firm to select the most appropriate case for itself. Linear programming based on closeness coefficient is coded in GAMS software programming. The cost and time data of the failure modes related to the iron works in the construction are presented in Table 14. When firms are thought to be composed of many processes, it is understandable the importance of examination of many different situations for each process. Closeness coefficient value based linear programming is solved in GAMS software program as we referenced the results in Table 15. The mathematical model and notations used are presented below:

Notations

x_i : Binary variable, equal to 1 when failure mode i is corrective
CC_i : Closeness coefficient value of failure mode i in Table 13
c_i : Total cost after corrective action
b_i : Total cost without corrective action
t_i : Necessary time to correct failure mode i

$$\max \sum_{i=1}^{n} (1 - CC_i).x_i \tag{22}$$

Eq. (22) indicates the objective function based on closeness coefficient values in Table 13.

st.

$$\sum_{i=1}^{n} (c_i.x_i + b_i.(1 - x_i)) \leq budget \tag{23}$$

$$\sum_{i=1}^{n} t_i.x_i \leq time \tag{24}$$

$$\sum_{i=1}^{n} x_i \leq number\ of\ correctable\ risk \tag{25}$$

$$x_3 + x_7 + x_8 \geq 2 \tag{26}$$

$$x_i = 0 - 1 \tag{27}$$

Eq.(23) show that the firm can allocate a certain budget for the occupational health and safety policy of the relevant process. It indicates the budget constraint in cases where failure modes are corrective events or not corrective events. The firm may encounter many failure modes in the working environment, but cannot correct all of them due to some limitations. After the corrective event occurs, the effect of the failure modes can be reduced or eliminated. If the firm continues without corrective events, it may face some problems in the long run. This can bring high costs to the firm. Eq.(24) shows the time, which the firm devotes to dealing with failure modes. Eq.(25) shows the maximum number of failure modes that the firm can handle due to its ability. Eq.(26)shows the limitation of association between failure modes. It indicates that at least two of the failure modes 3, 7 and 8 have to occur when the occupational health and safety policy is implemented. Eq.(27) shows the binary variable.

Table 14. The required cost and time for each failure mode.

Failure modes	c_i (TL)	b_i (TL)	t_i (Months)
FM1	500	800	0.5
FM2	700	1000	1
FM3	1000	1500	1
FM4	1100	2000	1.5
FM5	1000	1500	1
FM6	1500	3000	1.5
FM7	800	1500	0.5
FM8	1250	2500	1.5
FM9	1400	2750	2
FM10	1100	2250	1
FM11	1750	3000	2

The results obtained after solving the mathematical model are presented in Table 15. With the help of the results in Table 15, the company can determine the most appropriate occupational health and safety policy. Although failure modes 3, 7 and 8 are one of the first three of failure modes according to the IT2F number based TOPSIS method, they are considered together according to the results of none of the occupational health and safety policies based on the mathematical model.

Table 15. The results of linear programming

Budget	Time	NOCR	CR
20000	6	5	1,2,3,4,11
20000	6	6	1,2,3,4,5,7
20000	6	7	1,2,3,4,5,7
20000	6	8	1,2,3,4,5,7
20000	9	5	2,3,4,5,11
20000	9	6	2,3,4,5,8,11
20000	9	7	1,2,3,4,5,8,9
20000	9	8	1,2,3,4,5,7,8,9
20000	12	5	2,3,4,5,11
20000	12	6	2,3,4,5,8,11
20000	12	7	2,3,4,5,8,9,11
20000	12	8	2,3,4,5,7,8,9,11

*Corrected risks: CR, Number of corrective risk: NOCR

Managerial Implications

RPN is defined as the product of O, S and D in the traditional FMEA. This causes some problems to define the most important failure mode. This study proposes interval type-2 fuzzy number to define the weights of three parameters in order to overcome these problems. Interval type-2 fuzzy number provides us more information than crisp number and type-1 fuzzy number. This study presents TOPSIS approach with IT2F number for FMEA based risk assessment so that it contributes to the FMEA literature by presenting a novel method. The TOPSIS method uses crisp numbers while the fuzzy TOPSIS method uses fuzzy numbers to evaluate alternatives. In this study, TOPSIS method with IT2F number was used to model further uncertainty of subjective judgments. Multi-criteria decision making approach based FMEA deals with ranking the failure modes. It neglects the firm capacities such as budget, time, and resources. This may cause some problems in the long term. This study presents a linear programming approach that takes into account the firm capacities. This study is a leading study for risk assessment studies based on FMEA. This study presents 12 different situations in terms of an occupational health and safety policy. The construction firm can choose the most appropriate policy for itself. Since the firm consists of many processes, the flexibility to choose the most appropriate policy is very important.

CONCLUSION

One of the tasks that firms need to do in order to make their employees work efficiently is to provide a safe work environment for employees. Thus, employees can hire themselves more easily. Furthermore, the occurrence of a work accident may result in permanent or temporary job losses. The firms may face losses resulting from payment of compensation and downtime. In order to avoid these situations, firms should make risk assessments in the working environment. Risk assessment is crucial for an organization in a growing competitive environment in terms of occupational health and safety. Although it is not possible to make a completely safe work environment, the goal of a safer work environment can be achieved with the preventions to be taken.

This study is applied for the risk assessment of iron works in the construction sector. The benefit of the fuzzy TOPSIS method versus the crisp numbers in the TOPSIS method is that it handles fuzzy numbers to evaluate alternatives. IT2F number provides more information than type-1 fuzzy numbers to describe the uncertainty and vagueness in real-life applications. Therefore, IT2F number was preferred in this study. All opinions of decision makers are expressed on the basis of linguistic terms with IT2F number dealing with uncertainty.

The importance degrees of criteria and subjective opinions of decision makers about criteria based alternatives are combined. The weight of the criteria is used for the TOPSIS method with IT2F number, which is used to rank failure modes. This study examined approximate positive and negative ideal solutions with IT2F number to evaluate the alternatives due to their effectiveness and ease. This study uses possibility degrees instead of distance measurements in the current TOPSIS to determine a comparison index. The possibility based possibility coefficient of each alternative was calculated according to approximate positive and negative ideal solutions.

Risk assessment is crucial for an organization in a growing competitive environment in terms of occupational health and safety. The multi-criteria decision making method based on the possibility degree with IT2F number is a great chance for the risk assessment problem because the uncertain views of decision makers are examined. Another benefit of this study is that it overcomes the shortcomings of traditional FMEA. The RPN method, which is determined by multiplying O, S and D risk factors in traditional FMEA, has been criticized in the literature due to many deficiencies. Many papers focused on developing the drawbacks of RPN. This study expressed risk factors as fuzzy numbers with the help of linguistic terms rather than the crisp numbers in traditional FMEA. Decision makers expressed their views more easily. In order to determine the weights of these risk factors, the opinions of the decision makers were combined by using the importance of the decision makers and the group opinion was obtained. The proposed method is applied to assess the risks associated with ironwork in building a building using FMEA. The future papers can be handled different methods to improve the limitations of RPN.

REFERENCES

Chang, K. H., & Cheng, C. H. (2010). A risk assessment methodology using intuitionistic fuzzy set in FMEA. *International Journal of Systems Science*, *41*(12), 1457–1471. doi:10.1080/00207720903353633

Efe, B. (2019a). Website Evaluation Using Interval Type-2 Fuzzy-Number-Based TOPSIS Approach. In Multi-Criteria Decision-Making Models for Website Evaluation (pp. 166-185). IGI Global. doi:10.4018/978-1-5225-8238-0.ch009

Efe, B. (2019b). Analysis of operational safety risks in shipbuilding using failure mode and effect analysis approach. *Ocean Engineering*, *187*, 106214. doi:10.1016/j.oceaneng.2019.106214

Efe, B., & Efe, Ö. F. (2018). Intuitionistic Fuzzy Number Based Group Decision Making Approach For Personnel Selection. *Uludağ University Journal of The Faculty of Engineering*, *23*(3), 11–26. doi:10.17482/uumfd.338406

Efe, B., Kurt, M., & Efe, Ö. F. (2017). An Integrated Intuitionistic Fuzzy Set And Mathematical Programming Approach For An Occupational Health And Safety Policy. *Gazi University Journal of Science*, *30*(2), 73–95.

Efe, B., Yerlikaya, M. A., & Efe, Ö. F. (2016). İş Güvenliğinde Bulanık Promethee Yöntemiyle Hata Türleri ve Etkilerinin Analizi: Bir İnşaat Firmasında Uygulama. *Gümüşhane Üniversitesi Fen Bilimleri Enstitüsü Dergisi*, *6*(2), 126–137.

Efe, Ö. F. (2020). Hybrid Multi-Criteria Models: Joint Health and Safety Unit Selection on Hybrid Multi-Criteria Decision Making. In Multi-Criteria Decision Analysis in Management (pp. 62-84). IGI Global.

Fattahi, R., & Khalilzadeh, M. (2018). Risk evaluation using a novel hybrid method based on FMEA, extended MULTIMOORA, and AHP methods under fuzzy environment. *Safety Science*, *102*, 290–300. doi:10.1016/j.ssci.2017.10.018

Hwang, C. L., & Yoon, K. (1981). *Multiple attributes decision making methods and applications*. Springer. doi:10.1007/978-3-642-48318-9

Jee, T. L., Tay, K. M., & Lim, C. P. (2015). A New Two-Stage Fuzzy Inference System-Based Approach to Prioritize Failures in Failure Mode and Effect Analysis. *IEEE Transactions on Reliability*, *64*(3), 869–877. doi:10.1109/TR.2015.2420300

Kutlu, A. C., & Ekmekçioğlu, M. (2012). Fuzzy failure modes and effects analysis by using fuzzy TOPSIS-based fuzzy AHP. *Expert Systems with Applications*, *39*(1), 61–67. doi:10.1016/j.eswa.2011.06.044

Liu, H., You, J., Li, P., & Su, Q. (Article in Press). Failure Mode and Effect Analysis Under Uncertainty: An Integrated Multiple Criteria Decision Making Approach (2016). *IEEE Transactions on Reliability*.

Liu, H. C., Liu, L., & Li, P. (2014). Failure mode and effects analysis using intuitionistic fuzzy hybrid weighted Euclidean distance operator. *International Journal of Systems Science*, *45*(10), 2012–2030. doi:10.1080/00207721.2012.760669

Liu, H. C., Liu, L., & Lin, Q. L. (2013). Fuzzy failure mode and effects analysis using fuzzy evidential reasoning and belief rule-based methodology. *IEEE Transactions on Reliability*, *62*(1), 23–36. doi:10.1109/TR.2013.2241251

Liu, H. C., You, J. X., You, X. Y., & Shan, M. M. (2015). A novel approach for failure mode and effects analysis using combination weighting and fuzzy VIKOR method. *Applied Soft Computing*, *28*, 579–588. doi:10.1016/j.asoc.2014.11.036

Maity, S., De, S. K., & Mondal, S. P. (2020). A study of a backorder EOQ model for cloud-type intuitionistic dense fuzzy demand rate. *International Journal of Fuzzy Systems*, *22*(1), 201–211. doi:10.100740815-019-00756-1

Mohsen, O., & Fereshteh, N. (2017). An extended VIKOR method based on entropy measure for the failure modes risk assessment – A case study of the geothermal power plant (GPP). *Safety Science*, *92*, 160–172. doi:10.1016/j.ssci.2016.10.006

Qin, J., Xi, Y., & Pedrycz, W. (2020). Failure mode and effects analysis (FMEA) for risk assessment based on interval type-2 fuzzy evidential reasoning method. *Applied Soft Computing*, *89*, 106134. doi:10.1016/j.asoc.2020.106134

Shtovba, S., Rotshtein, A., & Pankevich, O. (2002). Fuzzy rule based system for diagnosis of stone construction cracks of buildings. In *Advances in Computational Intelligence and Learning* (pp. 401–411). Springer. doi:10.1007/978-94-010-0324-7_28

Tian, Z. P., Wang, J. Q., & Zhang, H. Y. (2018). An integrated approach for failure mode and effects analysis based on fuzzy best-worst, relative entropy, and VIKOR methods. *App. S. Comp.*, *72*, 636–646. doi:10.1016/j.asoc.2018.03.037

Wang, Y. M., Chin, K. S., Poon, G. K. K., & Yang, J. B. (2009). Risk evaluation in failure mode and effects analysis using fuzzy weighted geometric mean. *Expert Systems with Applications*, *36*(2), 1195–1207. doi:10.1016/j.eswa.2007.11.028

Yazdani, M., Abdi, M. R., Kumar, N., Keshavarz-Ghorabaee, M., & Chan, F. T. (2019). Improved decision model for evaluating risks in construction projects. *J. Cons. Eng. Man.*, *145*(5), 04019024. doi:10.1061/(ASCE)CO.1943-7862.0001640

Yazdi, M., Nedjati, A., Zarei, E., & Abbassi, R. (2020). A reliable risk analysis approach using an extension of best-worst method based on democratic-autocratic decision-making style. *Journal of Cleaner Production*, *256*, 120418. doi:10.1016/j.jclepro.2020.120418

Zhou, Q., & Thai, V. V. (2016). Fuzzy and grey theories in failure mode and effect analysis for tanker equipment failure prediction. *Safety Science*, *83*, 74–79. doi:10.1016/j.ssci.2015.11.013

Zhu, J., Shuai, B., Li, G., Chin, K. S., & Wang, R. (2020). Failure mode and effect analysis using regret theory and PROMETHEE under linguistic neutrosophic context. *Journal of Loss Prevention in the Process Industries*, *64*, 104048. doi:10.1016/j.jlp.2020.104048

Chapter 6
A Novel Entropy Measure for Interval-Valued Intuitionistic Fuzzy Set and Its Application to Failure Mode and Effect Analysis

Kamal Kumar
https://orcid.org/0000-0001-7903-4614
Amity University, Haryana, India

Naveen Mani
Sandip University, Nashik, India

Amit Sharma
Amity University, Haryana, India

Reeta Bhardwaj
Amity University, Haryana, India

ABSTRACT

The failure mode and effect analysis (FMEA) is widely used an effective pre-accident risk assessment tool to identify, eliminate, and assess potential failure modes in different industries for enhancing the safety and reliability of systems, process, services, and products. Therefore, this chapter presents a new approach to rank the failure modes under the interval-valued intuitionistic fuzzy set (IVIFS). For this, a novel measure to measure the fuzziness known as entropy measure is proposed. Some properties and axiom definition of the proposed entropy measure have been presented to show the validity of it. Afterwards, the proposed entropy measure is

DOI: 10.4018/978-1-7998-7564-2.ch006

utilized to obtain the weight of risk factor and developed an approach under the IVIFS environment to determine the risk priority order of failure modes. Finally, a real-life case of FMEA has been discussed to manifest the developed approach, and obtained results are compared with the results obtained by the existing methods for showing the feasibility and validity of the proposed approach.

INTRODUCTION

The Failure Mode and effect Analysis (FMEA) method is a risk analyzing method for identifying and removing possible faults, issues and errors to improve system, architecture, process and service efficiency and safety. FMEA first introduced by the United Nations in the 1960s as a formal and systematic approach in aerospace industry of countries with their obvious safety and reliability requirements(Bowles & Peláez, 1995). FMEA's main purpose is to correct key failure modes instead of fixing them after failures, they reach the customer before. In last decade, FMEA has become a powerful and useful tool that is commonly used in many sectors, including the power, mechanical aerospace, and healthcare industries(Chang & Cheng, 2011; Huang et al., 2017; Li et al., 2019; Song et al., 2013; Vahdani et al., 2015).In typical FMEA, failure modes (FMs) are evaluated with regard to risk factors: occurrence (O), severity (S), and detection (D). The risk priority setting of FMs is determined by calculating its risk priority numbers (RPNs), obtained by multiplying the risk factors O, S and D.

$$RPN = O \times S \times D$$

While traditional FMEA is a useful risk assessment tool, it is widely criticized for a number of inconveniences in the literature (H.-C. Liu, 2016; Zhao et al., 2017) which are summarized as follows:

1. The risk factors are di cult to quantify accurately and entirely, due to the uncertainty and vagueness of the decisions of FMEA team members.
2. In conventional FMEA, for the determination of FMs, only three risk factors O, S and D are used, which may ignore other important process or system factors.
3. The relative importance of risk factors, which is not fair in the real case, is not taken into account.
4. Using RPNs to assess the probability rating of modes of failure is controversial and debatable.

To overcome the disadvantages of conventional FMEA, scholars have implemented several theories of ambiguity and multi-attribute decision-making (MADM) approaches for tackling the FMEA issues. One of them,Zadeh(1965) introduced the concept of the fuzzy sets (FSs), and after that its extensions intuitionistic FSs (IFSs) (Atanassov, 1986) and interval-valued IFSs (IVIFSs)(Atanassov & Gargov, 1989) have presented as powerful tools to handling the uncertainty. Recently, these theories have been more applied to solve the real world FMEA issues. Song et al.(2013) introduced the FMEA based on the fuzzy TOPSIS. To describe the FMEA information,Vahdani et al.(2015) described a fuzzy structure.

During the decision-making (DM) process in MADM problems the attribute weights play crucial role because fluctuation in the weights can change the ranking of the alternatives. In some MADM problems criteria weights are given by the decision-makers but in some situations decision-makers cannot set the attribute weights due to the fuzziness of the data. Entropy measure is an important method to manage it, which is essentially known as the knowledge measure derived from C.E.Shannon(Shannon, 1948)'s fundamental paper "The Mathematical Theory of Communication." One of the trusted fields for calculating the degree of uncertainty in the data is information theory. Classical measurements of information, how-ever, deal with information that is detailed in nature. To resolve this, a collection of axioms for fuzzy entropy was suggested in (Deluca A, 1972). Zhang & Jiang(2008) generalized the logarithmic fuzzy entropy measure given in (De Luca & Termini, 1993) for IFSs. Wei et al.(2012) presented a new entropy by using the cosine function for IFS. Wang & Wang(2012) defined cotangent entropy measure for IFS. Verma & Sharma(2013) presented the exponential entropy measure of IFSs. Further, the all above existing entropy measures of IFSs do not contain the degree of hesitancy of IFS. M. Liu & Ren(2014) realized some drawbacks of above existing entropy measures of IFS and presented a new entropy measure by containing the degree of hesitance of IFS. Nguyen(2016) defined the knowledge measure based entropy for IVIFSs.Rashid et al.(2018) defined the distance based entropy measure in context of IVIFSs.Meng & Chen(2016) defined the similarity and entropy measure for IVIFS.

In the literature, authors found that the existing FMEA methods based on the interval-valued intuitionistic fuzzy values (IVIFVs) have various flaws in FMEA process. Most of existing FMEA methods are based on the interval-valued intuitionistic fuzzy weighted averaging (IVIFWA) aggregation operators (AOs). But, IVIFWA AOs have few limitations that if taking one value $\langle[1,1],[0,0]\rangle$ .out of n input values then get the aggregating value remains same $\langle[1,1],[0,0]\rangle$. Keeping the disadvantages in mind of the existing FMEA methods, authors introduce a novel FMEA method in context of IVIFVs environment. For this, the main contribution and aim of this chapter is given as follows:

1. To formulate the entropy measure for IVIFSs which help to evaluating the RF'sweight.
2. Some properties of the proposed entropy measure have been proved to show the legality and validity.
3. To develop a novel FMEA method based on the proposed entropy measure.
4. A real-life illustration has been taken to illustrate the proposed novel FMEA method, and obtained results are compared with the existing strategies to appear the preferences of the proposed FMEA method.

To achive above target, this chapter isarranged as follows.A literature review about the IVIFS is given in Section "**LITERATURE REVIEW**". Section "**PRELIMINARIES**" reviews basic concepts of IVIFSs theory. Section "**ENTROPY MEASURE FOR IVIFS**"introduce a new entropy measure for IVIFSs. In Section "**A NOVEL FMEA METHOD**", authors proposed a novel FMEA method. In Section "**APPLICATION OF THE PROPOSED FMEA METHOD**", an illustration is taken to demonastrate the proposed method The conclusions are presented in Section "**CONCLUSION**".

LITERATURE REVIEW

Recently, fuzzy theory and its extension widely used in di erent discipline like as computation intelligence, decision-making (DM), reliability analysis etc. In DM field, Kumar & Garg, (2018) defined the TOPSIS methods based on the SPA theory for IVIFSs environment. Joshi & Kumar(2018); Priyadharsini & Balasubramaniam(2019) defined a score and accuracy function for IVIFNs respectively. C.Y. Wang & Chen (2017) defined a score function and a DM approach for handling the MADM issues that overcomes the flaws of DM approach given byChen & Huang(2017). Garg & Kumar(2020)defined the power AOs based on the SPA theory under the IFS environment. In (Mishra et al., 2020), Mishra defined a new entropy by using the TODIM method for IVIFSs.Garg & Kumar(2019) introduced the concept of linguistic IVIFSs. Garg & Kumar, (2020a) defined the concept of possibility degree for ranking of IVIFSs. Garg & Kumar(2020a)introduced the novel exponential distance measure for IVIFSs.

Apart from this, Mondal & Roy(2015) solve the second order linear differential equations(DEq) with intuitionistic fuzzy boundary value. Mondal(2016)introduced the concept of the interval-valued fuzzy number to solve the firrst order DEqs. Mondal(2018) defined the application of the IVIFSs into DEqs. Mondal et al.(2018) utilized the non linear IVIFSs for solving the various form of difference equations. Mondal et al.(2019)formulate the arithmetic operations for non-linear IF number and its application for solving the linear integral equations. Maity et al.(2020) introduced the dense fuzzy set for handling inventory problem of backorder.Garg & Kumar(2018)provided the possibility degree and Einstein AOs for linguistic IFS environment.

For the reliability analysis, to solve the RPN model problem, in practice, the ranking can be regarded as MADM process, then ranking results can be obtained by using the different MADM methods. For instant, Liu et al. (2016) introduced the integrated method foe tackling the FMEA issues. Tooranloo & sadat Ayatollah (2016) introduced a novel IF approach for FMEA.Zhao et al.(2017) present the MULTIMOORA method for FMEA in IVIFS context. Wang et al.(2016) introduced a hybrid model for FMEA by using the IVIFSs theory. Tian et al. (2018) developed a approach for FMEA based on the entropy and VIKOR method. Liu et al. (2019) introduced an integrated approach for FMEA. Li et al. (2019) defined the FMEA method using the IVIFs theory. Qin et al.(2020) defined the FMEA method for type-2 fuzzy set. Boral et al.(2020) introduced an integrated method based on the AHP for FMEA. A systematic literature review for the FMEA is given by Liu et al.(2020).

PRELIMINARIES

Definition 2.1 (Atanassov & Gargov, 1989) An IVIFS A .is defined as

$$H=\left\{x,\left[\tau_H\left(x\right),\eta_H\left(x\right)\right],\left[\theta_H\left(x\right),\upsilon_H\left(x\right)\right]\mid x\in X\right\},. \tag{1}$$

where $[\tau_H(x),\eta_H(x)]$ and $[\theta_H(x),\upsilon_H(x)]$ represents the interval-valued "membership degree" and the interval-valued "non-membership" degree of x to H respectively, such that $0\le\tau_H(x)\le\eta_H(x)\le 1$, $0\le\theta_H(x)\le\upsilon_H(x)\le 1$, and $0\le\eta_H(x)=\upsilon_H(x)\le 1$, holds for $\forall x\in X$ The hesitation interval for H .is computed as $[\pi_H^L(x),\pi_H^U(x)]=[1-\eta_H(x)-\upsilon_H(x),1-\tau_H(x)-\theta_H(x)]$, $\forall x\in X$.

In [29], Xu called the the pair $\langle[\tau,\eta],[\theta,\upsilon]\rangle$ an IVIFV where $0\le\tau\le\eta\le 1$. $0\le\theta\le\upsilon\le 1$, and $0\le\eta+\upsilon\le 1$.

Definition 2.2 (Atanassov & Gargov, 1989)Let $A=\langle[\tau,\eta],[\theta,\upsilon]\rangle$, $A_1=\langle[\tau_1,\eta_1],[\theta_1,\upsilon_1]\rangle$ and $A_2=\langle[\tau_2,\eta_2],[\theta_2,\upsilon_2]\rangle$ be any three IVIFVs, then

1. $A_1\prec A_2$ if $\tau_1<\tau_2,\eta_1<\eta_2,\theta_1>\theta_2$ and $\upsilon_1>\upsilon_2$;
2. $A_1=A_2$ .ff $\tau_1=\tau_2,\eta_1=\eta_2,\theta_1=\theta_2$ and $\upsilon_1=\upsilon_2$;

3. $A^c = \langle[\theta, \upsilon], [\tau, \eta]\rangle$, where A^c .denotes the complement of the IVIFV A .

Definition 2.3 (Xu, 2007)For the collection of IVIFVs $\gamma_t = [\tau_t, \eta_t][\theta_t, \upsilon_t], (t = 1, 2, \ldots, n)$. the IVIFWA operator is defined as

$$IVIFWA(\gamma_1, \gamma_2, \ldots, \gamma_t) = \sum_{t=1}^{n} \omega_t \gamma_t$$

$$= \left(\left[1 - \prod_{t=1}^{n}(1-\tau_t)^{\omega_t}, 1 - \prod_{t=1}^{n}(1-\eta_t)^{\omega_t}\right], \left[\prod_{t=1}^{n}\theta_t^{\omega_t}, \prod_{t=1}^{n}\upsilon_t^{\omega_t}\right]\right), \tag{2}$$

Where $\omega_t > 0$ .and $\sum_{t=1}^{n} \omega_t = 1$.

Definition2.4. (Nguyen, 2016)The entropy measure is referred to areal-valuedfunction $E : \varphi(X) \to [0,1]$. if $H \in \varphi(X)$.ulfills the following properties:

(P1) $E(H) = 0 \Leftrightarrow$ H is a crisp set;

(P2) $E(H) = 1 \Leftrightarrow [\tau_H(x), \eta_H(x)] = [\theta_H(x), \upsilon_H(x)] \quad, \forall x \in X$.

(P3) $E(H) = E(H^c)$;. (P4) $If\ H_1 \le H_2\ then\ E(H_1) \le E(H_2)$..

ENTROPY MEASURE FOR IVIFS

Definition 3.1. For an IVIFS $H = \{x_t [\tau_H(x_t), \eta_H(x_t)], [\theta_H(x_t), \upsilon_H(x_t)] \mid x_t \in X\}, (t = 1, 2, \ldots, n)$.we define the following entropy measure:

$$E(H) = \frac{1}{6n}\sum^{n}(4\sqrt{(\tau_H(x_t) + \eta_H(x))(\theta_H(x_t) + \upsilon_H(x_t))} + \pi_H^L(x_t) + \pi_H^U(x_t) + 2\sqrt{(2 - \tau_H(x_t) - \eta_H(x))(2 - \theta_H(x_t) - \upsilon_H(x))}) \tag{3}$$

Theorem3.1. The entropy measure $\mathrm{E}\left(\mathrm{H}\right)$.orIVIFS H .defined in Eq. (3) satisfies the properties(P1)–(P4)as defined in

Definition2.4.

Proof.Let $H = \left\{x,\left[\tau_H\left(x_t\right),\eta_H\left(x_t\right)\right],\left[\theta_H\left(x_t\right),\upsilon_H\left(x_t\right)\right] \mid x_t \in X\right\}$.eanIVIFS.

(P1) We have

$\mathrm{E}\left(\mathrm{H}\right) = 0$

$$\Leftrightarrow \frac{1}{6n}\sum_{t=1}^{n}(4\sqrt{\left(\tau_H\left(x_t\right)+\eta_H\left(x\right)\right)\left(\theta_H\left(x_t\right)+\upsilon_H\left(x_t\right)\right)}+\pi_H^L\left(x_t\right)+\pi_H^U\left(x_t\right)+2\sqrt{\left(2-\tau_H\left(x_t\right)-\eta_H\left(x\right)\right)\left(2-\theta_H\left(x_t\right)-\upsilon_H\left(x\right)\right)})=0$$

$$\Leftrightarrow \left(\tau_H\left(x_t\right)+\eta_H\left(x\right)\right)=0\,or\left(\theta_H\left(x_t\right)+\upsilon_H\left(x_t\right)\right)=0,\ \ \pi_H^L\left(x_t\right)=\pi_H^U\left(x_t\right)=0\text{ and}$$
$\left(2-\tau_H\left(x_t\right)-\eta_H\left(x\right)\right)=0\,and\left(2-\theta_H\left(x_t\right)-\upsilon_H\left(x\right)\right)=0$.$\Leftrightarrow$*H* is a crisp set.

(P2) Since $\sqrt{xy}$ .achieves its maximum value $\frac{x+y}{2}$ .when $x = y$. therefore, we have

$$E\left(H\right)=1.$$

$$\Leftrightarrow\left(\tau_H\left(x_t\right)+\eta_H\left(x\right)\right)\left(\theta_H\left(x_t\right)+\upsilon_H\left(x_t\right)\right)+\pi_H^L\left(x_t\right)+\pi_H^U\left(x_t\right)+2\sqrt{\left(2-\tau_H\left(x_t\right)-\eta_H\left(x\right)\right)\left(2-\theta_H\left(x_t\right)-\upsilon_H\left(x\right)\right)}=6$$

$$\Leftrightarrow \tau_H\left(x_t\right)=\theta_H\left(x_t\right)\text{and } \eta_H\left(x_t\right)=\upsilon_H\left(x_t\right).$$

(P3)Since $H^c = \left\{x,\left[\theta_H\left(x_t\right),\upsilon_H\left(x_t\right)\right],\left[\tau_H\left(x_t\right),\eta_H\left(x_t\right)\right] \mid x_t \in X\right\}$,therefore, we have

$$E=\left(H\right)\frac{1}{6n}\sum_{t=1}^{n}(4\sqrt{\left(\tau_H\left(x_t\right)+\eta_H\left(x\right)\right)\left(\theta_H\left(x_t\right)+\upsilon_H\left(x_t\right)\right)}+\pi_H^L\left(x_t\right)+\pi_H^U\left(x_t\right)+\quad 2\sqrt{\left(2-\tau_H\left(x_t\right)-\eta_H\left(x\right)\right)\left(2-\theta_H\left(x_t\right)-\upsilon_H\left(x\right)\right)})$$
$$=\frac{1}{6n}\sum_{t=1}^{n}(4\sqrt{\left(\theta_H\left(x_t\right)+\upsilon_H\left(x\right)\right)\left(\tau_H\left(x_t\right)+\eta_H\left(x\right)\right)}+\pi_H^L\left(x_t\right)+\pi_H^U\left(x_t\right)+\quad 2\sqrt{\left(2-\theta_H\left(x_t\right)-\upsilon_H\left(x\right)\right)\left(2-\tau_H\left(x_t\right)-\eta_H\left(x\right)\right)})$$
$$=E\left(H^c\right)$$

(P4) Construct the function

$f\left(x,y\right)=4\sqrt{xy}+2-x-y+2\sqrt{\left(2-x\right)\left(2-y\right)}$, .here, $x, y \in \left[0,2\right]$, and $x+y\leq 2$. Now we will prove that the function $f\left(x,y\right)$is increasing with *x* and decreasing with *y* when $x \leq y$.

For this, we can derived the partial derivatives of $f(x,y)$.with respect to x and y.

$$\frac{\partial f(x,y)}{\partial x} = 2\sqrt{\frac{y}{x} - \sqrt{\frac{2-y}{2-x}} - 1}\,.\frac{\partial f(x,y)}{\partial y} = 2\sqrt{\frac{x}{y} - \sqrt{\frac{2-x}{2-y}} - 1}\,.$$

Since $\frac{\partial f(x,y)}{\partial x} \geq 0, \frac{\partial f(x,y)}{\partial y} \leq 0$, then $x \leq y$, .therefore, $f(x,y)$.is increasing with x, and decreasing with y. Thus,

$f((\tau_1(x_t) + \eta_1(x)), (\theta_1(x_t) + \upsilon_1(x_t))) \leq f((\tau_2(x_t) + \eta_2(x)), (\theta_2(x_t) + \upsilon_2(x_t)))$.

when $\tau_1(x_t) \leq \tau_2(x_t)$, $\eta_1(x_t) \leq \eta_2(x_t)$, $\theta_1(x_t) \geq \theta_2(x_t)$, and $\eta_1(x_t) \geq \eta_2(x_t)$.

Similarly, $\frac{\partial f(x,y)}{\partial x} \leq 0, \frac{\partial f(x,y)}{\partial y} \geq 0$, then $x \geq y$, .therefore, $f(x,y)$.is decreasing with x.and increasing with y. Thus,

$((\tau_1(x_t) + \eta_1(x)), (\theta_1(x_t) + \upsilon_1(x_t))) \leq f((\tau_2(x_t) + \eta_2(x)), (\theta_2(x_t) + \upsilon_2(x_t)))$,

when $\tau_1(x_t) \geq \tau_2(x_t)$, $\eta_1(x_t) \geq \eta_2(x_t)$, $\theta_1(x_t) \leq \theta_2(x_t)$, and $\eta_1(x_t) \leq \eta_2(x_t)$.

Therefore, If $H_1 \preceq H_2$.hen

$\frac{1}{6n}\sum_{t=1}^{n} f((\tau_1(x_t) + \eta_1(x)), (\theta_1(x_t) + \upsilon_1(x_t))) \leq \frac{1}{6n}\sum_{t=1}^{n} f((\tau_2(x_t) + \eta_2(x)), (\theta_2(x_t) + \upsilon_2(x_t)))$.

Hence $E(H_1) \leq E(H_2)$.

A NOVEL FMEA METHOD

Let us consider m failure modes (FMs) $A_1, A_2, \ldots,$ and A_m and n risk factors (RFs) $G_1, G_2, \ldots,$ and G_n, Assume that the DMk $D^{(i)}$, $i = 1, 2, \ldots, l$, .assesses RF G_t.of FM A_k.by using IVIFV $\tilde{\gamma}_{kt}^{(i)}$. where $\tilde{\gamma}_{kt}^{(i)} = \langle [\tilde{\tau}_{kt}^{(i)}, \tilde{\eta}_{kt}^{(i)}], [\tilde{\theta}_{kt}^{(i)}, \tilde{\upsilon}_{kt}^{(i)}] \rangle$. $k = 1, 2, \ldots, m$, .and $t = 1, 2, \ldots, n$. to construct the DMx $\tilde{R}^{(i)} = \left(\tilde{\gamma}_{kt}^{(i)}\right)_{m \times n}$. Each DMk $D^{(i)}$. has own weight λ_i, $i = 1, 2, \ldots, l$, .nd satisfy $\sum_{i=1}^{l} \lambda_i = 1$.to refelect the importance in FMEA process. Let the ω_t. $t = 1, 2, \ldots, n$, .be the weight of RFs will be deterimine by proposed entroy. The procedure steps of the proposed FMEA method are given as follows:

Step 1: Converte the DMx $\tilde{R}^{(i)} = \left(\tilde{\gamma}_{kt}^{(i)}\right)_{m\times n} = \left\langle\left[\tilde{\tau}_{kt}^{(i)}, \tilde{\eta}_{kt}^{(i)}\right], \left[\tilde{\theta}_{kt}^{(i)}, \tilde{\upsilon}_{kt}^{(i)}\right]\right\rangle$. $\left(i = 1, 2, \ldots, l\right)$ .nto the normalized DMx $R^{(i)} = \left(\gamma_{kt}^{(i)}\right)_{m\times n} = \left(\left[\tau_{kt}^{(i)}, \eta_{kt}^{(i)}\right], \left[\theta_{kt}^{(i)}, \upsilon_{kt}^{(i)}\right]\right)_{m\times n}$. where $k = 1,2,\ldots, m$ .and $t = 1,2,\ldots, n$ .shown as follows:

$$\gamma_{kt}^{(i)} = \begin{cases} \left[\tau_{kt}^{(i)}, \eta_{kt}^{(i)}\right], \left[\theta_{kt}^{(i)}, \upsilon_{kt}^{(i)}\right], & \textit{if } G_t \textit{ is a benifit type attribute} \\ \left[\theta_{kt}^{(i)}, \upsilon_{kt}^{(i)}\right], \left[\tau_{kt}^{(i)}, \eta_{kt}^{(i)}\right], & \textit{if } G_t \textit{ is a cost type attribute} \end{cases}. \tag{4}$$

Step 2: By using the Eq. (3), calculate the entropy measure $E\left(\gamma_{kt}^{(i)}\right)$.for each LIFV given in the DMx $R^{(i)} = \left(\gamma_{kt}^{(i)}\right)_{m\times n}$ .to obtain the group entropy matrix, $GE = \left(E_{kt}\right)_{m\times n}$ .s follows:

$$E_{kt} = \sum_{i=1}^{l} \lambda_i E\left(\gamma_{kt}^{(i)}\right). \tag{5}$$

Step 3: Calculate the weight ω_t .f the RF G_t,t=1,2, . . . ,n. as:

$$\omega_t = \frac{1 - e_t}{n - \sum_{t=n}^{n} e_t}. \tag{6}$$

Where $e_t = \frac{1}{m}\sum_{k=1}^{m} E_{kt}$.

Step4: Calculatetheoverallperformance $Q\left(H_k\right)$.feachalternative $H_k\left(k = 1, 2, ..., m\right)$.s follows:

$$Q\left(H_k\right) = \sum_{t=1}^{n} \omega_t E_{kt}. \tag{7}$$

Step5:Rankthealternatives $H_1, H_2, \ldots, H_m$ according to the descending sequence of overall performance $Q\left(H_k\right)$. $k = 1, 2, \ldots, m$.

APPLICATION OF THE PROPOSED FMEA METHOD: AN ILLUSTRATIVE EXAMPLE

In this section, a real-life practical implementation(Li et al., 2019; Vahdani et al., 2015) of steel manufacturing process risk management is used to illustrate the proposed FMEA method.

In this case,ten FMs of steel production process in a factory are evaluted by a team with respect to the three risk factor O (Occurance), S (Severity), D (Detection). The FMEA team consists of three experts $D^{(i)},(i=1,2,3)$.o perform the risk assessment and to recognize the most significant modes of failure to take critical steps. Due to their different domain knowledge and experience, they are associoated with the weighting vector $\lambda=(0.35,0.4,0.25)^{T}$. A total of ten potential failure modes have been identified by the FMEA team that included "non-acceptable formation" (A_1 ., "nipple thread pitted" (A_2 ., "arc formation loss" (A_3 ., "burn-out electrode" (A_4 ., "breaking of house of pipe" (A_5 ., "problem in moment of arm" (A_6 ., "refractory line damage" (A_7 ., "formation of steam" (A_8 ., "refractory line damage" (A_9 . and "moment of roof stop" (A_{10} .. FMEA team assessments for the ten FMs versus RFs are givren in Table 1.

The procedure steps of the proposed FMEA method for solving the above problem are given as follows:

Step 1: No need of normalizing process.

Step 2: By utilizing the Eq. (5), we calculate the group entropy matrix GE .which is given as:

Table 1. IVIFVs evaluation of the FMs

Experts	RFs	O	S	D
	FMs			
$D^{(1)}$	A_1	⟨[0.10,0.20],[0.70,0.80]⟩	⟨[0.10,0.20],[0.70,0.80]⟩	⟨[0.10,0.20],[0.70,0.80]⟩
	A_2	⟨[0.70,0.80],[0.10,0.20]⟩	⟨[0.70,0.80],[0.10,0.20]⟩	⟨[0.70,0.80],[0.10,0.20]⟩
	A_3	⟨[0.10,0.20],[0.70,0.80]⟩	⟨[0.10,0.20],[0.70,0.80]⟩	⟨[0.10,0.20],[0.70,0.80]⟩
	A_4	⟨[0.10,0.20],[0.70,0.80]⟩	⟨[0.10,0.20],[0.70,0.80]⟩	⟨[0.10,0.20],[0.70,0.80]⟩
	A_5	⟨[0.70,0.80],[0.10,0.20]⟩	⟨[0.70,0.80],[0.10,0.20]⟩	⟨[0.70,0.80],[0.10,0.20]⟩
	A_6	⟨[0.80,0.90],[0.05,0.10]⟩	⟨[0.80,0.90],[0.05,0.10]⟩	⟨[0.80,0.90],[0.05,0.10]⟩
	A_7	⟨[0.00,0.05],[0.90,0.90]⟩	⟨[0.00,0.05],[0.90,0.90]⟩	⟨[0.00,0.05],[0.90,0.90]⟩
	A_8	⟨[0.80,0.90],[0.05,0.10]⟩	⟨[0.80,0.90],[0.05,0.10]⟩	⟨[0.80,0.90],[0.05,0.10]⟩
	A_9	⟨[0.10,0.20],[0.70,0.80]⟩	⟨[0.10,0.20],[0.70,0.80]⟩	⟨[0.10,0.20],[0.70,0.80]⟩
	A_{10}	⟨[0.10,0.20],[0.70,0.80]⟩	⟨[0.10,0.20],[0.70,0.80]⟩	⟨[0.10,0.20],[0.70,0.80]⟩
$D^{(2)}$	A_1	⟨[0.70,0.80],[0.10,0.20]⟩	⟨[0.70,0.80],[0.10,0.20]⟩	⟨[0.70,0.80],[0.10,0.20]⟩
	A_2	⟨[0.70,0.80],[0.10,0.20]⟩	⟨[0.70,0.80],[0.10,0.20]⟩	⟨[0.70,0.80],[0.10,0.20]⟩
	A_3	⟨[0.00,0.05],[0.90,0.90]⟩	⟨[0.00,0.05],[0.90,0.90]⟩	⟨[0.00,0.05],[0.90,0.90]⟩
	A_4	⟨[0.80,0.90],[0.05,0.10]⟩	⟨[0.80,0.90],[0.05,0.10]⟩	⟨[0.80,0.90],[0.05,0.10]⟩
	A_5	⟨[0.30,0.40],[0.50,0.60]⟩	⟨[0.30,0.40],[0.50,0.60]⟩	⟨[0.30,0.40],[0.50,0.60]⟩
	A_6	⟨[0.00,0.05],[0.90,0.90]⟩	⟨[0.00,0.05],[0.90,0.90]⟩	⟨[0.00,0.05],[0.90,0.90]⟩
	A_7	⟨[0.10,0.20],[0.70,0.80]⟩	⟨[0.10,0.20],[0.70,0.80]⟩	⟨[0.10,0.20],[0.70,0.80]⟩
	A_8	⟨[0.50,0.50],[0.50,0.50]⟩	⟨[0.50,0.50],[0.50,0.50]⟩	⟨[0.50,0.50],[0.50,0.50]⟩
	A_9	⟨[0.00,0.05],[0.90,0.90]⟩	⟨[0.00,0.05],[0.90,0.90]⟩	⟨[0.00,0.05],[0.90,0.90]⟩
	A_{10}	⟨[0.00,0.05],[0.90,0.90]⟩	⟨[0.00,0.05],[0.90,0.90]⟩	⟨[0.00,0.05],[0.90,0.90]⟩
$D^{(3)}$	A_1	⟨[0.90,0.90],[0.00,0.05]⟩	⟨[0.90,0.90],[0.00,0.05]⟩	⟨[0.90,0.90],[0.00,0.05]⟩
	A_2	⟨[0.30,0.40],[0.50,0.60]⟩	⟨[0.30,0.40],[0.50,0.60]⟩	⟨[0.30,0.40],[0.50,0.60]⟩
	A_3	⟨[0.10,0.20],[0.70,0.80]⟩	⟨[0.10,0.20],[0.70,0.80]⟩	⟨[0.10,0.20],[0.70,0.80]⟩
	A_4	⟨[0.30,0.40],[0.50,0.60]⟩	⟨[0.30,0.40],[0.50,0.60]⟩	⟨[0.30,0.40],[0.50,0.60]⟩
	A_5	⟨[0.80,0.90],[0.05,0.10]⟩	⟨[0.80,0.90],[0.05,0.10]⟩	⟨[0.80,0.90],[0.05,0.10]⟩
	A_6	⟨[0.30,0.40],[0.50,0.60]⟩	⟨[0.30,0.40],[0.50,0.60]⟩	⟨[0.30,0.40],[0.50,0.60]⟩
	A_7	⟨[0.80,0.90],[0.05,0.10]⟩	⟨[0.80,0.90],[0.05,0.10]⟩	⟨[0.80,0.90],[0.05,0.10]⟩
	A_8	⟨[0.30,0.40],[0.50,0.60]⟩	⟨[0.30,0.40],[0.50,0.60]⟩	⟨[0.30,0.40],[0.50,0.60]⟩
	A_9	⟨[0.10,0.20],[0.70,0.80]⟩	⟨[0.10,0.20],[0.70,0.80]⟩	⟨[0.10,0.20],[0.70,0.80]⟩
	A_{10}	⟨[0.10,0.20],[0.70,0.80]⟩	⟨[0.10,0.20],[0.70,0.80]⟩	⟨[0.10,0.20],[0.70,0.80]⟩

$$GE = \begin{matrix} A_1 \\ A_2 \\ A_3 \\ A_4 \\ A_5 \\ A_6 \\ A_7 \\ A_8 \\ A_9 \\ A_{10} \end{matrix} \overset{\begin{matrix} O & \quad S & \quad D \end{matrix}}{\begin{pmatrix} 0.2153 & 0.2389 & 0.2881 \\ 0.2153 & 0.2785 & 0.2270 \\ 0.1591 & 0.1946 & 0.3040 \\ 0.1591 & 0.3263 & 0.2005 \\ 0.2626 & 0.3263 & 0.2548 \\ 0.2626 & 0.3263 & 0.2548 \\ 0.2626 & 0.3291 & 0.2153 \\ 0.1591 & 0.1946 & 0.3263 \\ 0.2626 & 0.3263 & 0.2548 \\ 0.3040 & 0.2548 & 0.3040 \end{pmatrix}}.$$

Step 3: Calculate the weights of RFs by using the Eq.(6) and obatined results are $\omega_1 = 0.3468$, $\omega_2 = 0.3229$, and $\omega_3 = 0.3303$.

Step 4: Overallperformance $Q(H_k).(k = 1,2,\ldots,10)$.for the FM H_k.by using the Eq. (7) as:

$Q(H_1)$=0.2470, $Q(H_2) =$ 0.2396, $Q(H_3) =$ 0.2184, $Q(H_4) =$ 0.2268, $Q(H_5) =$ 0.2806, $Q(H_6) =$ 0.2806, $Q(H_7) =$ 0.2685, $Q(H_8) =$ 0.2258 , $Q(H_9) =$ 0.2806 and $Q(H_{10}) = 0.2881$.

Step 5: Risk priority rank of FMs H_k.according to the $Q(H_k) H_{10} \succ H_5 = H_6 = H_9 \succ H_7 \succ H_1 \succ H_2 \succ H_4 \succ H_8 \succ H_3$.

From the results, it is clear that H_{10}.is the best FMs. In order to show the advantages and feasibility of the proposed FMEA method, authors conduct a comparative study with the existing methods and a sensitivity analysis of the FMs ranking.

COMPARISON AND DISCUSSION

To further demonstrate the effectiveness of the proposed FMEA method, authors utilize the different existings FMEA method(Li et al., 2019; Ye, 2010; Zavadskas et al., 2014; Zhao et al., 2017)on the above consider the illustrative illustration. Ranking orders by different existing methods are given in Table 2. From the Table 2, it is clear that the proposed FMEA method effectively solve the illustration and give a reasonable results.

Table 2. Ranking the FMs applying the different methods

FMs	IVIFV-MULTIMOORA (Zhao et al., 2017)	FMEA method (Li et al., 2019)	IVIFV-TOPSIS (Ye, 2010)	IVIFV-WASPAS (Zavadskas et al., 2014)	Proposed FMEA method
A_1	8	6	6	8	6
A_2	2	5	2	2	7
A_3	4	8	8	7	10
A_4	10	10	10	10	8
A_5	5	2	3	3	2
A_6	5	2	3	3	2
A_7	9	9	9	9	5
A_8	3	7	7	6	9
A_9	5	2	3	3	2
A_{10}	1	1	1	1	1

SENSITIVITY ANALYSIS

Here, authors conduct a sensitivity analysis by changing the weigts of the RFs as given in Table 3. Exp. 0 shows the RFs weight s that are evaluted by using the proposed entropy measures in this chapter. Other changes in the RFs weight s are shown by the the Exp. 1-4. Ranking order of the FMs under the different situations are given in Table 4.

Table 3. Weight information of sensitivity analysis

RFs	Exp 0	Exp 1	Exp 2	Exp 3	Exp 4
O	0.3468	0.3333	0.60	0.20	0.20
S	0.3229	0.3333	0.20	0.20	0.60
D	0.3303	0.3334	0.20	0.60	0.20

Table 4. Ranking orders of the sensitivity analysis

	Ranking order of the FMs
Exp 0	$H_{10} \succ H_5 = H_6 = H_9 \succ H_7 \succ H_1 \succ H_2 \succ H_4 \succ H_8 \succ H_3$
Exp 1	$H_{10} \succ H_5 = H_6 = H_9 \succ H_7 \succ H_1 \succ H_2 \succ H_4 \succ H_8 \succ H_3$
Exp 2	$H_{10} \succ H_5 = H_6 = H_9 \succ H_7 \succ H_1 \succ H_2 \succ H_4 \succ H_8 \succ H_3$
Exp 3	$H_{10} \succ H_5 = H_6 = H_9 \succ H_8 \succ H_1 \succ H_3 \succ H_7 \succ H_2 \succ H_4$
Exp 4	$H_5 = H_6 = H_9 \succ H_7 \succ H_{10} \succ H_4 \succ H_2 \succ H_1 \succ H_8 \succ H_3$

As according to the sensitivity analysis, change in the weight of RFs the different ranking of the FMs obtained. For the Exp 0,1,2,3 the best FMs is H_{10} .while for the Exp 4 the best FMs are H_5, H_6 .and H_9. Thus the sensitivity analysis shows that the RFs weights play a crucial role in finding the best FMs. Therefore, we should choose a desire method for finding the weights of the RFs. Hence, authors are use entropy measure for finding the RFs weights which is the powerfull tool for determing the uncertainty.

FUTURE RESEARCH DIRECTIONS

Fuzzy reliability analysis receiving more attention by the recerchers. This book provide a plateform to the researchers to give the new ides and techiniques about the fuzzy relaibility in the field of the different real-life issues. Therefore, in this chapter, authors developed a novel entropy measure and a novel FMEA method under the IVIFV environment based on the proposed entropy measure. In future, authors can apply this method in different extensions such as IFS, picture fuzzy set, dual hesitant fuzzy set. Type-2 fuzzy set, linguistic IFS, complex IFS, cubic fuzzy set, etc. Other than this, authors can add more risk factors to make method more effective according to the situations and time demand. In future, proposed FMEA method can be apply to handle the MADM problems and decision-making problems of real-life. Proposed FMEA method is an alternate solution to handling the decision-making issues.

CONCLUSION

This chapter mainly introduces the a novel FMEA method in context of IVIFVs environment which is an integral part of the reliability analysis. For this, authors propose a new entropy measure for IVIFSs which helps to depict the uncertainty of the decision-makers. Few axioms are illustrated to show the vailidty of the proposed entropy measure. Proposed entropy measure helps to find the RFs weights. Afterwards, anovel FMEA method has been developed based on entropy measure. A real-life case of FMEA analysis has been consider to illustrate the proposed novel FMEA method, and obtained results are compared with the existing FMEA methods. Compare outcomes show the validity and effectiveness of the proposed FMEA method. From the results, authors conclude that the proposed FMEA method is an effective and powerfull tool when the incorrect and extream evaluation information of decision makers exits. The proposed novel FMEA method overcomes the drawbacks of the tradetional FMEA methods and methods which include the aggregation operators. In future other rsiak factor can be added to improve the rationality of FMEA method. Otherthan this, authors will be extend this method for other environment and use the latest techniques and environment. This novel FMEA method can be use in the other field of the reliability analysis.

REFERENCES

Atanassov, K., & Gargov, G. (1989). Interval valued intuitionistic fuzzy sets. *Fuzzy Sets and Systems*, *31*(3), 343–349. doi:10.1016/0165-0114(89)90205-4

Atanassov, K. T. (1986). Intuitionistic fuzzy sets. *Fuzzy Sets and Systems*, *20*(1), 87–96. doi:10.1016/S0165-0114(86)80034-3

Boral, S., Howard, I., Chaturvedi, S. K., McKee, K., & Naikan, V. N. A. (2020). An integrated approach for fuzzy failure modes and effects analysis using fuzzy AHP and fuzzy MAIRCA. *Engineering Failure Analysis*, *108*, 104195. doi:10.1016/j.engfailanal.2019.104195

Bowles, J. B., & Peláez, C. E. (1995). Fuzzy logic prioritization of failures in a system failure mode, effects and criticality analysis. *Reliability Engineering & System Safety*, *50*(2), 203–213. doi:10.1016/0951-8320(95)00068-D

Chang, K. H., & Cheng, C.-H. (2011). Evaluating the risk of failure using the fuzzy OWA and DEMATEL method. *Journal of Intelligent Manufacturing*, *22*(2), 113–129. doi:10.100710845-009-0266-x

Chen, S. M., & Huang, Z. C. (2017). Multiattribute decision making based on interval-valued intuitionistic fuzzy values and linear programming methodology. *Information Sciences*, *381*, 341–351. doi:10.1016/j.ins.2016.11.010

De Luca, A., & Termini, S. (1993). A definition of a nonprobabilistic entropy in the setting of fuzzy sets theory. In *Readings in Fuzzy Sets for Intelligent Systems* (pp. 197–202). Elsevier. doi:10.1016/B978-1-4832-1450-4.50020-1

Deluca, A. T. S., & Termini, S. (1972). A definition of a non probabilistic entropy in the setting of fuzzy set theory. *Information and Control*, *20*(4), 301–312. doi:10.1016/S0019-9958(72)90199-4

Garg, H., & Kumar, K. (2018). Group decision making approach based on possibility degree measure under linguistic interval-valued intuitionistic fuzzy set environment. *Journal of Industrial and Management Optimization*, *13*(5), 1–23. doi:10.3934/jimo.2018162

Garg, H., & Kumar, K. (2019). Linguistic interval-valued Atanassov intuitionistic fuzzy sets and their applications to group decision-making problems. *IEEE Transactions on Fuzzy Systems*, *27*(12), 2302–2311. doi:10.1109/TFUZZ.2019.2897961

Garg, H., & Kumar, K. (2020a). A novel exponential distance and its based TOPSIS method for interval-valued intuitionistic fuzzy sets using connection number of SPA theory. *Artificial Intelligence Review*, *53*(1), 595–624. doi:10.100710462-018-9668-5

Garg, H., & Kumar, K. (2020b). A novel possibility measure to interval-valued intuitionistic fuzzy set using connection number of set pair analysis and its applications. *Neural Computing & Applications*, *32*(8), 3337–3348. doi:10.100700521-019-04291-w

Garg, H., & Kumar, K. (2020c). Power Geometric Aggregation Operators Based on Connection Number of Set Pair Analysis Under Intuitionistic Fuzzy Environment. *Arabian Journal for Science and Engineering*, *45*(3), 2049–2063. doi:10.100713369-019-03961-0

Huang, J., Li, Z. S., & Liu, H. C. (2017). New approach for failure mode and effect analysis using linguistic distribution assessments and TODIM method. *Reliability Engineering & System Safety*, *167*, 302–309. doi:10.1016/j.ress.2017.06.014

Joshi, D., & Kumar, S. (2018). Improved accuracy function for interval-valued intuitionistic fuzzy sets and its application to multi—Attributes group decision making. *Cybernetics and Systems*, *49*(1), 64–76. doi:10.1080/01969722.2017.1412890

Kumar, K., & Garg, H. (2018). {TOPSIS} method based on the connection number of set pair analysis under interval-valued intuitionistic fuzzy set environment. *Computational & Applied Mathematics*, *37*(2), 1319–1329. doi:10.100740314-016-0402-0

Li, Y. L., Wang, R., & Chin, K.-S. (2019). New failure mode and effect analysis approach considering consensus under interval-valued intuitionistic fuzzy environment. *Soft Computing*, *23*(22), 11611–11626. doi:10.100700500-018-03706-5

Liu, H. C. (2016). FMEA using uncertainty theories and MCDM methods. In *FMEA using uncertainty theories and MCDM methods* (pp. 13–27). Springer.

Liu, H. C., You, J. X., Chen, S., & Chen, Y. Z. (2016). An integrated failure mode and effect analysis approach for accurate risk assessment under uncertainty. *IIE Transactions*, *48*(11), 1027–1042. doi:10.1080/0740817X.2016.1172742

Liu, H. C., You, J. X., & Duan, C. Y. (2019). An integrated approach for failure mode and effect analysis under interval-valued intuitionistic fuzzy environment. *International Journal of Production Economics*, *207*, 163–172. doi:10.1016/j.ijpe.2017.03.008

Liu, H. C., Zhang, L. J., Ping, Y. J., & Wang, L. (2020). Failure mode and effects analysis for proactive healthcare risk evaluation: A systematic literature review. *Journal of Evaluation in Clinical Practice*, *26*(4), 1320–1337. doi:10.1111/jep.13317 PMID:31849153

Liu, M., & Ren, H. (2014). A new intuitionistic fuzzy entropy and application in multi-attribute decision making. *Information (Basel)*, *5*(4), 587–601. doi:10.3390/info5040587

Maity, S., De, S. K., & Mondal, S. P. (2020). A study of a backorder EOQ model for cloud-type intuitionistic dense fuzzy demand rate. *International Journal of Fuzzy Systems*, *22*(1), 201–211. doi:10.100740815-019-00756-1

Meng, F., & Chen, X. (2016). Entropy and similarity measure for Atannasov's interval-valued intuitionistic fuzzy sets and their application. *Fuzzy Optimization and Decision Making*, *15*(1), 75–101. doi:10.100710700-015-9215-7

Mishra, A. R., Rani, P., Pardasani, K. R., Mardani, A., Stević, Ž., & Pamučar, D. (2020). A novel entropy and divergence measures with multi-criteria service quality assessment using interval-valued intuitionistic fuzzy TODIM method. *Soft Computing*, *24*(15), 1–21. doi:10.100700500-019-04627-7

Mondal, S. P., & Roy, T. K. (2015). Second order linear differential equations with generalized trapezoidal intuitionistic Fuzzy boundary value. *Journal of Linear and Topological Algebra*, *4*(2), 115–129.

Mondal, S. P. (2016). Differential equation with interval valued fuzzy number and its applications. *International Journal of System Assurance Engineering and Management*, *7*(3), 370–386. doi:10.100713198-016-0474-7

Mondal, S. P. (2018). Interval valued intuitionistic fuzzy number and its application in differential equation. *Journal of Intelligent & Fuzzy Systems*, *34*(1), 677–687. doi:10.3233/JIFS-161898

Mondal, S. P., Goswami, A., & Kumar De, S. (2019). Nonlinear Triangular Intuitionistic Fuzzy Number and Its Application in Linear Integral Equation. *Advances in Fuzzy Systems*, *2019*, 2019. doi:10.1155/2019/4142382

Mondal, S. P., Mandal, M., & Bhattacharya, D. (2018). Non-linear interval-valued fuzzy numbers and their application in difference equations. *Granular Computing*, *3*(2), 177–189. doi:10.100741066-017-0063-4

Nguyen, H. (2016). A new interval-valued knowledge measure for interval-valued intuitionistic fuzzy sets and application in decision making. *Expert Systems with Applications*, *56*, 143–155. doi:10.1016/j.eswa.2016.03.007

Priyadharsini, J., & Balasubramaniam, P. (2019). Multi-criteria decision making method based on interval-valued intuitionistic fuzzy sets. *The Journal of Analysis*, *27*(1), 259–276. doi:10.100741478-018-0122-5

Qin, J., Xi, Y., & Pedrycz, W. (2020). Failure mode and effects analysis (FMEA) for risk assessment based on interval type-2 fuzzy evidential reasoning method. *Applied Soft Computing*, *89*, 106134. doi:10.1016/j.asoc.2020.106134

Rashid, T., Faizi, S., & Zafar, S. (2018). Distance Based Entropy Measure of Interval-Valued Intuitionistic Fuzzy Sets and Its Application in Multicriteria Decision Making. *Advances in Fuzzy Systems*. doi:10.1155/2018/3637897

Shannon, C. E. (1948). A mathematical theory of communication. *The Bell System Technical Journal*, *27*(3), 379–423. doi:10.1002/j.1538-7305.1948.tb01338.x

Song, W., Ming, X., Wu, Z., & Zhu, B. (2013). Failure modes and effects analysis using integrated weight-based fuzzy TOPSIS. *International Journal of Computer Integrated Manufacturing*, *26*(12), 1172–1186. doi:10.1080/0951192X.2013.785027

Tian, Z., Wang, J., & Zhang, H. (2018). An integrated approach for failure mode and effects analysis based on fuzzy best-worst, relative entropy, and VIKOR methods. *Applied Soft Computing*, *72*, 636–646. doi:10.1016/j.asoc.2018.03.037

Tooranloo, H. S. (2016). A model for failure mode and effects analysis based on intuitionistic fuzzy approach. *Applied Soft Computing*, *49*, 238–247. doi:10.1016/j.asoc.2016.07.047

Vahdani, B., Salimi, M., & Charkhchian, M. (2015). A new FMEA method by integrating fuzzy belief structure and TOPSIS to improve risk evaluation process. *International Journal of Advanced Manufacturing Technology*, *77*(1–4), 357–368. doi:10.100700170-014-6466-3

Verma, R., & Sharma, B. D. (2013). Exponential entropy on intuitionistic fuzzy sets. *Kybernetika*, *49*(1), 114–127.

Wang, C. Y., & Chen, S.-M. (2017). An improved multiattribute decision making method based on new score function of interval-valued intuitionistic fuzzy values and linear programming methodology. *Information Sciences*, *411*, 176–184. doi:10.1016/j.ins.2017.05.022

Wang, J. Q., & Wang, P. (2012). Intuitionistic linguistic fuzzy multi-criteria decision-making method based on intuitionistic fuzzy entropy. *Control and Decision*, *27*(11), 1694–1698.

Wang, L. E., Liu, H. C., & Quan, M. Y. (2016). Evaluating the risk of failure modes with a hybrid MCDM model under interval-valued intuitionistic fuzzy environments. *Computers & Industrial Engineering*, *102*, 175–185. doi:10.1016/j.cie.2016.11.003

Wei, C. P., Gao, Z. H., & Guo, T. T. (2012). An intuitionistic fuzzy entropy measure based on trigonometric function. *Control and Decision*, *27*(4), 571–574.

Xu, Z. S. (2007). Models for multiple attribute decision making with intuitionistic fuzzy information. *International Journal of Uncertainty, Fuzziness and Knowledge-based Systems*, *15*(03), 285–297. doi:10.1142/S0218488507004686

Ye, F. (2010). An extended {TOPSIS} method with interval-valued intuitionistic fuzzy numbers for virtual enterprise partner selection. *Expert Systems with Applications*, *37*(10), 7050–7055. doi:10.1016/j.eswa.2010.03.013

Zadeh, L. A. (1965). Fuzzy sets. *Information and Control*, *8*(3), 338–353. doi:10.1016/S0019-9958(65)90241-X

Zavadskas, E. K., Antucheviciene, J., Hajiagha, S. H. R., & Hashemi, S. S. (2014). Extension of weighted aggregated sum product assessment with interval-valued intuitionistic fuzzy numbers (WASPAS-IVIF). *Applied Soft Computing*, *24*, 1013–1021. doi:10.1016/j.asoc.2014.08.031

Zhang, Q. S., & Jiang, S. Y. (2008). A note on information entropy measures for vague sets and its applications. *Information Sciences*, *178*(21), 4184–4191. doi:10.1016/j.ins.2008.07.003

Zhao, H., You, J. X., & Liu, H. C. (2017). Failure mode and effect analysis using MULTIMOORA method with continuous weighted entropy under interval-valued intuitionistic fuzzy environment. *Soft Computing*, *21*(18), 5355–5367. doi:10.100700500-016-2118-x

Chapter 7
Solving Interval Type-2 Fuzzy Reliability-Redundancy Allocation Systems With Efficient PSO Algorithm

Zubair Ashraf
https://orcid.org/0000-0001-7122-2856
South Asian University, India

Md. Gulzarul Hasan
Madanapalle Institute of Technology and Science, India

Mohd Shoaib Khan
South Asian University, India

ABSTRACT

A new type-2 fuzzy reliability-redundancy allocation problem (IT2FRRAP) is introduced in this study, which maximizes system reliability underneath the type-2 fuzzy uncertainty. Throughout the design process, estimation of system parameters with type-2 fuzzy sets is very rational, as many experts may have varying views on the numerous risks involved with the multi-state framework. The authors consider fuzzy membership functions of interval type-2 to fuzzy set to reflect the reliability of the sub-system. By using the extension theory, they evaluate the total system reliability as per respective configurations (i.e., series, parallel, non-parallel, and bridge). They developed a novel solution method focused on particle swarm optimization to address the IT2FRRAP. They carried out the computational simulations on various numerical databases to show the gravity of the proposed formulations. To compare its productiveness, a detailed analysis of the proposed solution was made with the genetic algorithm (GA).

DOI: 10.4018/978-1-7998-7564-2.ch007

INTRODUCTION

As the environment is more and more tech-savvy, a large volume of international investment goes into the generation of the complex-standard and quite a reliable system. Consequently, the reliability calculation is deemed an inherently valuable method that can be applied to determine the lifespan of numerous systems in specific industries, such as robots, power plants, satellites, and so on. Seeing that individual of particular industries has distinct requirements, they adapt efficiency testing schemes to their preferences. Over the last few decades, this field of study has increasingly been the science community's priority.

Reliability of a system is interpreted as the capability of the system to fulfill the required action within a specific operating environment for a particular timeframe (Rausand, 2014). Here the component might have been an aspect or a subsystem. The manufacturer's job may be one process or a mixture of tasks, while the timeframe can differ from days to years. The system's experts use reliability testing to make critical choices on the architecture of a device. Such required judgments are meant to cope with issues that may occur before or after the program begins service. The problems may be everything except design, specification management, retrieval, device maintenance, etc.

A system's efficiency can be improved in two main ways. First, by increasing the functionality of the independent units, this approach isn't suitable because of the implausibility of providing components with high efficiency present at the very same time as needed. Second, though redundancy including its modules, this strategy includes using many expendable or redundant modules which influence the functionality of the system by having its majority of the redundancy if any particular component were to malfunction. Nevertheless, other concerns like weight, cost, and volume become adversely impacted mostly with the inclusion of all these redundant elements to a design. Simply because of that, a pay-off between the reliability of the structure as well as its redundancies within unique design restrictions is essential. The issue is labeled as the reliability-redundancy allocation problem (RRAP).

Several mathematical models have been developed to formulate the RRAP (Kuo & Prasad, 2000), (Ramirez-Marquez & Coit, 2004), (Tian, Zuo, & Huang, 2008), (Wang, Tianshi, Tang, & Yao, 2009), (Wei-Chang Yeh, Yi-Cheng Lin, Yuk Ying Chung, & Mingchang Chih, 2010); amongst them the furthermost common one was proposed by (Kuo & Prasad, 2000). The traditional RRAP aims to optimize the efficiency of the system underneath the system's cost, volume, and weight limitations. Mathematically, the RRAP model is defined as follows:

$$\text{Maximize } R_s(r,n). \tag{1}$$

Subject to $C_s(r,n) \le C$. $W_s(w,n) \le W$. $V_s(v,n) \le V$. $r = (r_1,\ldots,r_m)$. $n = (n_1,\ldots,n_m)$. $r_i \in [0,1] \subset \mathbb{R}$. $n_i \subset -^{+}$. In above model, the objective functions is the total reliability of a system, represented by $R_s(r,n)$.while the restraints cost, weight and volume, are represented by $C_s(r,n)$. $W_s(w,n)$.and $V_s(v,n)$. respectively.

The traditional RRAP systems are designed underneath the premise that all the knowledge surrounding the system's efficiency and expense parameters is precise or deterministic. However, as we think about systems architecture in actual life, various complexities occur at all stages of implementation. Such uncertainties contribute to critical parts of reliability of the system and may emerge for a variety of reasons (Nannapaneni & Mahadevan, 2016), (Zio, 2009), (Muhuri, Ashraf, & Danish Lohani, 2018): planners, parametric, production and environmental possibility. Incorporating these complexities into the device design phase dramatically affects the quality of the final program. In detailed, fuzzy reliability theory has been incorporated to promise through certain hesitations with wonderful advancement (Mahapatra & Roy, 2006), (Klir & Yuan, 2008), (Garg & Sharma, 2013), (Soltani, 2014), (Ashraf, Khan, & Lohani, 2019).

Many studies were performed on reliability modelling focused on problem categories, process frameworks, and methodology classification, as described in (Tzafestas, 1980), and (Soltani, 2014). W. Kuo offered an outline of conventional optimization of reliability coupled with evolving solutions methods (Kuo & Prasad, 2004; Kuo & Wan, 2007). In 2014, R. Soltani (Soltani, 2014) provided several lists to analyze the studies from the multiple viewpoints of modelling uncertainty on the RRAP. Cheng and Mon (1993) suggested an approach in order to achieve fuzzy reliability through alpha-cutting using interval algebra which requires the use of fuzzy numbers (FNs) across a confidence interval. The (Mon & Cheng, 1994) have discussed the issue of providing differing probabilities of malfunction in specific system components then proposed fuzzy distributions rather than likelihood functions to compensate for such a discrepancy. Chen (1994) described and the system's reliability through triangular FNs in creating a framework for evaluating a fuzzy reliability scheme that uses FN arithmetic and logical operations. Huang (1997) suggested an ambiguous decision-making approach for the multi-goal optimization process to maximize the reliability of a device with much more than one goal. Ravi et al. (2000) developed a fuzzy multi-goal optimization process for complicated structures, whereby not just the functionality as well as the price, weights and volume of design is regarded as vague goals. The researchers have looked at the effects of multiple aggregate operations upon this solving problem.

Mahapatra and Roy (2006) have addressed the multiple objective strategic decisions of series and complex system reliability through the use of the FNs to cope mostly with ambiguity around the knowledge gathered from the system analysts. Ebrahimipour and Sheikhalishahi (2011) developed the fuzzy RRAP problem utilizing triangular FNs for the inconsistency throughout the element specifications. Garg and Sharma (2013) have formulated the multi-objective RRAP of the fuzzy purposes and have overcome it by using PSO employing weight preferences approach. Komal and Sharma (2014) built a Lambda-Tau framework with genetic algorithm methodology that sequenced learning algorithm as well as genetic algorithm to evaluate the fuzzy reliability of the paper plant's washing process.

Thought, fuzzy sets theory have been used in RRAP, type-1 fuzzy sets (T1 FS) are not sufficient enough to accommodate the decision-makers' diverse viewpoints. Therefore, Type-2 fuzzy sets (T2 FS) initially proposed by (Zadeh, 1975) seem to be the most effective method to modeling the numerous uncertainties arising from various perspectives as well as the diverse views of framework expertise (Bustince et al., 2016). The IT2 FSs have even been implemented in a variety of real-life implementations, such as (Gaxiola, Melin, Valdez, Castro, & Castillo, 2016), (Ashraf, Malhotra, Muhuri, & Danish Lohani, 2017), (Valdez, Vazquez, Melin, & Castillo, 2017), (Ashraf, Roy, Muhuri, & Danish Lohani, 2018), (Ashraf, Roy, Muhuri, & Danish Lohani, 2020) and so on. Moreover, (Ashraf, Muhuri, Danish Lohani, & Nath, 2014) developed a fuzzy multi-objective reliability redundancy allocation optimization with fuzzy type-2 uncertainty and solve it using NSGA-II (non- dominated sorting genetic algorithm-II). (Ashraf, Muhuri, & Danish Lohani, 2015) reported the reliability-redundancy allocation problem through a particle swam optimization based procedure in which the parameters are as type-2 fuzzy quantities.

Recently, (Muhuri et al., 2018) suggested an interval type-2 fuzzy multi-objective RRAP, for simulate higher-order uncertainties within system element specifications utilizing IT2 FNs, and use NSGA-II to achieve the best solution. Using the interval type-2 fuzzy sets for the maximization of the system IT2 fuzzy reliability simultaneously minimization of the system IT2 fuzzy cost, Ashraf et al. in (Ashraf, Muhuri, Lohani, & Roy, 2019) proposed a fuzzy multi-objective reliability–redundancy allocation problem. Centered on the description earlier, while modeling the reliability of the modules, we proposed type-2 fuzzy sets (T2 FSs). Owing to the higher computational workload of T2 FS, we employed type-2 fuzzy sets of intervals.

A newly interval type-2 fuzzy reliability-redundancy allocation problem (IT2FRRAP) is introduced in this chapter, which maximizes system reliability in the type-2 fuzzy condition under cost, weight, and volume restrictions. IT2FRRAP's purpose is designed as interval type-2 fuzzy sets (IT2 FSs) with the associated type-2 fuzzy membership functions (IT2 MFs). Rendering to respective structures, i.e. series, parallel, series-parallel and bridge, the IT2 MFs of entire system reliability is determined by consolidating the sub-system reliabilities utilizing the Zadeh's extension theory. We propose a new solution strategy focused on PSO (particle swarm optimization) to solve optimization problems for all four formulations of IT2FRRAP. Experimental simulations for the formulations were carried out using computational instances, including series, parallel, series-parallel and bridge IT2FRRAPs. We carried out experimental computations on real-life numeric sets of data to illustrate the usefulness of our presented formulations.

The rest of the chapter is as follows. The conceptual preliminary findings that would be used in this work are stated in Section 2. The detailed mathematical description of the theoretical problem is set out in Section 3. Section 4 discusses the full solution to these problems. In Section 5, we implemented experimental simulations, including computational measurements, and illustrated the effectiveness of our proposed strategy by comparisons. After this, we shall conclude in Section 6.

PRELIMINARIES

Type-1 fuzzy set (T1 FS) (Zadeh, 1965)**:** The T1 FS (or fuzzy set) A .should be defined by the basic functions called type-1 membership function (T1 MF) of the component over the universal set X .as described in the following:

$$A = \{x, \mu_A(x) \mid \forall x \in X\}. \tag{1}$$

where $\mu_A(x)$ .symbolizes the T1 MF of A. such that $0 \leq \mu_A(x) \leq 1$.

Interval type-2 fuzzy set (IT2 FS) (John, 1998): A type-2 fuzzy set (T2 FS) $\tilde{A}$ .is described through a T2 MF (type-2 membership function) that himself is T1 FS. $\tilde{A}$ may be described in the following:

$$\tilde{A} = \{(x,u), \mu_{\tilde{A}}(x,u) \mid \forall x \in X, \forall u \in J_x \subseteq [0,1]\}. \tag{2}$$

here, x .is a primary variable occupied within the universe of discourse X. u .is a secondary variable in such a way the $u \in J_x$ .for every $x \in X$ .and J_x .represents the primary membership grade of x.

Once entirely secondary membership grades of T2 FS will equal to 1, that is, $\mu_{\tilde{A}}(x,u) = 1$. T2 FS converts to IT2 FS (interval type-2 fuzzy set) (J.M. Mendel & John, 2002) as follows:

$$\tilde{A} = \{(x,u), 1 \mid \forall x \in X, \forall u \in J_x \subseteq [0,1]\}. \tag{3}$$

Footprint of uncertainty (Jerry M. Mendel & Wu, 2010): Doubtfulness in an IT2 FS $\tilde{A}$ .involves is a closed area named the FOU (footprint of uncertainty). It is a 2-dimensional support, i.e.

$$FOU(\tilde{A}) = \{(x,u) \in X \times U \mid \mu_{\tilde{A}}(x,u) \rangle 0\}. \tag{4}$$

The UMF (upper membership function) as well as LMF (lower membership function) of $\tilde{A}$ .are T1 MFs which unavoidable by FOU ($\tilde{A}$. as displayed in Fig 1 (Jerry M. Mendel & Wu, 2010). The UMF and LMF are related through the superior certain represented by $\bar{\mu}_{\tilde{A}}(x)$. $\forall x \in X$ .and the inferior certain represented by $\underline{\mu}_{\tilde{A}}(x)$. $\forall x \in X$ .of FOU($\tilde{A}$., correspondingly.

$$\underline{\mu}_{\tilde{A}}(x) = LMF(\tilde{A}) = \inf\{u \mid u \in [0,1], \mu_{\tilde{A}}(x) \rangle 0\} \ \forall x \in X. \tag{5}$$

$$\bar{\mu}_{\tilde{A}}(x) = UMF(\tilde{A}) = \sup\{u \mid u \in [0,1], \mu_{\tilde{A}}(x) \rangle 0\} \qquad \forall x \in X. \tag{6}$$

Hence, the IT2 MF $\mu_{\tilde{A}}(x)$ can be said as follows:

$$\mu_{\tilde{A}}(x) = [\underline{\mu}_{\tilde{A}}(x), \bar{\mu}_{\tilde{A}}(x)]$$

Figure 1. Interval type-2 fuzzy set (IT2 FS)

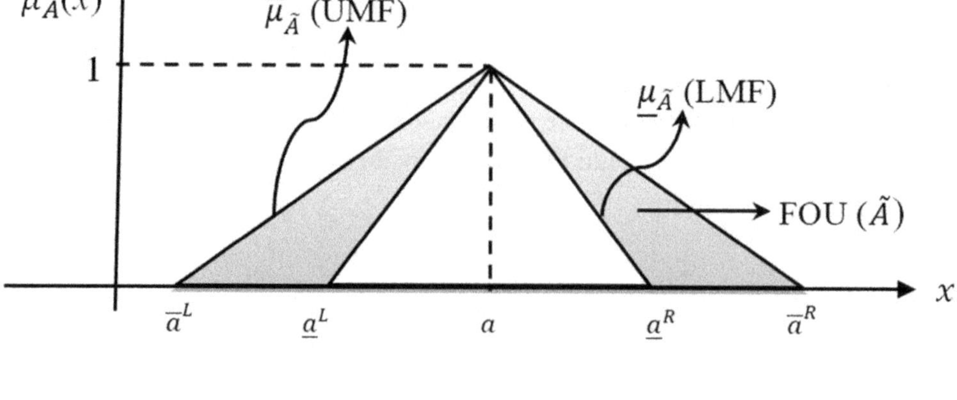

A pictorial demonstration of triangular IT2 FS is shown in Fig. 1 where triangular T1 MFs represents both the LMF as well as UMF. In this figure, $\overline{a}^L . \overline{a}^R$ .are the left and right upper ends of UMF and $\underline{a}^L . \underline{a}^R$ .are the left and right lower ends of LMF, which generates the triangular fuzzy number corresponding to $\overline{\mu}_{\tilde{A}}\left(x;\overline{a}^L, a, \overline{a}^R\right)$ .and $\underline{\mu}_{\tilde{A}}\left(x;\underline{a}^L, a, \underline{a}^R\right)$. respectively.

Let us consider, $\tilde{A}$ .and $\tilde{B}$ .are two IT2 FSs and $\mu_{\tilde{A}}(x)$ .and $\mu_{\tilde{B}}(x)$ .represents their respective IT2 MF, then the three basic set operations can be considered as follows:

Union: $\mu_{\tilde{A}\oplus\tilde{B}}(x) = \mu_{\tilde{A}}(x) \vee \mu_{\tilde{B}}(x) = \left[\underline{\mu}_{\tilde{A}}(x) \vee \underline{\mu}_{\tilde{B}}(x), \overline{\mu}_{\tilde{A}}(x) \vee \overline{\mu}_{\tilde{B}}(x)\right]$.(8)

Intersection: $\mu_{\tilde{A}\wedge\tilde{B}}(x) = \mu_{\tilde{A}}(x) \wedge \mu_{\tilde{B}}(x) = \left[\underline{\mu}_{\tilde{A}}(x) \wedge \underline{\mu}_{\tilde{B}}(x), \overline{\mu}_{\tilde{A}}(x) \wedge \overline{\mu}_{\tilde{B}}(x)\right]$.(9)

Complement: $1 \ominus \mu_{\tilde{A}}(x) = \left[1 - \underline{\mu}_{\tilde{A}}(x), 1 - \overline{\mu}_{\tilde{A}}(x)\right]$.(10)

Figure 2. Layout of system configurations

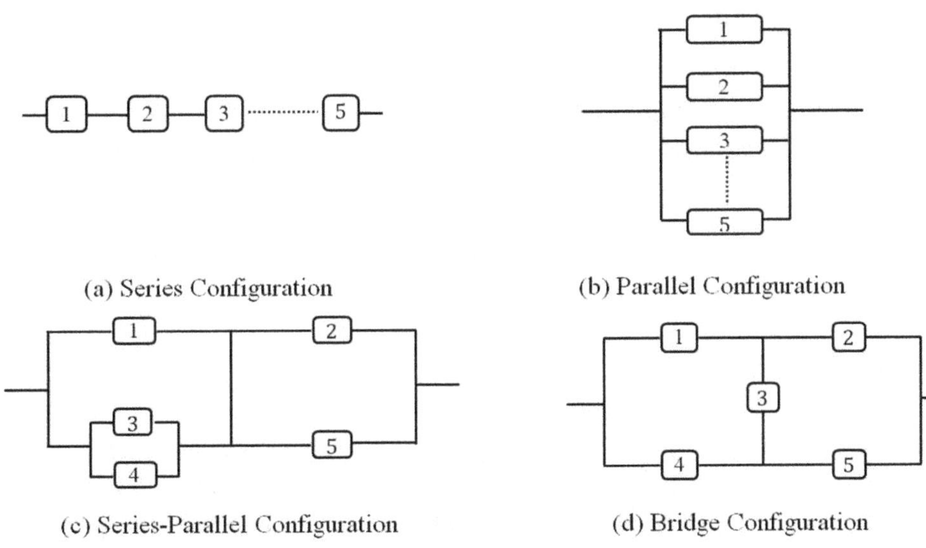

PROPOSED PROBLEM FORMULATIONS

Assumptions

1. Underlying characteristics are deterministic for all the components.

2. Components include sequential (active/deactivated) operational conditions and are anti-repairable.
3. Redundant components within each module are identical.
4. Any component faults are time-independent, and so don't harm the systems.

Notations

m Total number of component in the system.
n_i Redundancy in the i-th component.
r_i Component reliability.
w_i Component weight.
v_i Component volume.
α_i, β_i Shaping and scaling parameters of components.
R_S Total reliability of the system.
C_S Total cost of the system.
V_S Total volume of the system.
W_S Total weight of the system.
$\tilde{R}_i$ Interval type-2 fuzzy reliability of the redundant module.
$\tilde{R}_s$ Total IT2 fuzzy reliability function of system.
$\mu_{\tilde{R}_S}$ Interval type-2 fuzzy membership function (IT2 MFs) of total reliability.
$\underline{\mu}_{\tilde{R}_S}, \overline{\mu}_{\tilde{R}_S}$ LMF (lower membership function) and UMF (upper membership function).
$\underline{R}_i^L, \underline{R}_i^R$ Left and right ends of LMF in the i-th sub-systems' reliability.
$\overline{R}_i^L, \overline{R}_i^R$ Left and right ends of UMF in the i-th sub-systems' reliability.

System Configurations

In the present work, we have considered a RRAP on the series, parallel, series-parallel systems and bridge system configurations as displayed in Fig. 2. For all the four configurations, m components are of redundant that is $n_i; i = 1,\ldots,m$. The IT2 fuzzy optimization model is established through correlating a grade of gratification to the reliability of all systems._

Suppose, $\tilde{R}_1, \tilde{R}_2, \ldots, \tilde{R}_m$ symbolizes the IT2 fuzzy reliabilities m redundant components. The total system IT2 fuzzy reliability $\tilde{R}_s$ is characterized by the structure function φ (Chen, 1994), (Mon & Cheng, 1994) as follows:

$$\tilde{R}_s = \varphi\left(\tilde{R}_1, \tilde{R}_2, \ldots, \tilde{R}_m\right) \tag{11}$$

Where φ represents the structural combination of different configurations for system. For all the four systems the goal is the maximization of systems' reliability restricted to the non-linear constraints and the reliabilities of each redundant component as well as allocations of redundancy are to be decision variables. Hence, the IT2 fuzzy optimization models are expressed as below.

Series System

In series system (Fig. 2(a)), the n_m redundant components are linked in series. The IT2 fuzzy reliability of series system is thus calculated by the following equations:

$$\tilde{R}_S(r,n) = \tilde{R}_1(r_1,n_1) \otimes \tilde{R}_2(r_2,n_2) \otimes \ldots \otimes \tilde{R}_m(r_m,n_m) \tag{12}$$

So, the IT2 MF of $\tilde{R}_S$, represented by $\mu_{\tilde{R}_S}(x) = \left[\underline{\mu}_{\tilde{R}_S}(x), \bar{\mu}_{\tilde{R}_S}(x)\right]$, may be calculated using the follows:

$$\mu_{\tilde{R}_S}(x) = \mu_{\tilde{R}_1}(x_1) \otimes \mu_{\tilde{R}_2}(x_2) \otimes \ldots \otimes \mu_{\tilde{R}_m}(x_m) \tag{13}$$

$$\left[\underline{\mu}_{\tilde{R}_S}(x), \bar{\mu}_{\tilde{R}_S}(x)\right] = \left[\begin{matrix} \underline{\mu}_{\tilde{R}_1}(x_1) \otimes \underline{\mu}_{\tilde{R}_2}(x_2) \otimes \ldots \otimes \underline{\mu}_{\tilde{R}_m}(x_m), \\ \bar{\mu}_{\tilde{R}_1}(x_1) \otimes \bar{\mu}_{\tilde{R}_2}(x_2) \otimes \ldots \otimes \bar{\mu}_{\tilde{R}_m}(x_m) \end{matrix}\right]$$

$$= \left[\wedge_{i=1}^{m} \underline{\mu}_{\tilde{R}_i}(x_i), \wedge_{i=1}^{m} \bar{\mu}_{\tilde{R}_i}(x_i)\right] \tag{14}$$

Parallel System

For parallel system (Fig. 2(b)), the m redundant components are linked in parallel. The IT2 fuzzy reliability of parallel system is thus calculated by the following equations:

$$\tilde{R}_S(r,n)=1\ominus\left(\left(1\ominus\tilde{R}_1(r_1,n_1)\right)\otimes\left(1\ominus\tilde{R}_2(r_2,n_2)\right)\otimes\ldots\otimes\left(1\ominus\tilde{R}_m(r_m,n_m)\right)\right) \tag{15}$$

So, the IT2 MF of $\tilde{R}_S$, represented by $\mu_{\tilde{R}_S}(x)=\left[\underline{\mu}_{\tilde{R}_S}(x),\bar{\mu}_{\tilde{R}_S}(x)\right]$, may be calculated using the follows:

$$\mu_{\tilde{R}_S}(x)=1\ominus\left(\left(1\ominus\mu_{\tilde{R}_1}(x_1)\right)\otimes\left(1\ominus\mu_{\tilde{R}_2}(x_2)\right)\otimes\ldots\otimes(1\ominus\mu_{\tilde{R}_m}(x_m)\right) \tag{16}$$

$$\left[\underline{\mu}_{\tilde{R}_S}(x),\bar{\mu}_{\tilde{R}_S}(x)\right]=\begin{bmatrix}1\ominus\left(\left(1\ominus\underline{\mu}_{\tilde{R}_1}(x_1)\right)\otimes\left(1\ominus\underline{\mu}_{\tilde{R}_2}(x_2)\right)\otimes\ldots\otimes\left(1\ominus\underline{\mu}_{\tilde{R}_m}(x_m)\right)\right),\\ 1\ominus\left(\left(1\ominus\bar{\mu}_{\tilde{R}_1}(x_1)\right)\otimes\left(1\ominus\bar{\mu}_{\tilde{R}_2}(x_2)\right)\otimes\ldots\otimes\left(1\ominus\bar{\mu}_{\tilde{R}_m}(x_m)\right)\right)\end{bmatrix}$$

$$=\left[1\ominus\wedge_{i=1}^{m}\left(1\ominus\underline{\mu}_{\tilde{R}_i}(x_i)\right),1\ominus\wedge_{i=1}^{m}\left(1\ominus\bar{\mu}_{\tilde{R}_i}(x_i)\right)\right] \tag{17}$$

Series-Parallel System

In series-parallel system (Fig. 2(c)), the five redundant components are linked in series-parallel structure. The IT2 fuzzy reliability of series-parallel system is thus calculated by the following equations:

$$\tilde{R}_S(r,n)=1\ominus\left(1\ominus\tilde{R}_1\otimes\tilde{R}_2\right)\left[1\ominus\left(\tilde{R}_3\oplus\tilde{R}_4\ominus\tilde{R}_3\otimes\tilde{R}_4\right)\otimes\tilde{R}_3\right] \tag{18}$$

So, the IT2 MF of $\tilde{R}_S$, represented by $\mu_{\tilde{R}_S}(x)=\left[\underline{\mu}_{\tilde{R}_S}(x),\bar{\mu}_{\tilde{R}_S}(x)\right]$, may be calculated using the follows:

$$\mu_{\tilde{R}_S}(x)=1\ominus\left(1\ominus\mu_{\tilde{R}_1}(x_1)\otimes\mu_{\tilde{R}_2}(x_2)\right)\left[1\ominus\left(\mu_{\tilde{R}_3}(x_3)\oplus\mu_{\tilde{R}_4}(x_4)\ominus\mu_{\tilde{R}_3}(x_3)\otimes\mu_{\tilde{R}_4}(x_4)\right)\otimes\mu_{\tilde{R}_5}(x_5)\right] \tag{19}$$

$$\left[\underline{\mu}_{\tilde{R}_S}(x), \overline{\mu}_{\tilde{R}_S}(x)\right] =$$

$$1 \ominus \left(1 \ominus \underline{\mu}_{\tilde{R}_1}(x_1) \otimes \underline{\mu}_{\tilde{R}_2}(x_2)\right)\left[1 \ominus \left(\underline{\mu}_{\tilde{R}_3}(x_3) \oplus \underline{\mu}_{\tilde{R}_4}(x_4) \ominus \underline{\mu}_{\tilde{R}_3}(x_3) \otimes \underline{\mu}_{\tilde{R}_4}(x_4)\right) \otimes \underline{\mu}_{\tilde{R}_5}(x_5)\right]$$

$$,$$

$$\left[1 \ominus \left(1 \ominus \overline{\mu}_{\tilde{R}_1}(x_1) \otimes \overline{\mu}_{\tilde{R}_2}(x_2)\right)\left[1 \ominus \left(\overline{\mu}_{\tilde{R}_3}(x_3) \oplus \overline{\mu}_{\tilde{R}_4}(x_4) \ominus \overline{\mu}_{\tilde{R}_3}(x_3) \otimes \overline{\mu}_{\tilde{R}_4}(x_4)\right) \otimes \overline{\mu}_{\tilde{R}_5}(x_5)\right]\right] \tag{20}$$

Bridge (Complex) System

In bridge system (Fig. 2(d)), the five redundant components are linked in complex structure. The IT2 fuzzy reliability of bridge system is thus calculated by the following equations:

$$\tilde{R}_S(r,n) = \tilde{R}_5 \otimes \left(1 \ominus (1 \ominus \tilde{R}_1) \otimes (1 \ominus \tilde{R}_3)\right) \otimes \left(1 \ominus (1 \ominus \tilde{R}_2) \otimes (1 \ominus \tilde{R}_4)\right)$$

$$\oplus \left(1 \ominus \tilde{R}_5\right)\left[1 \ominus \left(1 \ominus \tilde{R}_1 \otimes \tilde{R}_2\right) \otimes \left(1 \ominus \tilde{R}_3 \otimes \tilde{R}_4\right)\right] \tag{21}$$

So, the IT2 MF of $\tilde{R}_S$, represented by $\mu_{\tilde{R}_S}(x) = \left[\underline{\mu}_{\tilde{R}_S}(x), \overline{\mu}_{\tilde{R}_S}(x)\right]$, may be calculated using the follows:

$$\mu_{\tilde{R}_S}(x) = \mu_{\tilde{R}_5}(x_5) \otimes \left\{1 \ominus \left(1 \ominus \mu_{\tilde{R}_1}(x_1)\right) \otimes \left(1 \ominus \mu_{\tilde{R}_3}(x_3)\right)\right\} \otimes \left\{1 \ominus \left(1 \ominus \mu_{\tilde{R}_2}(x_2)\right) \otimes \left(1 \ominus \mu_{\tilde{R}_4}(x_4)\right)\right\} \oplus \left(1 \ominus \mu_{\tilde{R}_5}(x_5)\right)\left[1 \ominus \left\{1 \ominus \mu_{\tilde{R}_1}(x_1) \otimes \mu_{\tilde{R}_2}(x_2)\right\} \otimes \left\{1 \ominus \mu_{\tilde{R}_3}(x_3) \otimes \mu_{\tilde{R}_4}(x_4)\right\}\right] \tag{22}$$

$$\left[\underline{\mu}_{\tilde{R}_S}(x), \overline{\mu}_{\tilde{R}_S}(x)\right]$$

$$\underline{\mu}_{\tilde{R}_5}(x_5) \otimes \left\{1 \ominus \left(1 \ominus \underline{\mu}_{\tilde{R}_1}(x_1)\right) \otimes \left(1 \ominus \underline{\mu}_{\tilde{R}_3}(x_3)\right)\right\} \otimes \left\{1 \ominus \left(1 \ominus \underline{\mu}_{\tilde{R}_2}(x_2)\right) \otimes \left(1 \ominus \underline{\mu}_{\tilde{R}_4}(x_4)\right)\right\}$$

$$\oplus\left(1\ominus\underline{\mu}_{\tilde{R}_5}(x_5)\right)\left[1\ominus\left\{1\ominus\underline{\mu}_{\tilde{R}_1}(x_1)\otimes\underline{\mu}_{\tilde{R}_2}(x_2)\right\}\otimes\left\{1\ominus\underline{\mu}_{\tilde{R}_3}(x_3)\otimes\underline{\mu}_{\tilde{R}_4}(x_4)\right\}\right],$$

$$\overline{\mu}_{\tilde{R}_5}(x_5)\otimes\left\{1\ominus\left(1\ominus\overline{\mu}_{\tilde{R}_1}(x_1)\right)\otimes\left(1\ominus\overline{\mu}_{\tilde{R}_3}(x_3)\right)\right\}\otimes\left\{1\ominus\left(1\ominus\overline{\mu}_{\tilde{R}_2}(x_2)\right)\otimes\left(1\ominus\overline{\mu}_{\tilde{R}_4}(x_4)\right)\right\}$$

$$\oplus\left(1\ominus\overline{\mu}_{\tilde{R}_5}(x_5)\right)\left[1\ominus\left\{1\ominus\overline{\mu}_{\tilde{R}_1}(x_1)\otimes\overline{\mu}_{\tilde{R}_2}(x_2)\right\}\otimes\left\{1\ominus\overline{\mu}_{\tilde{R}_3}(x_3)\otimes\overline{\mu}_{\tilde{R}_4}(x_4)\right\}\right] \tag{23}$$

We create the interval type-2 fuzzy reliability-redundancy allocation problem (IT2FRRAP) for n_m-redundant components systems as follows:

$$\text{Maximize } \widetilde{R_S}(r,n) \tag{24}$$

$$\text{Subject to } C_s(r,n)=\sum_{i=1}^{m}\left[\alpha_i\left(-\frac{t}{\ln(r_i)}\right)^{\beta_i}\cdot\left(n_i+\exp(n_i/4)\right)\right]\le C\cdot$$

$$W_s(r,n)=\sum_{i=1}^{m}w_i.\left(n_i+\exp(n_i/4)\right)\le W.$$

$$V_s(r,n)=\sum_{i=1}^{m}w_iv_i^2n_i^2\le V.$$

$$r_i\in[0,1]\subset\mathbb{R},\ n_i\subset-^{+};\ i=1,2,\ldots,m.$$

Here, $C_s(r,n)$ is the cost constraint, α_i and β_i characterize inherent properties (shaping and scaling factor) of the system. The total weight constrain is given by $W_s(w,n)$ with the weights of the distinct components (w_i) and the redundant components (n_i). Likewise, the total volume constrain of the system is given in $V_s(v,n)$ in relations of the distinct volumes (v_i) and redundant components (n_i).

In the IT2FRRAP, as given in Eq. (24), the $\widetilde{R_S}(r,n)$ is the IT2 fuzzy objective function corresponding to various configurations of system as series, parallel, series-parallel systems and bridge configuration. The system specialists offer perspectives on the system's functionality throughout the context of IT2 MFs which appear to differ yet remain within a similar range. Therefore by amidst IT2 MFs, we could indeed approximate the ambiguity and vagueness associated with the reliability of the system. There are particular distinctions (Muhuri et al., 2018) that need to be taken as follows to consider an appropriate form of IT2 MF from several existing one such as trapezoidal, Gaussian, triangular, etc. The system specialists offer perspectives on the system's functionality throughout the context of IT2 MFs which appear to differ yet remain within a similar range. Therefore by amidst IT2 MFs, we could indeed approximate the ambiguity and vagueness associated with the reliability of the system. There are particular distinctions (Muhuri et al., 2018) that need to be taken as follows to consider an appropriate form of IT2 MF from several existing one such as trapezoidal, Gaussian, triangular, etc. Firstly, the reliability of a system characterized as the likelihood that such a system will be able to service properly without malfunction for a specified period of time (T). Every time the manufacturers will promise, ensuring the importance of the time's durability is unambiguous. Furthermore, it is also not feasible to construct a system that operates over an indefinite length of time; having a long tail over a particular stage in the IT2 MF is not useful.

Hence, an interval type-2 triangular membership function (IT2 TMF) is the finest selection for demonstrating the IT2 fuzzy reliability of the systems in IT2FRRAP. The IT2 TMF $\mu_{\tilde{R}_i} = \left[\underline{\mu}_{\tilde{R}_i}, \overline{\mu}_{\tilde{R}_i}\right]$ for the IT2 fuzzy reliability of the i-th sub-system might be considered as:

$$\left[\underline{\mu}_{\tilde{R}_i}\left(x_i;\underline{R}_i^L,R_i,\underline{R}_i^R\right),\overline{\mu}_{\tilde{R}_i}\left(x_i;\overline{R}_i^L,R_i,\overline{R}_i^R\right)\right] = [\quad 0, \quad x_i \le \underline{R}_i^L \quad 0, \quad x_i \le \overline{R}_i^L \; \frac{x_i-\underline{R}_i^L}{R_i-\underline{R}_i^L}, \underline{R}_i^L \le x_i \le R_i \; \frac{x_i-\overline{R}_i^L}{R_i-\overline{R}_i^L}, \overline{R}_i^L \le x_i \le R_i \; \frac{\underline{R}_i^R-x_i}{\underline{R}_i^R-R_i}, R_i \le x_i \le \underline{R}_i^R \; \frac{\overline{R}_i^R-x_i}{\overline{R}_i^R-R_i}, R_i \le x_i \le \overline{R}_i^R \; 0, \quad x_i \le \underline{R}_i^R \, 0, \quad x_i \le \overline{R}_i^R \tag{25}$$

In Eq. (25), $\underline{R}_i^L$, $\underline{R}_i^R$ represents the left and right end-points of LMF; $\overline{R}_i^L$, $\overline{R}_i^R$ represents the left and right end-points of UMF, required in the making of triangular fuzzymembershipfunctionsmatchingto $\underline{\mu}_{\tilde{R}_i}\left(x_i;\underline{R}_i^L,R_i,\underline{R}_i^R\right)$ and $\overline{\mu}_{\tilde{R}_i}\left(x_i;\overline{R}_i^L,R_i,\overline{R}_i^R\right)$, correspondingly. Algorithm 1 provides the processes of producing the IT2 TMF and various features of IT2 MF are made by Algorithm 1 are displayed in Fig. 3.

Figure 3. IT2 fuzzy triangular membership function for $\tilde{R}_i$.

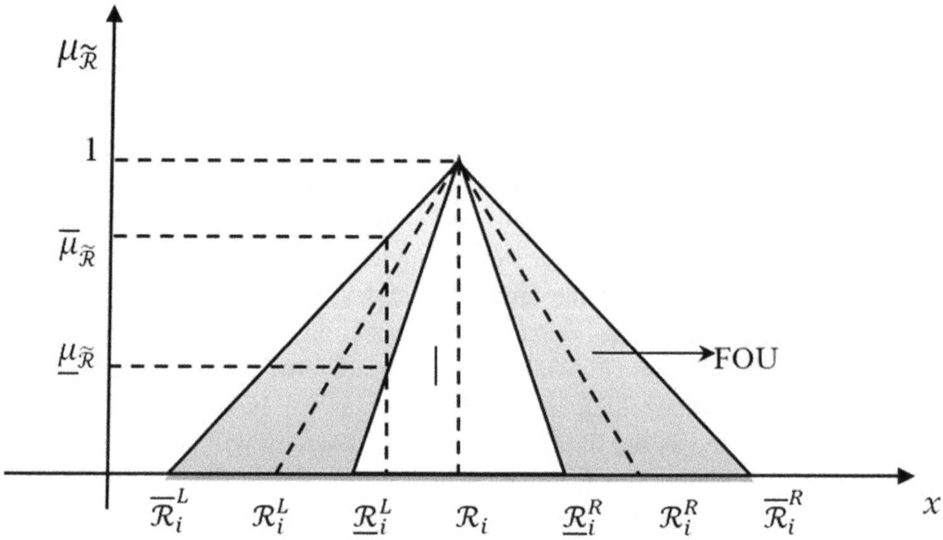

Algorithm 1: Interval Type-2 Triangular Membership Function

Input: Provided the $\mathcal{R}_i; \mathcal{R}_i^L$ and $\mathcal{R}_i^R$ such that $\mathcal{R}_i^L \leq \mathcal{R}_i \leq \mathcal{R}_i^R$ and $\mathcal{R}_i \in [a,b]$.

Output: Interval Type-2 Triangular Membership Function $\mu_{\tilde{\mathcal{R}}_i}(x) = \left[\underline{\mu}_{\mathcal{R}_i}(x), \bar{\mu}_{\mathcal{R}_i}(x)\right]$.

1: Calculate the lower end-points $\underline{\mathcal{R}}_i^L$ of $\mathcal{R}_i^L$ using the following:

$$\underline{\mathcal{R}}_i^L = \mathcal{R}_i^L + \left(\mathcal{R}_i - \mathcal{R}_i^L\right) * rand\ ;$$

2: Calculate the upper end-points $\bar{\mathcal{R}}_i^L$ of $\mathcal{R}_i^L$ using the following:

$$\bar{\mathcal{R}}_i^L = \mathcal{R}_i^L - \left(\mathcal{R}_i^L - a\right) * rand\ ;$$

3: Calculate the lower end-points $\underline{\mathcal{R}}_i^R$ of $\mathcal{R}_i^R$ using the following:

$$\underline{\mathcal{R}}_i^R = \mathcal{R}_1^R - \left(\mathcal{R}_1^R - \mathcal{R}_i\right) * rand\ ;$$

4: Calculate the upper end-points $\bar{\mathcal{R}}_i^R$ of $\mathcal{R}_i^R$ using the following:

$$\bar{\mathcal{R}}_i^R = \mathcal{R}_1^R + \left(b - \mathcal{R}_1^R\right) * rand\,;$$

5: Evaluate the LMF $\underline{\mu}_{\mathcal{R}_i}\left(x; \underline{\mathcal{R}}_i^L, \mathcal{R}_1^{\mathrm{L}}, \underline{\mathcal{R}}_i^L\right)$ using Eq. (25).
6: Evaluate the UMF $\bar{\mu}_{\mathcal{R}_i}\left(x; \bar{\mathcal{R}}^L, \mathcal{R}, \bar{\mathcal{R}}^R\right)$ using Eq. (25).

The crisp values of total system reliability are indeed the outcomes of the optimization model, therefore the type-reduction and defuzzification method incorporates to the IT2 MFs to calculate numerical values. The prominent Enhanced Karnik and Mendel (EKM) approach (Jerry M. Mendel & Wu, 2010) would be used for type-reduction procedure as well as the centroid procedure would be used for defuzzification (Jerry M. Mendel, 2007; Jerry M. Mendel & Liu, 2013). For a given IT2 FS $\tilde{A}$, the centroid $C_{\tilde{A}}(x)$ is the collective centroid of all its embedded T1 FSs. This would be to say,

$$C_{\tilde{A}}(x) = \left\{c_l\left(\tilde{A}\right), \ldots, c_r\left(\tilde{A}\right)\right\} = \left[c_l\left(\tilde{A}\right), c_r\left(\tilde{A}\right)\right] \tag{26}$$

Algorithm 2: EKM Procedure

Procedure for $c_l\left(\tilde{A}\right)$

Procedure for $c_r\left(\tilde{A}\right)$

$$c_l\left(\tilde{A}\right) = \min_{\forall \theta_i \in [\underline{\mu}, \bar{\mu}]} \left(\sum_{i-1}^{N} x_i \theta_i \Big/ \sum_{i-1}^{N} \theta_i \right)$$

$$c_r\left(\tilde{A}\right) = \max_{\forall \theta_i \in [\underline{\mu}, \bar{\mu}]} \left(\sum_{i-1}^{N} x_i \theta_i \Big/ \sum_{i-1}^{N} \theta_i \right)$$

1.Established $k = \left[N / 2.4\right]$ (the adjacent integer to $N / 2.4$) and calculate:

$$a = \sum_{i=1}^{k} x_i \bar{\mu}_{\tilde{A}}(x_i) + \sum_{i=k+1}^{k} x_i \underline{\mu}_{\tilde{A}}(x_i)$$

$$b = \sum_{i=1}^{k} \bar{\mu}_{\tilde{A}}(x_i) + \sum_{i=k+1}^{k} \underline{\mu}_{\tilde{A}}(x_i)$$

Established $k = [N/1.7]$ (the adjacent integer to $N/1.7$) and calculate:

$$a = \sum_{i=1}^{k} x_i \underline{\mu}_{\tilde{A}}(x_i) + \sum_{i=k+1}^{k} x_i \bar{\mu}_{\tilde{A}}(x_i)$$

$$b = \sum_{i=1}^{k} \underline{\mu}_{\tilde{A}}(x_i) + \sum_{i=k+1}^{k} \bar{\mu}_{\tilde{A}}(x_i)$$

2.Evaluate $c' = a / b$
3.Discovery $k' \in [1, N-1]$ so that $x_{k'} \le c' x_{k'+1}$
4.Check if $k' = k$. If yes, stop and set $c' = c_l(L)$, and $k = L$. If no, go to step 5.

Check if $k' = k$. If yes, stop and set $c' = c_r(R)$, and $k = R$. If no, go to step 5.

5.Compute and

$$b' = b + s \sum_{i=\min(k,k')+1}^{\max(k,k')} \left[\bar{\mu}_{\tilde{A}}(x_i) - \underline{\mu}_{\tilde{A}}(x_i)\right]$$

Computeand

$$b' = b - s \sum_{i=\min(k,k')+1}^{\max(k,k')} \left[\bar{\mu}_{\tilde{A}}(x_i) - \underline{\mu}_{\tilde{A}}(x_i)\right]$$

6.Calculate $c''(k') = a' / b'$

7.Set $c' = c''(k'), a = a', b = b'$ and go to Step 2.

In Algorithm 2, we have included essential stages of a EKM methodology. By employing left and right centroids that is, (c_l) and (c_r), in the defuzzification technique, the ultimate compact value (y_d) is the sole average of centroid left and centroid right.

$$y_d = \frac{c_l + c_r}{2} \tag{27}$$

Finally, we create the crisp model for IT2FRRAP for n_m-redundant components systems as follows:

$$\text{Maximize } \xi_t \tag{28}$$

Subject to $C_s(r,n) - C \le 0$, $W_s(r,n) - W \le 0$, $V_s(r,n) - V \le 0$

$$\xi_t = \mu_{\tilde{R}_S}(r,n) = \left[\underline{\mu}_{\tilde{R}_S}(r,n), \bar{\mu}_{\tilde{R}_S}(r,n)\right]$$

$$\left[\underline{\mu}_{\tilde{R}_i}(r,n), \bar{\mu}_{\tilde{R}_i}(r,n)\right] = \left[\underline{\mu}_{\tilde{R}_i}\left(x_i; \underline{R}_i^L, R_i, \underline{R}_i^R\right), \bar{\mu}_{\tilde{R}_i}\left(x_i; \bar{R}_i^L, R_i, \bar{R}_i^R\right)\right];$$

$$R_i = 1 - \prod_{j=1}^{n_i}\left(1 - r_j\right)$$

$$r_{i,min} \le r_i \le r_{i,max},\ n_{i,min} \le n_i \le n_{i,max},\ R_{i,min} \le R_i^L, R_i^R \le R_{i,max},$$

$$r_i \in [0,1] \subset \mathbb{R},\ n_i \subset -^{+};\ i = 1,2,\ldots,m.$$

Solution Approach

This section explains the solution procedure castoff to resolve the IT2FRRAPs, expressed in Eq. (28). We utilized a well-established meta-heuristic optimization technique named as the particle swarm optimization (PSO) introduced by Kennedy

(2011) within our solution method. In pursuit of feed, PSO reproduces the social interactions of flocks of birds or schools of fish. It is a search algorithm focused mostly on a community that finds behaviour of individuals in a swarm for sequential and fast motions within the same flock (Engelbrecht, 2007), (Clerc & Kennedy, 2002). Numerous variations of PSO have been established and is used in so many applications ever since the original incarnation, e.g. (Clerc, 2006), (Poli, Kennedy, & Blackwell, 2007), (Olivas, Valdez, Castillo, & Melin, 2016), (Valdez et al., 2017), (Olivas et al., 2017) and (Muhuri, Ashraf, & Goel, 2020). Because the IT2FRRAPs is a non-linear issue of optimization, proven being an NP-hard problem by (Chern, 1992), evolutionary methods are primarily adapted to its solutions. PSO is thus one of the ideal evolutionary optimization algorithms that might be used to determine the best reliability of the modules and the number of dependencies which relate to the optimization problem. The entire methodology for implementing our solution procedure is given in Algorithm 3. The descriptions of each and every phase associated with the proposed solution algorithm will be addressed in sub-sections below.

Algorthim 3: Solution Algorthim

Input: Set the specifications of PSO and input parameters of IT2FRRAP.
Output: Global best particle ($gbest$) and Optimal results.

1: For each particle $i = 1 \ to \ N_p$
2: For each dimension $d = 1 \ to \ \dim$ // Initialize all particles' position and velocity
3: Evaluate the positions $X_{i,d} \leftarrow Rand\left(X_{min}, X_{max}\right)$
4: Evaluate the velocities $V_{i,d} \leftarrow Rand\left(V_{min}, V_{max}\right)$
5: end For
6: $pbest \leftarrow X_i$ // Initialize particles' personal best position
7: $gbest \leftarrow \varnothing$ // Initialize particles' global best position
8: end For
9: Fitness function evaluation
10: IT2 Fuzzification \\ Reliability
11: Type-reduction \\ EKM Algorithm
12: Defuzzification \\ Centroid method
13: While Maximum iterations or stopping criteria is not attained
14: For each particle $i = 1 \ to \ N_p$
15: If $f\left(X_i\right) < f\left(pbest_i\right)$
16: Update the personal best $pbest_i \leftarrow X_i$
17: end If

18: If $f\left(pbest_i\right) < f\left(gbest\right)$

19: Update the global best $gbest \leftarrow pbest_i$

20: end If

21: end For

22: For each particle $i = 1 \ to \ N_p$

23: For each dimension $d = 1 \ to \ \dim$

24: Update the velocity $V'_{i,d} \leftarrow \left(V_{i,d}, pbest_i, gbest\right)$ // Using Eq. (29)

25: Update the position $X'_{i,d} \leftarrow \left(X_{i,d}, V_{i,d}\right)$ // Using Eq. (30)

26: end For

27: end For

28: Fitness function evaluation

29: End While

30: Global best particle $gbest$ and Optimal results.

Initialization Process

In this phase the following factors of PSO are constructed: (a) the maximum size of individual particles in the population (N_p) (b) dimension (dim) of individual particles, (b) the initial position (X) and velocity (V) of all particle in population, referring to variable (r, n), (c) the initial values are applied to the parameters (set of constriction coefficient) involved. The swarm population has arbitrarily configured locations $\left(X\right)$ and velocities $\left(V\right)$ that lead to increasing vector choice, and the population's personal best ($pbest$), and global best ($gbest$).

Fitness Function Evaluation

The IT2 fuzzy reliability (objective functions) of IT2FRRAPs, as represented in Eq. (28), is assessed in this fitness function evaluation phase. It requires three phases: (I) Fuzzification; (II) Accurate assessment; and (III) Type-Reduction and Defuzzification. The fuzzification algorithm defines IT2 MFs to the sub-system functionality and is included in the objective assessment. Furthermore, the frameworks of defuzzified objective function equations are expressed by the type-reduction and defuzzification techniques (as they suggested in Section 3).

Finding gbest and pbest

Within the feasible space the particles of the population travel by using a precise algebraic principle regarding momentum and location at each instance. The motions of each particle are influenced by their corresponding personal best ($pbest$) or best local position. At about the same time, each particles is guided toward the space 's best established optimal location, recognized as that of the global best $(gbest)$. All $pbest$ and $gbest$ aid push colony into the ideal option feasible. Algorithm 3 provides the method for identifying the community with $pbest$ and $gbest$.

Updating the Position and Velocity

Let t symbolizes the instance of iteration, the measure of the i-th particle at the iteration $t+1$ is $X_i(t+1)$, intended by accumulation its preceding position $X_i(t)$ to its velocity $V_i(t)$ at the t, as follows:

$$X_i(t+1) = X_i(t) + V_i(t) \tag{29}$$

Additional, the velocity $V_i(t+1)$ of the i-th particle at the iteration $t+1$ is rationalized by the earlier velocity $V_i(t)$, position $X_i(t)$, $pbest_i(t)$, and $gbest(t)$, as follows:

$$V_i(t+1) = \chi * \left[V_i(t) + c_1 \times r_1 \times (pbest_i(t) - X_i(t)) + c_2 \times r_2 \times (gbest(t) - X_i(t))\right] \tag{30}$$

In above equation, c_1, c_2 and k characterize the construction coefficient set of the PSO and χ is the construction coefficient considered as $\chi = \frac{2*k}{\left|2 - \phi - \sqrt{\phi(\phi - 4)}\right|}$, $k \in [0,1]$, $\phi_1 = c_1 r_1$, $\phi_2 = c_2 r_2$, and $\phi = \phi_1 + \phi_2$. r_1 & r_2 are random variables with uniform distribution between 0 and 1 (Clerc, 2006).

Constraint Handling

To solve a constrained problem of optimization, it is necessary to manage the constraints involved to the target functions. Throughout the presence of different kinds of inequalities, seeking the feasible solutions to an optimization method, be it sole or multilevel, could become quite challenging. Several approaches have been suggested for addressing these cases. Among such techniques, the approach incorporating penalty functions is by far the most acceptable. We used the penalty driven dependent method in our optimization method to handle the PSO restrictions

(Coello Coello, 2002), (Marler & Arora, 2010). It is achieved using a updated function J in solution space X to address a problem as described in the following:

$$J(x) = \begin{cases} F + \sum_i g_i(x), & if \quad x \notin X \\ F(x), & if \quad x \in X \end{cases} \tag{31}$$

Here, F characterizes the functional values of the poorest feasible solution x and g_i is the restraint function.

Experimental Simulations

We utilized the suggested solution method in this segment to address the IT2FRRAPs for the four configuration systems, i.e. the series, parallel, series-parallel and bridge system. These all IT2FRRAPs are addressed by using the proposed PSO centered solution method. Input parameters are given in Table 1 for the series, parallel, and bridge IT2FRRAP formulations. However, the input information presented in Table 2 is for series-parallel IT2FRRAP formulation. With each used redundant component of five the system input parameters shaping element (α), the scale parameter (β), the element weights (w) and the element volumes (v) are listed in Table 1 and Table 2. Further, Table 1 and Table 2 provides the extreme range of overall cost of the system (C), weight (W) and volume (V) that can offer the feasible zone. System analysts give the range metrics for determining the IT2 fuzzy goal for enhancement according to each configuration. The objective function, IT2 MFs represent approximately the IT2 Fuzzy satisfaction area (as the goal), that is developed by system experts. Furthermore, the system utilizes appropriate input parameters to determine the optimal output for different combinations. The left and right ideal values of the sub-system reliability are given in Table 3.

Table 1. Input parameters: for Series, Parallel and Bridge system.

i	$10^5 * \pm_i$	$^2{}_i$	w_i	v_i	V	C	W	$T(h)$
1	2.330	1.5	7.0	1.0				
2	1.450	1.5	8.0	2.0				
3	0.541	1.5	8.0	3.0	110	175	200	1000
4	8.050	1.5	6.0	4.0				
5	1.950	1.5	9.0	2.0				

Table 2. Input Data: for Series-Parallel System.

i	$10^5 * \pm_i$	$^{2}{}_{i}$	w_i	v_i	V	C	W	$T(h)$
1	2.500	1.5	3.5	2				
2	1.450	1.5	4.0	4				
3	0.541	1.5	4.0	5	180	175	100	1000
4	0.541	1.5	3.5	8				
5	2.100	1.5	3.5	4				

Table 3. The ideal value of sub-systems for all system.

	Series System	**Parallel System**	**Series-Parallel System**	**Bridge System**
R_i^L	0.6452566	0.6108074	0.6065163	0.6307274
R_i^R	0.8199358	0.8471696	0.9227051	0.8144518

PSO's standard parameters are as: the size of the population (= 100), maximum iteration (= 100), acceleration factors $c_1 = c_2 = 1.5$ and k is randomly selected from the uniform distribution, as indicated in (Shi & Eberhart, 1998), (Beielstein, 2002), (Trelea, 2003). We also introduced Goldberg's suggested real-coded Genetic Algorithm (GA) (Golberg, 1989) for comparison. The thorough description for GA

Table 4. **Optimal result for Series IT2FRRAP with GA and Proposed method.**

Method	**GA**	**Proposed method**
$(n_1, n_2, n_3, n_4, n_5)$	(3,2,2,3,3)	(3,2,2,3,3)
r_1	0.779427	0.7794057356
r_2	0.869482	0.8718339654
r_3	0.902674	0.9028816561
r_4	0.714038	0.7114011564
r_5	0.786896	0.7878058325
$\boldsymbol{R_s}$	0.931578	**0.9316823357**
Slacks of C_S	0.121454	0.000000004
Slacks of V_S	27	27
Slacks of W_S	7.518918	7.518918

is available in Deb (2001). GA characteristics areas: the size of the population (= 100), maximum iteration (= 100), crossover probability of one-point crossover and mutation probability of one-point mutation, respectively, are taken as 0.6 and 0.4. The method of selecting the tournament is used to produce the playback. We had already done 50 distinct spins to get the better IT2FRRAP optimum result.

For series configuration, the IT2FRRAP is formulated using the parameters provided in Table 1 and the upper and lower ideal values of the subsystems are taken from Table 3. Now, the formulated model is solved using the proposed solution approach, and the results are compared with GA. The best reliability system, the correspondent reliabilities of each component and the number of redundant components are given in Table 4, obtained using the proposed solution approach and GA. In Table 4, the slack in constraints function that is system cost, weight and volume are also presented. Moreover, the maximum achievable system reliability using the PSO based solution method is 0.93168234, whereas, for GA, it is 0.931578. Hence, the total reliability of the system significantly improved via the proposed IT2FRRAP framework.

Table 5. **Optimal result for Parallel IT2FRRAP with GA and Proposed method.**

Method	GA	Proposed method
$(n_1, n_2, n_3, n_4, n_5)$	(3,2,2,4,2)	(3,2,2,4,2)
r_1	0.583327	0.5833962547
r_2	0.697626	0.6976823645
r_3	0.562849	0.5629269678
r_4	0.698137	0.6981395787
r_5	0.553977	0.5540595425
R_s	0.851456	**0.8515354254**
Slacks of C_S	71.9442	71.9353354
Slacks of V_S	9	9
Slacks of W_S	7.86824	7.8682357

Further, the IT2FRRAP is established for parallel configuration by using specifications provided in Table 1, and the superior and inferior acceptable ideals of the subsystems are obtained from Table 3. By implementing the proposed solving methodology, the IT2FRRAP system is solved, and the results have been compared with GA. Table 5 provides the best reliability of the system, the corresponding reliability of every component as well as the number of redundant components achieved when using the PSO and GA. Table 5 provides the slack in constraints function, which is device cost, weight, and volume. In addition, the overall achievable system reliability using the PSO-based solution approach is 0.85153543, while it is 0.851456 for the GA. Consequently, the system's overall reliability improved dramatically through the proposed IT2FRRAP design.

For series-parallel configuration, the IT2FRRAP is formulated using the parameters provided in Table 2 and the upper and lower ideal values of the subsystems are taken from Table 3. Now, the formulated model is solved using the proposed solution approach, and the results are compared with GA. The best reliability system, the correspondent reliabilities of each component and the number of redundant components are given in Table 6, obtained using the proposed solution approach and GA. In Table 6, the slack in constraints function that is system cost, weight and volume are also presented. Moreover, the maximum achievable system reliability using the PSO based solution method is 0.99997669, whereas, for GA, it is 0.99997418. Hence, the total reliability of the system significantly improved via the proposed IT2FRRAP framework.

Similar to series and parallel, for complex configuration, the IT2FRRAP is formulated using the parameters provided in Table 1 and the upper and lower ideal values of the subsystems are taken from Table 3. Now, the IT2FRRAP model is solved using the PSO based solution approach and the results are compared with GA. The best reliability system, the correspondent reliabilities of each component

Table 6. **Optimal result for Series- Parallel IT2FRRAP with GA and Proposed method.**

Method	GA	Proposed method
$(n_1, n_2, n_3, n_4, n_5)$	(2,2,2,2,4)	(2,2,2,2,4)
r_1	0.785452	0.8196583365
r_2	0.842998	0.8449101524
r_3	0.885333	0.8954871214
r_4	0.917958	0.8955148254
r_5	0.870318	0.8684681325
R_s	0.99997418	**0.99997669687**
Slacks of C_S	1.194440	0.00001753564
Slacks of V_S	40	40
Slacks of W_S	1.609289	1.609289

Table 7. **Optimal result for Bridge IT2FRRAP with GA and Proposed method.**

Method	GA	Proposed method
$(n_1, n_2, n_3, n_4, n_5)$	(3,3,3,3,1)	(3,3,2,4,1)
r_1	0.814090	0.8280606187
r_2	0.864614	0.8580404545
r_3	0.890291	0.9141487365
r_4	0.701190	0.6479689653
r_5	0.734731	0.70420485956
$\boldsymbol{R_s}$	0.99987916	**0.99988963546**
Slacks of C_S	0.376347	0.0001468178
Slacks of V_S	18	5
Slacks of W_S	4.26477	1.560466

and the number of redundant components are given in Table 7, obtained using the proposed solution approach and GA. In Table 7, the slack in constraints function that is system cost, weight and volume are also presented. Moreover, the maximum achievable system reliability using the PSO method is 0.0.999889, whereas, for GA, it is 0.9998791. Hence, the total reliability of the system significantly improved via the proposed IT2FRRAP framework. Finally, it has been observed from the solutions that the proposed ones can dominate other methods for the four test problems discussed. From the solutions, it was suggested that the proposed ones could overtake other methodologies for the four different issues mentioned.

CONCLUSION

The chapter mainly utilized PSO (particle swarm optimization) and GA (genetic algorithm) dependent approach techniques to address the interval type-2 fuzzy reliability-redundancy allocation problem (IT2FRRAP) on series, parallel, series-parallel and bridge structures. The primary objective of the IT2FRRAP is to enhance reliability that's been constructed underneath the dynamic environments of type-2 fuzzy. Towards this purpose, the reliability of the sub-components was designed by interval type-2 fuzzy membership functions (IT2 MF), as well as the system 's overall IT2 fuzzy reliability functions, are assessed on the basis of each specification using the Zadeh extension theory. For appropriate datasets, the

IT2FRRAP frameworks are optimized, and the findings are reported. As shown by our suggested PSO-based solution strategy, the best solutions are much higher than the solutions that GA obtains.

In the future, the current modelling of type 2 uncertainty might also be investigated for complex systems, k-out-of-n systems, etc., and the methodology to the solution could've been enhanced utilizing different optimization techniques. Also, the very same development approach may have been used in many other real-life applications like supply chain, strategic inventory level, forecasting etc.

REFERENCES

Ashraf, Z., Khan, M. S., & Lohani, Q. M. D. (2019). New bounded variation based similarity measures between Atanassov intuitionistic fuzzy sets for clustering and pattern recognition. *Applied Soft Computing*, *85*, 105529. Advance online publication. doi:10.1016/j.asoc.2019.105529

Ashraf, Z., Malhotra, D., Muhuri, P. K., & Danish Lohani, Q. M. (2017). Interval type-2 fuzzy demand based vendor managed inventory model. *IEEE International Conference on Fuzzy Systems*. 10.1109/FUZZ-IEEE.2017.8015738

Ashraf, Z., Muhuri, P. K., & Danish Lohani, Q. M. (2015). Particle swam optimization based reliability-redundancy allocation in a type-2 fuzzy environment. *2015 IEEE Congress on Evolutionary Computation, CEC 2015 - Proceedings*, 1212–1219. 10.1109/CEC.2015.7257027

Ashraf, Z., & Muhuri, P. K., Danish Lohani, Q. M., & Nath, R. (2014). Fuzzy multi-objective reliability-redundancy allocation problem. *2014 IEEE International Conference on Fuzzy Systems (FUZZ-IEEE)*, 2580–2587. 10.1109/FUZZ-IEEE.2014.6891889

Ashraf, Z., Muhuri, P. K., Lohani, Q. M. D., & Roy, M. L. (2019). Type-2 fuzzy reliability–redundancy allocation problem and its solution using particle-swarm optimization algorithm. *Granular Computing*, *4*(2), 145–166. doi:10.100741066-018-0106-5

Ashraf, Z., Roy, M. L., Muhuri, P. K., & Danish Lohani, Q. M. (2018). A Novel Image Steganography Approach Based on Interval Type-2 Fuzzy Similarity. *2018 IEEE International Conference on Fuzzy Systems (FUZZ-IEEE)*, 1–8. 10.1109/FUZZ-IEEE.2018.8491582

Ashraf, Z., Roy, M. L., Muhuri, P. K., & Danish Lohani, Q. M. (2020). Interval type-2 fuzzy logic system based similarity evaluation for image steganography. *Heliyon*, *6*(5), e03771. doi:10.1016/j.heliyon.2020.e03771 PMID:32420466

Beielstein, T. (2002). Tuning PSO parameters through sensitivity analysis. *Technical Report, Reihe Computational Intelligence CI 124/02. 2002. Collaborative Research Cent*, (3). Advance online publication. doi:10.13140/2.1.1466.2080

Bustince, H., Barrenechea, E., Pagola, M., Fernandez, J., Xu, Z., Bedregal, B., Montero, J., Hagras, H., Herrera, F., & De Baets, B. (2016). A Historical Account of Types of Fuzzy Sets and Their Relationships. *IEEE Transactions on Fuzzy Systems*, *24*(1), 179–194. doi:10.1109/TFUZZ.2015.2451692

Chen, S. M. (1994). Fuzzy system reliability analysis using fuzzy number arithmetic operations. *Fuzzy Sets and Systems*, *64*(1), 31–38. doi:10.1016/0165-0114(94)90004-3

Cheng, C. H., & Mon, D. L. (1993). Fuzzy system reliability analysis by interval of confidence. *Fuzzy Sets and Systems*, *56*(1), 29–35. doi:10.1016/0165-0114(93)90182-H

Chern, M. S. (1992). On the computational complexity of reliability redundancy allocation in a series system. *Operations Research Letters*, *11*(5), 309–315. doi:10.1016/0167-6377(92)90008-Q

Clerc, M. (2006). Particle Swarm Optimization. Particle Swarm Optimization. doi:10.1002/9780470612163

Clerc, M., & Kennedy, J. (2002). The particle swarm - explosion, stability, and convergence in a multidimensional complex space. *IEEE Transactions on Evolutionary Computation*, *6*(1), 58–73. doi:10.1109/4235.985692

Coello Coello, C. A. (2002). Theoretical and numerical constraint-handling techniques used with evolutionary algorithms: A survey of the state of the art. *Computer Methods in Applied Mechanics and Engineering*, *191*(11–12), 1245–1287. doi:10.1016/S0045-7825(01)00323-1

Deb, K. (2001). *Multi-objective optimization using evolutionary algorithms*. John Wiley & Sons, Inc.

Ebrahimipour, V., & Sheikhalishahi, M. (2011). Application of multi-objective particle swarm optimization to solve a fuzzy multi-objective reliability redundancy allocation problem. *2011 IEEE International Systems Conference*, 326–333. 10.1109/SYSCON.2011.5929085

Engelbrecht, A. P. (2007). Computational Intelligence. *Computational Intelligence*. Advance online publication. doi:10.1002/9780470512517

Garg, H., & Sharma, S. P. (2013). Multi-objective reliability-redundancy allocation problem using particle swarm optimization. *Computers & Industrial Engineering*, *64*(1), 247–255. doi:10.1016/j.cie.2012.09.015

Gaxiola, F., Melin, P., Valdez, F., Castro, J. R., & Castillo, O. (2016). Optimization of type-2 fuzzy weights in backpropagation learning for neural networks using GAs and PSO. *Applied Soft Computing*, *38*, 860–871. doi:10.1016/j.asoc.2015.10.027

Golberg, D. E. (1989). *Machine Learning and Its Applications* (G. Paliouras, V. Karkaletsis, & C. D. Spyropoulos, Eds.). doi:10.1007/3-540-44673-7

Huang, H.-Z. (1997). Fuzzy multi-objective optimization decision-making of reliability of series system. *Microelectronics and Reliability*, *37*(3), 447–449. doi:10.1016/S0026-2714(96)00040-6

John, R. (1998). Type 2 Fuzzy Sets: An Appraisal of Theory and Applications. *International Journal of Uncertainty, Fuzziness and Knowledge-based Systems*, *06*(06), 563–576. doi:10.1142/S0218488598000434

Kennedy, J. (2011). Particle Swarm Optimization. In C. Sammut & G. I. Webb (Eds.), *Encyclopedia of Machine Learning* (Vol. 4, pp. 760–766). doi:10.1007/978-0-387-30164-8_630

Klir, G. J., & Yuan, B. (2008). *Fuzzy sets and fuzzy logic, theory and applications*. Retrieved from http://digilib.uin-suka.ac.id/7049/

Komal, & Sharma, S. P. (2014). Fuzzy reliability analysis of repairable industrial systems using soft-computing based hybridized techniques. *Applied Soft Computing Journal, 24*, 264–276. doi:10.1016/j.asoc.2014.06.054

Kuo, W., & Prasad, R. (2004). *System Reliability Optimization: An Overview*. doi:10.1007/978-1-4419-9021-1_2

Kuo, W., & Prasad, V. R. (2000). An annotated overview of system-reliability optimization. *IEEE Transactions on Reliability*, *49*(2), 176–187. doi:10.1109/24.877336

Mahapatra, G. S., & Roy, T. K. (2006). Fuzzy multi-objective mathematical programming on reliability optimization model. *Applied Mathematics and Computation*, *174*(1), 643–659. doi:10.1016/j.amc.2005.04.105

Marler, R. T., & Arora, J. S. (2010). The weighted sum method for multi-objective optimization: New insights. *Structural and Multidisciplinary Optimization, 41*(6), 853–862. doi:10.100700158-009-0460-7

Mendel, J. M., & John, R. I. B. (2002). Type-2 fuzzy sets made simple. *IEEE Transactions on Fuzzy Systems, 10*(2), 117–127. doi:10.1109/91.995115

Mendel, J. M. (2007). Advances in type-2 fuzzy sets and systems. *Information Sciences, 177*(1), 84–110. doi:10.1016/j.ins.2006.05.003

Mendel, J. M., & Xinwang Liu. (2013). Simplified interval type-2 fuzzy logic systems. *IEEE Transactions on Fuzzy Systems, 21*(6), 1056–1069. doi:10.1109/TFUZZ.2013.2241771

Mendel, J. M. (2010). *Perceptual Computing*. doi:10.1002/9780470599655

Mon, D.-L., & Cheng, C.-H. (1994). Fuzzy system reliability analysis for components with different membership functions. *Fuzzy Sets and Systems, 64*(2), 145–157. doi:10.1016/0165-0114(94)90330-1

Muhuri, P. K., Ashraf, Z., & Lohani, Q. M. D. (2018). Multiobjective Reliability Redundancy Allocation Problem With Interval Type-2 Fuzzy Uncertainty. *IEEE Transactions on Fuzzy Systems, 26*(3), 1339–1355. doi:10.1109/TFUZZ.2017.2722422

Muhuri, P. K., Ashraf, Z., & Goel, S. (2020). A Novel Image Steganographic Method based on Integer Wavelet Transformation and Particle Swarm Optimization. *Applied Soft Computing, 92*, 106257. Advance online publication. doi:10.1016/j.asoc.2020.106257

Nannapaneni, S., & Mahadevan, S. (2016). Reliability analysis under epistemic uncertainty. *Reliability Engineering & System Safety, 155*, 9–20. doi:10.1016/j.ress.2016.06.005

Olivas, F., Amador-Angulo, L., Perez, J., Caraveo, C., Valdez, F., & Castillo, O. (2017). Comparative study of type-2 fuzzy Particle swarm, Bee Colony and Bat Algorithms in optimization of fuzzy controllers. *Algorithms, 10*(3), 101. Advance online publication. doi:10.3390/a10030101

Olivas, F., Valdez, F., Castillo, O., & Melin, P. (2016). Dynamic parameter adaptation in particle swarm optimization using interval type-2 fuzzy logic. *Soft Computing, 20*(3), 1057–1070. doi:10.100700500-014-1567-3

Poli, R., Kennedy, J., & Blackwell, T. (2007). Particle swarm optimization. *Swarm Intelligence, 1*(1), 33–57. doi:10.100711721-007-0002-0

Ramirez-Marquez, J. E., & Coit, D. W. (2004). A heuristic for solving the redundancy allocation problem for multi-state series-parallel systems. *Reliability Engineering & System Safety*, *83*(3), 341–349. doi:10.1016/j.ress.2003.10.010

Rausand, M. (2014). Reliability of Safety-Critical Systems. Reliability of Safety-Critical Systems. doi:10.1002/9781118776353

Ravi, V., Reddy, P. J., & Zimmermann, H. J. (2000). Fuzzy global optimization of complex system reliability. *IEEE Transactions on Fuzzy Systems*, *8*(3), 241–248. doi:10.1109/91.855914

Shi, Y., & Eberhart, R. C. (1998). *Parameter selection in particle swarm optimization*. doi:10.1007/BFb0040810

Soltani, R. (2014). Reliability optimization of binary state non-repairable systems: A state of the art survey. *International Journal of Industrial Engineering Computations*, *5*(3), 339–364. doi:10.5267/j.ijiec.2014.5.001

Tian, Z., Zuo, M. J., & Huang, H. (2008). Reliability-redundancy allocation for multi-state series-parallel systems. *IEEE Transactions on Reliability*, *57*(2), 303–310. doi:10.1109/TR.2008.920871

Trelea, I. C. (2003). The particle swarm optimization algorithm: Convergence analysis and parameter selection. *Information Processing Letters*, *85*(6), 317–325. doi:10.1016/S0020-0190(02)00447-7

Tzafestas, S. G. (1980). Optimization of system reliability: A survey of problems and techniques†. *International Journal of Systems Science*, *11*(4), 455–486. doi:10.1080/00207728008967030

Valdez, F., Vazquez, J. C., Melin, P., & Castillo, O. (2017). Comparative study of the use of fuzzy logic in improving particle swarm optimization variants for mathematical functions using co-evolution. *Applied Soft Computing*, *52*, 1070–1083. doi:10.1016/j.asoc.2016.09.024

Wang, Z., Tianshi, C., Tang, K., & Yao, X. (2009). A multi-objective approach to redundancy allocation problem in parallel-series systems. *2009 IEEE Congress on Evolutionary Computation, CEC 2009*, 582–589. 10.1109/CEC.2009.4982998

Yeh, W.-C., Lin, Y.-C., Chung, Y. Y., & Chih, M. (2010). A Particle Swarm Optimization Approach Based on Monte Carlo Simulation for Solving the Complex Network Reliability Problem. *IEEE Transactions on Reliability*, *59*(1), 212–221. doi:10.1109/TR.2009.2035796

Zadeh, L. A. (1965). Fuzzy sets. *Information and Control*, *8*(3), 338–353. doi:10.1016/S0019-9958(65)90241-X

Zadeh, L. A. (1975). The concept of a linguistic variable and its application to approximate reasoning-I. *Information Sciences*, *8*(3), 199–249. doi:10.1016/0020-0255(75)90036-5

Zio, E. (2009). Reliability engineering: Old problems and new challenges. *Reliability Engineering & System Safety*, *94*(2), 125–141. doi:10.1016/j.ress.2008.06.002

Chapter 8
Fuzzy Reliability Evaluation of Complex Systems Using Intuitionistic Fuzzy Sets

Ashok Singh Bhandari
Graphic Era (Deemed), India

Akshay Kumar
Graphic Era Hill University, India

Mangey Ram
https://orcid.org/0000-0002-8221-092X
Graphic Era (Deemed), India

ABSTRACT

In a system where it cannot be reduced to a pure series and parallel form (i.e., complex), reliability is obtained by universal generating function (UGF) technique. The intuitionistic fuzzy sets (IFS) concept with triangular fuzzy number (TFN) and Weibull lifetime distribution is introduced to find the fuzzy reliability of the same system. Also, averaging operator with given equal weights is used with the set of three triangular intuitionistic fuzzy number. A numerical example is solved for demonstration.

INTRODUCTION

Reliability is one of the most important characteristics of any system. When faced with an unusual failure mode, we need to learn what is causing the failure in order to solve the problem. To solve all these kinds of problems reliability modelling

DOI: 10.4018/978-1-7998-7564-2.ch008

and assessment are critical for pursuing high reliability or availability level. In the real-world problem, the data or the system parameters are often imprecise and for conventional reliability analysis use of probabilistic approach is inadequate. Therefore, IFS technique is introduced to analyse the reliability of a complex system.

In context of UGF technique, Levitin et al. (2000) solved the redundancy optimization problem for power systems using genetic algorithms (GA), where the elements of the system had different capacities. Also, UGF method was used to estimate the reliability of multi-state power system with series-parallel structure. Levitin and Lisnianski (2001) presented a new technique for reliability optimization problems for the family of multistate system. Authors combined UGF for quick reliability evaluation and GA for optimization. Levitin (2002) had used modified UGF technique on correct classification probability of weighted voting classifiers. Author suggested a method to make the classification decisions with both plurality and threshold voting without imposing constraints on unit weights. Levitin (2005) proposed a new multi-state model of linear consecutive *k*-out-of-*r*-from-*n*: F system with multiple failure criteria and calculated the reliability of the proposed system with the help of UGF technique. Lisnianski and Ding (2009) proposed a redundant model that is typically in a multi-state system. Authors had considered two interconnected multi-state systems where one of them was a redundant system and calculated the reliability of the considered system using UGF and random process methods. Ding et al. (2010) studied the impact of high wind power on the windmill for long term planning point of view. In the model windmill and conventional generators were combined and the reliability of the system was evaluated with the help of UGF technique. Yingkui and Jing (2012) presented a systematic review on the reliability evaluation of multi-state systems. Authors also summarized optimization and maintenance of the complex multi-state system. Tillman et al. (1970) considered a complex system and estimated it's reliability using Bayes theorem and then the reliability of the system was optimized with the help of sequential unconstrained minimization technique.

Binary state assumption in probist (*i.e.*, conventional) reliability theory has shown earlier. This Concept of presenting a system completely failed or working was not extensively acceptable, and thus binary state assumption were replaced by the fuzzy state. Cai et al. (1993) introduced the conceptual framework of profust reliability which was developed based on the fuzzy state assumption and probability assumption. Cheng and Mon (1993) proposed a more general and simplified approach of evaluating fuzzy system reliability by arithmetic and α- cuts interval. Jiang and Chen (2003) has given a numerical algorithm to compute fuzzy reliability of systems such as mechanical components, sensors, electronic units etc. This algorithm grounds the reliability analysis of the system where components are with fuzzy reliability. Wu (2004) proposed the Bayesian reliability estimation under fuzzy environment.

Invoking the Resolution Identity theorem fuzzy Bayes point estimation was created. Fuzzy reliability for series and parallel systems using intuitionistic fuzzy set theory was analysed with a new developed approach by Kumar et al. (2011). Authors introduced trapezoidal intuitionistic fuzzy set number and arithmetic operations between two trapezoidal intuitionistic fuzzy set numbers. Authors have also shown that triangular intuitionistic fuzzy set number is a special case of trapezoidal intuitionistic fuzzy set number. Garg et al. (2014) presented multi-objective reliability optimization methodology with triangular interval data. Authors resolved a conflict between the objectives with the help of intuitionistic fuzzy programming technique. Kumar and Ram (2018 & 2019) evaluated the fuzzy reliability of binary and complex system on the basis of hesitant and dual hesitant fuzzy set theory using Weibull distribution and recursive methods. Kumar et al. (2019) discussed the fuzzy reliability of complex system having dissimilar components with IFS theory, Weibull distribution and Markov process techniques.

As discussed above, in this chapter an example of complex system is used to get the reliability of the system by UGF technique and IFS. Section 2 describes the model in detail i.e. the arrangement of components and the key component which is taken to solve the model. Section 3 contains the definition of some of the important terms as IFS, TIFN and Weibull Distribution and their properties. In section 4 the taken complex system is solved in order to find it's reliability by UGF technique and IFS theory.

DEFINITIONS

Intuitionistic Fuzzy Sets

Suppose we have a unique space E. Let A be a subset of E. Let us construct the set

$$A' = \{x, \mu_A(x), v_A(x) | x \in E\}$$

where, $0 \leq \mu_A(x) + v_A(x) \leq 1$, then A' is said to be IFS. Mostly we work with $A's$ concept instead of the fixed set A for simplification purpose. Mathematically, a formal definition of the IFS can be stated as follows:

$$A' = \{x, \mu_A(x), v_A(x) | x \in E \ \& 0 \leq \mu_A(x) + v_A(x) \leq 1\}$$

Here $\mu_A(x) : E \rightarrow [0,1]$ and $v_A(x) : E \rightarrow [0,1]$ are membership (validity, etc.) and non-membership (non-validity, etc.) degrees respectively. Also, uncertainty degreeisdefinedbythefunction $\pi_A(x) : E \rightarrow [0,1]$ through $\pi(x) = 1 - \mu(x) - v(x)$.

Properties of IFS

Various relations and operations have been defined for every two intuitionistic fuzzy sets. From these the most important relations and operations are:

- Inclusion: $A \subset B \, iff\,(\forall x \in E)(\mu_A(x) \leq \mu_B(x) \;\&\; v_A(x) \geq v_B(x))$.
- Equality: $A = B \, iff\,(\forall x \in E)(\mu_A(x) = \mu_B(x) \;\& v_A(x) = v_B(x))$.
- Classical negation: $\bar{A} = \{x, \mu_A(x), v_A(x) | x \in E\}$.
- Conjunction:
  $A \cap B = \{x, \min(\mu_A(x), \mu_B(x)), \max(v_A(x), v_B(x)) \, | x \in E\}$.
- Disjunction:
  $A \cup B = \{x, \max(\mu_A(x), \mu_B(x)), \min(v_A(x), v_B(x)) \, | x \in E\}$.

Triangular Fuzzy Number

Let α be a TIFN, where the $\mu_A(x)$ and $v_A(x)$ have the following definitions for α:

Membership Function:

$$\mu_A(x) = \begin{cases} \dfrac{(x-a)\mu_\alpha}{(b-a)} & (a \leq x < b), \\ \mu_\alpha & (x = b), \\ \dfrac{(c-x)\mu_\alpha}{(c-b)} & (b < x \leq c) \end{cases}$$

For non-membership function, it is given as:

$$\nu_A(x) = \begin{cases} \dfrac{\left(b - x + \nu_\alpha\left(x - a'\right)\right)}{\left(b - a'\right)} & \left(a' \le x < b\right), \\ \nu_\alpha & \left(x = b\right), \\ \dfrac{\left(x - b + \nu_\alpha\left(c' - x\right)\right)}{\left(c' - b\right)} & \left(b < x \le c'\right) \end{cases}$$

where, $0 \le \mu_A(x) \le 1; 0 \le \nu_A(x) \le 1; 0 \le \mu_A(x) + \nu_A(x) \le 1; a, b, c, a', c' \in R$

The TIFN is therefore denoted as $\tilde{\alpha} = \left(\left[a,b,c\right]; \mu_\alpha\right), \left(\left[a,b,c\right]; \nu_\alpha\right)$, when $\mu_\alpha = 1$, and $\nu_\alpha = 0$, and $\tilde{\alpha}$ will change into the traditional triangular fuzzy number (TFN).

Weibull Distribution

A random variable X follows a Weibull distribution with parameters α and β where $\alpha > 0, \beta > 0$, if the probability density function (Pdf) of X is

$$f\left(X; \alpha, \beta\right) = \frac{\alpha}{\beta^\alpha} X^{(\alpha-1)} e^{-(X/\beta)^\alpha} \text{ for } X \ge 0 \text{, otherwise zero.}$$

MODEL DESCRIPTION

A complex system with five elements is proposed in this chapter, where it is cannot be modelled as series, parallel, series- parallel, parallel- series or *k*-out-of-*n* system. The reliability of such system can be determined using UGF and IFS theory. In this model component 3 is selected as keystone component and component 1 and 2 are in series and similarly 4 and 5 are in series as shown in Figure 1. Reliability for the two possible cases is evaluated, one is when element 3 is working and other is when element 3 fails discuss in Figure 1.

Figure 1. Complex System

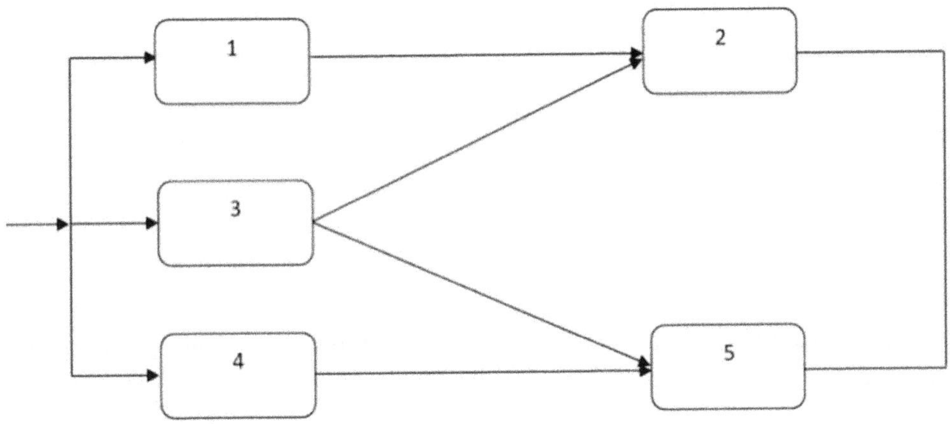

The system structure function is

$$X = \varphi\left(X_1, X_2, X_3, X_4, X_5\right)$$

$$X = min.\ \{max.\left(X_2, X_5\right), X_3 + min.\left[max.\left\{min.\left(X_1, X_2\right), min.\left(X_4, X_5\right)\right\}, \bar{X}_3\right\}\right]$$

The form of elements of u-functions is

$$u_j(z) = \left(1 - p_j\right) z^0 + p_j z^1$$

$$\bar{u}_j(z) = \left(1 - p_j\right) z^1 + p_j z^0$$

Recursively defined system structure is:

$$X_6 = max.\left(X_2, X_5\right)$$

$$X_7 = min.\left(X_6, X_3\right)$$

$$X_8 = min.\left(X_1, X_2\right)$$

$$X_9 = min.\left(X_4, X_5\right)$$

$$X_{10} = max.\left(X_8, X_9\right)$$

$$X_{11} = min.\left(X_{10}, \bar{X}_3\right)$$

$$X = X_7 + X_{11}$$

Now we use the recursive approach to find the reliability of the system, which is as follows

$$U_6\left(z\right) = u_2\left(z\right) \otimes_{max} u_5\left(z\right)$$

$$= \left[\left(1 - p_2\right)z^0 + p_2 z^1\right] \otimes_{max} \left[\left(1 - p_5\right)z^0 + p_5 z^1\right]$$

$$= \left(p_2 + p_5 - p_2 p_5\right)z^1 + \left[\left(1 - p_2\right)\left(1 - p_5\right)z^0\right].$$

$$U_7\left(z\right) = U_6\left(z\right) \otimes_{min} u_3\left(z\right)$$

$$= \left[\left(p_2 + p_5 - p_2 p_5\right)z^1 + \left[\left(1 - p_2\right)\left(1 - p_5\right)z^0\right]\right] \otimes_{min} \left[\left(1 - p_3\right)z^0 + p_3 z^1\right]$$

$$= \left(p_2 + p_5 - p_2 p_5\right)p_3 z^1 + \left(1 - p_2 p_3 - p_3 p_5 + p_2 p_3 p_5\right)z^0.$$

$$U_8\left(z\right) = u_1\left(z\right) \otimes_{min} u_2\left(z\right)$$

$$= \left[\left(1 - p_1\right)z^0 + p_1 z^1\right] \otimes_{min} \left[\left(1 - p_2\right)z^0 + p_2 z^1\right]$$

$$= p_1 p_2 z^1 + \left(1 - p_1 p_2\right)z^0.$$

$$U_9\left(z\right) = u_4\left(z\right) \otimes_{min} u_5\left(z\right)$$

$$= \left[\left(1 - p_4\right) z^0 + p_4 z^1\right] \otimes_{min} \left[\left(1 - p_5\right) z^0 + p_5 z^1\right]$$

$$= p_4 p_5 z^1 + \left(1 - p_4 p_5\right) z^0,$$

$$U_{10}\left(z\right) = U_8\left(z\right) \otimes_{max} U_9\left(z\right) = \left[p_1 p_2 z^1 + \left(1 - p_1 p_2\right) z^0\right] \otimes_{max} \left[p_4 p_5 z^1 + \left(1 - p_4 p_5\right) z^0\right] = (p_1 p_2 + p_4 p_5 - p_1 p_2 p_4 p_5) z^1 + \left(1 - p_1 p_2\right)\left(1 - p_4 p_5\right) z^0.$$

$$U_{11}\left(z\right) = U_{10}\left(z\right) \otimes_{min} \bar{u}_3\left(z\right) = \left[\left(1 - p_{10}\right) z^0 + p_{10} z^1\right] \otimes_{min} \left[\left(1 - p_3\right) z^1 + p_3 z^0\right]$$

(p1p2 + p4p5 – p1p2p4p5) (1-p3) z^1

$$+\left[\left(1 - p_1 p_2\right)\left(1 - p_4 p_5\right)\left(1 - p_3\right) + \left(1 - p_1 p_2\right)\left(1 - p_4 p_5\right) p_3 + + \left(p_1 p_2 + p_4 p_5 - p_1 p_2 p_4 p_5\right) p_3\right] z^0.$$

Now,

$$U\left(z\right) = U_7\left(z\right) + U_{11}\left(z\right)$$

$$= [\left(p_2 + p_5 - p_2 p_5\right) p_3 + (p_1 p_2 + p_4 p_5 - p_1 p_2 p_4 p_5)\left(1 - p_3\right)] z^1$$

$$+\left[\left\{\left(1 - p_1 p_2\right)\left(1 - p_4 p_5\right)\left(1 - p_3\right) + \left(1 - p_1 p_2\right)\left(1 - p_4 p_5\right) p_3 + \left(1 - p_2 p_3 - p_3 p_5 + p_2 p_3 p_5\right) + \left(p_1 p_2 + p_4 p_5 - p_1 p_2 p_4 p_5\right) p_3\right\}\right] z^0.$$

The u-function $U\left(z\right)$ is presentation of probability mass function of the variable *X*. The expected value of *X* is the reliability of the system. The system reliability can now be obtained as $U'\left(z\right)$ at $z = 1$,

$$R = E\left(X\right) = U'\left(1\right) = \left(p_2 + p_5 - p_2 p_5\right) p_3 + (p_1 p_2 + p_4 p_5 - p_1 p_2 p_4 p_5)\left(1 - p_3\right). \quad (1)$$

If all the components of the system are independent identically distributed, then we can have

$$p_1 = p_2 = p_3 = p_4 = p_5 = p.$$

Then the reliability of the system is,

$$R = p^5 - p^4 - 3p^3 + 4p^2. \tag{2}$$

NUMERICAL EXAMPLE

Considered the lifetime component of a system with Weibull distribution to the triangular membership function having intuitionistic fuzzy parameter φ are obtained if

$$\varphi = \varphi_1;(3.00,\ 4.50,5.00;2.50,4.50,6.50),$$

$\varphi_2;(1.25,\ 1.30,2.50;1.00,1.30,3.00), \varphi_3;(2.50,3.00,4.75;2.25,3.00,5.00)$, then the expression for α and β - cut in terms of the membership and non-membership function is

$$\varphi_1 = (3.00 + 1.50\alpha,5.00 - 0.50\alpha;4.50 - 2.00\beta,4.50 + 1.50\beta),$$

$$\varphi_2 = (1.25 + 0.05\alpha,2.50 - 1.20\alpha;1.30 - 0.30\beta,1.30 + 0.50\beta\ ),$$

$$\varphi_3 = (2.5 + 0.5\alpha,4.75 - 1.75\alpha;3.00 - 0.75\beta,3.00 + 0.25\beta).$$

Now using all these above values in equation (2), we get

$$R(t)[\alpha] = \exp(-(t / 3.00 + 1.50\alpha)^{\beta})^5 -$$

$$\exp(-(t / 3.00 + 1.50\alpha)^{\beta})^4 - 3\exp(-(t / 3.00 + 1.50\alpha)^{\beta})^3 + 4\exp(-(t / 3.00 + 1.50\alpha)^{\beta})^2 \tag{3}$$

$$R(t)[\alpha] = \exp(-(t/5.00-0.50\alpha)^{\beta})^5 - \exp(-(t/5.00-0.50\alpha)^{\beta})^4 - 3\exp(-(t/5.00-0.50\alpha)^{\beta})^3 + 4\exp(-(t/5.00-0.50\alpha)^{\beta})^2 \quad (4)$$

$$R(t)[\beta] = \exp(-(t/4.50-2.00(1-\alpha))^{\beta})^5 - \exp(-(t/4.50-2.00(1-\alpha))^{\beta})^4 - 3\exp(-(t/4.50-2.00(1-\alpha))^{\beta})^3 + 4\exp(-(t/4.50-2.00(1-\alpha))^{\beta})^2 \quad (5)$$

$$R(t)[\beta] = \exp(-(t/4.50+1.50(1-\alpha))^{\beta})^5 - \exp(-(t/4.50+1.50(1-\alpha))^{\beta})^4 - 3\exp(-(t/4.50+1.50(1-\alpha))^{\beta})^3 + 4\exp(-(t/4.50+1.50(1-\alpha))^{\beta})^2. \quad (6)$$

For all α, where taking $\beta = 0.5$ and $t = 5$.

In Table 1, shown the fuzzy reliability of triangular fuzzy number using Weibull distribution in form of $\alpha\, and \beta$ cut from Eqs.(3)-(6).

Table 1. $\alpha\, and \beta$ cut of the Intuitionistic fuzzy membership function.

(α, β)- cut	$R[\alpha]$	$R[\beta]$
0.0	(0.2360, 0.3804)	(0.1907, 0.4588)
0.1	(0.2488, 0.3774)	(0.2093, 0.4495)
0.2	(0.2613, 0.3744)	(0.2272, 0.4398)
0.3	(0.2734, 0.3714)	(0.2445, 0.4298)
0.4	(0.2852, 0.3683)	(0.2613, 0.4195)
0.5	(0.2966, 0.3652)	(0.2773, 0.4088)
0.6	(0.3078, 0.3620)	(0.2928, 0.3978)
0.7	(0.3186, 0.3589)	(0.3078, 0.3863)
0.8	(0.3291, 0.3557)	(0.3221, 0.3744)
0.9	(0.3393, 0.3595)	(0.3359, 0.3620)
1.0	(0.3492, 0.3492)	(0.3492, 0.3492)

$$R(t)[\alpha] = \exp(-(t/1.25 + 0.05\alpha)^{\beta})^{5} -$$

$$\exp(-(t/1.25 + 0.05\alpha)^{\beta})^{4} - 3\exp(-(t/1.25 + 0.05\alpha)^{\beta})^{3} + 4\exp(-(t/1.25 + 0.05\alpha)^{\beta})^{2}. \tag{7}$$

$$R(t)[\alpha] = \exp(-(t/2.50 - 1.20\alpha)^{\beta})^{5} -$$

$$\exp(-(t/2.50 - 1.20\alpha)^{\beta})^{4} - 3\exp(-(t/2.50 - 1.20\alpha)^{\beta})^{3} + 4\exp(-(t/2.50 - 1.20\alpha)^{\beta})^{2}. \tag{8}$$

$$R(t)[\beta] = \exp(-(t/1.30 - 0.30(1-\alpha))^{\beta})^{5} -$$

$$\exp(-(t/1.30 - 0.30(1-\alpha))^{\beta})^{4} - 3\exp(-(t/1.30 - 0.30(1-\alpha))^{\beta})^{3} + 4\exp(-(t/1.30 - 0.30(1-\alpha))^{\beta})^{2}. \tag{9}$$

$$R(t)[\beta] = \exp(-(t/1.30 + 0.50(1-\alpha))^{\beta})^{5} -$$

$$\exp(-(t/1.30 + 0.50(1-\alpha))^{\beta})^{4} - 3\exp(-(t/1.30 + 0.50(1-\alpha))^{\beta})^{3} + 4\exp(-(t/1.30 + 0.50(1-\alpha))^{\beta})^{2}. \tag{10}$$

For all α, where taking $\beta = 0.5$ and $t = 5$.

In Table 2, also shown fuzzy reliability of membership and non-membership α, β - cut using Eqs.(7)-(10).

$$R(t)[\alpha] = \exp(-(t/2.5 + 0.5\alpha)^{\beta})^{5} -$$

$$\exp(-(t/2.5 + 0.5\alpha)^{\beta})^{4} - 3\exp(-(t/2.5 + 0.5\alpha)^{\beta})^{3} + 4\exp(-(t/2.5 + 0.5\alpha)^{\beta})^{2}. \tag{11}$$

Table 2. Fuzzy reliability of α, β - cut from Intuitionistic fuzzy set

(α, β)- cut	$R[\alpha]$	$R[\beta]$
0.0	(0.0655, 0.1907)	(0.0419, 0.1214)
0.1	(0.0660, 0.1792)	(0.0446, 0.1163)
0.2	(0.0665, 0.1676)	(0.0474, 0.1112)
0.3	(0.0670, 0.1558)	(0.0501, 0.1061)
0.4	(0.0675,0.1438)	(0.0530, 0.1009)
0.5	(0.0680, 0.1316)	(0.0558, 0.0958)
0.6	(0.0685, 0.1194)	(0.0587, 0.0907)
0.7	(0.0690, 0.1071)	(0.0616, 0.0856)
0.8	(0.0695, 0.0948)	(0.0646, 0.0805)
0.9	(0.0700, 0.0826)	(0.0675, 0.0755)
1.0	(0.0705, 0.0705)	(0.0705, 0.0705)

$$R(t)[\alpha] = \exp(-(t / 4.75 - 1.75\alpha)^{\beta})^{5} -$$

$$\exp(-(t / 4.75 - 1.75\alpha)^{\beta})^{4} - 3\exp(-(t / 4.75 - 1.75\alpha)^{\beta})^{3} + 4\exp(-(t / 4.75 - 1.75\alpha)^{\beta})^{2}. \tag{12}$$

$$R(t)[\beta] = \exp(-(t / 3.00 - 0.75(1-\alpha))^{\beta})^{5} -$$

$$\exp(-(t / 3.00 - 0.75(1-\alpha))^{\beta})^{4} - 3\exp(-(t / 3.00 - 0.75(1-\alpha))^{\beta})^{3} + 4\exp(-(t / 3.00 - 0.75(1-\alpha))^{\beta})^{2}. \tag{13}$$

$$R(t)[\beta] = \exp(-(t / 3.00 + 0.25(1-\alpha))^{\beta})^{5} -$$

$$\exp(-(t / 3.00 + 0.25(1-\alpha))^{\beta})^{4} - 3\exp(-(t / 3.00 + 0.25(1-\alpha))^{\beta})^{3} + 4\exp(-(t / 3.00 + 0.25(1-\alpha))^{\beta})^{2}. \tag{14}$$

For all α, where taking $\beta = 0.5$ and $t = 5$.

In Table 3, discuss the α, β - cut intuitionistic fuzzy reliability in form membership and non-membership function of Eqs.(11)-(14).

Table 3. Intuitionistic fuzzy reliability in form of α, β - cut.

(α, β)- cut	$R[\alpha]$	$R[\beta]$
0.0	(0.1907, 0.3652)	(0.1666, 0.2571)
0.1	(0.1954, 0.3541)	(0.1739, 0.2551)
0.2	(0.2000, 0.3426)	(0.1812, 0.2530)
0.3	(.2047, 0.3308)	(0.1883, 0.2509)
0.4	(0.2093, 0.3186)	(0.1954, 0.2488)
0.5	(0.2138, 0.3059)	(0.2024, 0.2467)
0.6	(0.2183, 0.2928)	(0.2093, 0.2445)
0.7	(0.2228, 0.2793)	(0.2161, 0.2424)
0.8	(0.2272, 0.2653)	(0.2228, 0.2403)
0.9	(0.2316, 0.2509)	()0.2294, 0.2381)
1.0	(0.2360, 0.2360)	(0.2360, 0.2360)

CONCLUSION

In this chapter, we have evaluated the reliability function for a complex system with the help of UGF technique. Also, on the basis of Intuitionistic Fuzzy Set theory and (α, β) – cut method arithmetic operations of proposed triangular fuzzy numbers are evaluated. The result of considered system and techniques shown in Table 1, Table 2 and Table 3 in form of fuzzy reliability and membership and non-membership function.

REFERENCES

Cai, K. Y., Wen, C. Y., & Zhang, M. L. (1993). Fuzzy states as a basis for a theory of fuzzy reliability. *Microelectronics and Reliability*, *33*(15), 2253–2263.

Cheng, C. H., & Mon, D. L. (1993). Fuzzy system reliability analysis by interval of confidence. *Fuzzy Sets and Systems*, *56*(1), 29–35.

Ding, Y., Wang, P., Goel, L., Loh, P. C., & Wu, Q. (2010). Long-term reserve expansion of power systems with high wind power penetration using universal generating function methods. *IEEE Transactions on Power Systems*, *26*(2), 766–774.

Garg, H., Rani, M., Sharma, S. P., & Vishwakarma, Y. (2014). Intuitionistic fuzzy optimization technique for solving multi-objective reliability optimization problems in interval environment. *Expert Systems with Applications*, *41*(7), 3157–3167.

Jiang, Q., & Chen, C. H. (2003). A numerical algorithm of fuzzy reliability. *Reliability Engineering & System Safety*, *80*(3), 299–307.

Kumar, A., & Ram, M. (2018). Reliability of the system based on Weibull distribution and hesitant fuzzy set. *International Journal of Mathematical. Engineering and Management Sciences*, *3*(4), 513–521.

Kumar, A., & Ram, M. (2019). Reliability Analysis for Environment Systems Using Dual Hesitant Fuzzy Set. In *Advanced Fuzzy Logic Approaches in Engineering Science* (pp. 162–173). IGI Global.

Kumar, A., Singh, S. B., & Ram, M. (2019). Reliability appraisal for consecutive-*k*-out-of-*n*:F system of non-identical components with intuitionistic fuzzy set. *Int. J. Operational Research.*, *36*(3), 362–374.

Kumar, M., Yadav, S. P., & Kumar, S. (2011). A new approach for analysing the fuzzy system reliability using intuitionistic fuzzy number. *International Journal of Industrial and Systems Engineering*, *8*(2), 135–156.

Levitin, G. (2002). Evaluating correct classification probability for weighted voting classifiers with plurality voting. *European Journal of Operational Research*, *141*(3), 596–607.

Levitin, G. (2005). Reliability of linear multistate multiple sliding window systems. *Naval Research Logistics*, *52*(3), 212–223.

Levitin, G., & Lisnianski, A. (2001). A new approach to solving problems of multi-state system reliability optimization. *Quality and Reliability Engineering International*, *17*(2), 93–104.

Levitin, G., Lisnianski, A., Haim, H. B., & Elmakis, D. (2000, April). Genetic algorithm and universal generating function technique for solving problems of power system reliability optimization. In *DRPT2000. International Conference on Electric Utility Deregulation and Restructuring and Power Technologies. Proceedings (Cat. No. 00EX382)* (pp. 582-586). IEEE.

Lisnianski, A., & Ding, Y. (2009). Redundancy analysis for repairable multi-state system by using combined stochastic processes methods and universal generating function technique. *Reliability Engineering & System Safety*, *94*(11), 1788–1795.

Tillman, F. A., Hwang, C. L., Fan, L. T., & Lai, K. C. (1970). Optimal reliability of a complex system. *IEEE Transactions on Reliability*, *19*(3), 95–100.

Wu, H. C. (2004). Fuzzy reliability estimation using Bayesian approach. *Computers & Industrial Engineering*, *46*(3), 467–493.

Yingkui, G., & Jing, L. (2012). *Multi-state system reliability: A new and systematic review*. Procedia Engineeri.

Chapter 9
Fuzzy Reliability Assessment of Consecutive r Out of k From n G System Using UGF

KM. Meenakshi
M. B. G. P. G. College, Haldwani, India

Akshay Kumar
Graphic Era Hill University, India

S. B. Singh
G. B. Pant University of Agriculture and Technology, Pantnagar, India

ABSTRACT

This chapter considers a consecutive r-out of-k-from n: G system taken for the study. The considered system works if at least r elements out of k consecutive elements within n elements work. The consecutive r-out of-k-from n: G system is special case of linear consecutive k-out-of-r-from-n:G system. A technique is applied for analyzing fuzzy reliability of the system using universal generating function and fuzzy set theory. At the end of the chapter, the proposed technique is demonstrated from a numerical example.

INTRODUCTION

Consecutive *r*-out of -*k*-from *n*: G is one of the several system used in engineering. The concept of an consecutive *r*-out of -*k*-from *n*: G system was made known by Griffith(1986) but initially mentioned by Tong (1986) in his research, In consecutive *r*-out of -*k*-from *n*: G system, the system consist of *n* elements and is working if

DOI: 10.4018/978-1-7998-7564-2.ch009

at least r elements out of k consecutive elements working or at least consecutive k elements working. Reliability/Availability analysis of the system completely depends on reliability/availability of its elements. During the reliability analysis of the system it is not always necessary to have complete information about the system which may lead to uncertainty. Earlier, there are many theories introduced to handle it, fuzzy set theory is one of them. In the theory of fuzzy sets the membership of an element is a single value between 0 and 1. Cai et al. (1993) evaluated the fuzzy reliability of binary system and discussed relation between failure rate and fuzzy reliability.

Most of the researchers Sun and Liao (1990), Levitin (2005); Habib *et al.* (2007) have done work in binary and multi-state consecutive systems from different approaches with the help of crisp data which is not applicable for the case where the data available has uncertainty that arises due to insufficiency in data or incompleteness in information. To overcome this type of problem the fuzzy universal generating function (FUGF) technique was introduced and developed to extend the universal generating function (UGF) with crisp sets by Ding and Lisnianski (2008). Chongshan (2009) discussed a repairable linear consecutive 2-out-of-3 failure system having repair elements and evaluated the fuzzy reliability from exponential distribution. Ding *et al.* (2010) studied a multi-state k-out-of-n system which is generalized form of multi-state weighted k-out-of-n system and computed fuzzy reliability with the help of UGF technique. Baziuk *et al.* (2016) reviewed the application of fuzzy reliability theory in real world and mathematically formulation. Authors introducing new concept of human reliability analysis to discussed some methodologies. Kumar and Goel (2017) considered a pulping system which used in industry having four subsystem and calculated fuzzy reliability, failure and repair rate using trapezoidal fuzzy number. Song *et al.* (2017) discussed a multi-state system weighted k-out-of-n system and also find fuzzy reliability using stochastic multiple-valued approach and FUGF technique. Kumar and Ram (2018) and Kumar *et al* (2020) assessment the fuzzy reliability of binary and multi-state systems such as series-parallel and consecutive linear and circular k-out-of-n systems using well known Weibull distribution and hesitant fuzzy set.

In the above discussion, we have studied consecutive r-out of -k-from n: G system using FUGF technique using UGF and triangular fuzzy number as probability. It is worth mentioning here that with the incorporation of above mentioned points the reliability of system using FUGF technique is never obtained before. In this chapter section 1 discuss review of literature based on fuzzy reliability analysis and in section 2 define some basic definition of UGF, fuzzy, triangular fuzzy number and FUGF. Section 3 describe the system functioning, section 4 given an example based on proposed technique and last section 5 define result and discussion of consider system.

DEFINITIONS

Universal Generating Function

UGF first discussed by Ushakov [1986]. Let a discrete random variable X has M possible values ($x_1, x_2, \ldots, x_M$) and ($p_1, p_2, \ldots, p_M$) be the corresponding probabilities. Then UGF of discrete random variable X= ($x_1, x_2, \ldots, x_M$) is defined by a polynomial

$$U(z) = \sum_{j=1}^{M} p_j z^{(X=x_j)}, j = 1,2,3,\ldots,M \tag{1}$$

Fuzzy Set

Zadeh (1965) discussed fuzzy set theory to represent data analysis and information under uncertainties. Fuzzy set was explicitly intended to mathematically represent vagueness and uncertainty. Different authors like Dubois and Prade (1988), Ji *et al.* (2014) discussed the work with application of fuzzy set theory.

Let T is denoted by collection of objects assigned by t where T is known as universe of discourse. A fuzzy set K in the universe of discourse T is identified by membership function $\eta_K : T \to [0,1]$ and $\eta_K(t)$ is degree of membership of element t in fuzzy set K for each $t \in T$ and fuzzy set k is represented by order pair as the follows

$$K = \{(t, \eta_K(t)) \mid t \in T\}$$

Triangular Fuzzy Number

Triangular fuzzy number (TFN) is one of the well known fuzzy numbers. A triangular fuzzy number K can be defined as a triplet $K = [k_1, k_2, k_3]$, where

1. k_1 to k_2 function is increasing
2. k_2 to k_3 function is decreasing
3. $k_1 \le k_2 \le k_3$
4. 4. Its membership function (Figure 1) is defined as:

$$\eta_K(t) = \begin{cases} \dfrac{t-k_1}{k_2-k_1}, & \text{for } k_1 \le t \le k_2 \\ 1, & \text{for } t = k_2 \\ \dfrac{k_3-t}{k_3-k_2}, & \text{for } k_2 \le t \le k_3 \\ 0, & \text{otherwise} \end{cases} \tag{2}$$

Figure 1. Triangular fuzzy number

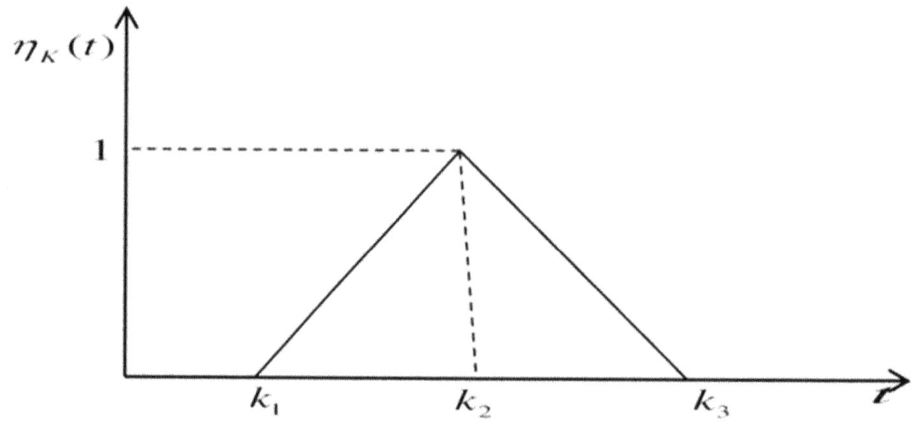

Crisp interval can be calculated by applying α-cut of the TFN, which is denoted by K_α and calculated as

$$\frac{K^L_{(\alpha)} - k_1}{k_2 - k_1} = \alpha \text{ and } \frac{k_3 - K^U_{(\alpha)}}{k_3 - k_2} = \alpha$$

$$\Rightarrow K^L_{(\alpha)} = k_1 + \alpha(k_2 - k_1) \text{ and } K^U_{(\alpha)} = k_3 - \alpha(k_3 - k_2)$$

$$\text{Then } K_\alpha = [K^L_{(\alpha)}, K^U_{(\alpha)}] = [k_1 + \alpha(k_2 - k_1), k_3 - \alpha(k_3 - k_2)]. \tag{3}$$

Fuzzy Universal Generating Function

FUGF the extended form of UGF where the state probabilities of the elements can be measured as fuzzy values. Here the fuzzy number probability for representing the uncertainty in the system probability distribution corresponding to different states. Fuzzy probability of ith element at (X_{j_i}) state is denoted by $(\widehat{p}^{i}_{j_i})$ UGF with fuzzy probability is follows:

$$\widehat{u}_{(i)}(z) = \sum_{j_i=1}^{n} (\widehat{P}^{(i)}_{j_i}) z^{X_{j_i}} \text{ where i= 1,2,3,...n} \tag{4}$$

$$\widehat{U}(z) = \underset{f}{\otimes}(\widehat{u}_{(1)}(z), \widehat{u}_{(2)}(z), ..., \widehat{u}_{(n)}(z))$$

$$\widehat{U}(z) = \sum_{j_1=1}^{l_1} \sum_{j_2=1}^{l_2} ... \sum_{j_n=1}^{l_n} (\prod_{i=1}^{n} (\widehat{P}^{(i)}_{j_i})) z^{f(X_{j_1},...,X_{j_n})} \tag{5}$$

Fuzzy Reliability of Consecutive *r*-out of -*k*-from *n*: G System

The consider consecutive *r*-out of -*k*-from *n*: G system has *n* binary state elements. Let X_{ij_i} be state variable of element *i* where *i* =1,2,..,*n*. FUGF of each element *i* can be determined using equation (4). Considered system has *n*-*k*+1 possible group of *k* consecutive elements. h^{th} group of *k* consecutive elements has elements from *h* to *h*+*k*-1. To obtained fuzzy u-function (FUGF) corresponding to all groups of *k* consecutive elements following procedure is taken

Step 1. Defined FUGF of $\widehat{U}_{1-k}(z)$ is

$$\widehat{U}_{1-k}(z) = z^{X_0} \tag{6}$$

where, x_0 vector consists *k* zeros

Step 2. Defined following shift operator over FUGF $\widehat{U}_h(z) = \sum_{t=1}^{N} \widehat{P}^{\circ}_{h,t} \; z^{Y_{h,t}}$

$$\widehat{U}_h(z) \underset{\leftarrow}{\otimes} \widehat{u}_{h+k}(z) = (\widehat{U}_h(z) \underset{\leftarrow}{\otimes} (\widehat{P}^{(h+k)}_1 z^1 + \widehat{q}^{(h+k)}_2 z^0)$$

$$= \widehat{P}_1^{(h+k)} \sum_{t=1}^{N} \widehat{P}_{h,t}^{\circ} \; z^{Y_{h,t} \leftarrow 1} + \widehat{q}_2^{(h+k)} \sum_{t=1}^{N} \widehat{P}_{h,t}^{\circ} \; z^{Y_{h,t} \leftarrow 0} \tag{7}$$

Here, $\widehat{P}_{h,t}^{\circ}$ is probability of element corresponding to t^{th} state

Let $Y \underset{\leftarrow}{\otimes} x$ shift operator over vector Y and value x shift all the vector elements of Y one place left i.e. remove the 1st element of state Y and adds new state of successive element then FUGF of $\widehat{U}_h(z) \underset{\leftarrow}{\otimes} \widehat{u}_{h+k}(z)$ is re-write as

$$\widehat{U}_h(z) \underset{\leftarrow}{\otimes} \widehat{u}_{h+k}(z) = \sum_{m=1}^{2N} \widehat{P}_{h+1,m}^{*} \; z^{Y_{h+1,m}}$$

where $\widehat{P}_{h+1,m}^{*}$ is probability of group corresponding to m^{th} state and $Y_{h+1,m} = (X_{j_{h+1}}, X_{j_{h+2}}, \ldots, X_{j_{h+k}})$ Failure probability of group is obtained as follow

$$\text{if } z = \begin{cases} 1, \text{ if } S_{h+1} \leq k - r \\ 0, \text{ otherwise} \end{cases} \tag{8}$$

where, $S_{h+1} = X_{j_{h+1}} + X_{j_{h+2}} + \ldots + X_{j_{h+k}}$.

Remove the terms from equation (8) having state vectors containing k or more than k zeros then the remaining fuzzy u-function of equation (8) is denoted by $U_h(Z)$and continue the process given in equation (7) and (8) until we get the fuzzy u-function of $(n$-k+1)th group. The fuzzy u-function of $(n$-k+1) th group is written as

$$\widehat{U}_{n-k+1}(z) = \sum_{m=1}^{N^*} \widehat{P}_{n-k+1,m}^{\bullet} \; z^{Y_{n-k+1,m}} \tag{9}$$

where, $Y_{n-k+1,m} = (X_{j_{n-k+1}}, \ldots, X_{j_n})$

Reliability of reliability of consecutive r-out of -k-from n: G system is obtained

$$\widehat{R}_s = \sum_{m=1}^{N^*} \widehat{P}_{n-k+1,m}^{\bullet} \; z^{Y_{n-k+1,m}} \tag{10}$$

$$\text{if } z = \begin{cases} 0, \text{ if } \sum_{j_{n-k+1}=n-k+1}^{n} X_{j_{n-k+1}} \leq k-r \\ 1, \text{ otherwise} \end{cases}$$

ILLUSTRATIVE EXAMPLE

Consider a consecutive 2-out of-3-from 4: G system have four binary state elements of considered system are C_1, C_2, C_3 and C_4 with triangular fuzzy probabilities $\widehat{P}_1^{(1)} = (.70, .72, .74)$, $\widehat{q}_2^{(1)} = (.2, .25, .3)$, $\widehat{P}_1^{(2)} = (.6, .65, .7)$, $\widehat{q}_2^{(2)} = (.3, .32, .34)$, $\widehat{P}_1^{(3)} = (.8, .85, .9)$, $\widehat{q}_2^{(3)} = (.09,.10,.11)$, $\widehat{P}_1^{(4)} = (.9, .93, .95)$, $\widehat{q}_2^{(4)} = (.08, .09, .10)$ and $X_{j_1}, X_{j_2}, X_{j_3}, X_{j_4}$ are corresponding states there values are 0 or 1. There are n-k+1=2 groups G_1 with elements states $(X_{j_1}, X_{j_2}, X_{j_3})$ and G_2 with elements states $(X_{j_2}, X_{j_3}, X_{j_4})$ of k=3 consecutive elements. FUGF of C_1, C_2, C_3, C_4 elements are $\widehat{u}_1(z), \widehat{u}_2(z), \widehat{u}_3(z), \widehat{u}_4(z)$ respectively given as follows

$$\widehat{u}_1(z) = \widehat{P}_1^{(1)} z^1 + \widehat{q}_2^{(1)} z^0$$

$$\widehat{u}_2(z) = \widehat{P}_1^{(2)} z^1 + \widehat{q}_2^{(2)} z^0$$

$$\widehat{u}_4(z) = \widehat{P}_1^{(3)} z^1 + \widehat{q}_2^{(3)} z^0$$

$$\widehat{u}_4(z) = \widehat{P}_1^{(4)} z^1 + \widehat{q}_2^{(4)} z^0$$

FUGF of group is calculated with the help of equations (6)-(10) are

$$\widehat{U}_{-2}(z) = z^{(0,0,0)}$$

$$\widehat{U}_{-1}(z) = z^{(0,0,0)} \underset{\leftarrow}{\otimes} (\widehat{P}_1^{(1)} z^1 + \widehat{q}_2^{(1)} z^0)$$
$$= \widehat{P}_1^{(1)} z^{(0,0,1)} + \widehat{q}_2^{(1)} z^{(0,0,0)}$$

$$\hat{U}_0(z) = (\hat{P}_1^{(1)} z^{(0,0,1)} + \hat{q}_2^{(1)} z^{(0,0,0)}) \underset{\leftarrow}{\otimes} (\hat{P}_1^{(2)} z^1 + \hat{q}_2^{(2)} z^0)$$
$$= (\hat{P}_1^{(1)} \hat{P}_1^{(2)}) z^{(0,1,1)} + (\hat{q}_2^{(1)} \hat{P}_1^{(2)}) z^{(0,0,1)} + (\hat{P}_1^{(1)} \hat{q}_2^{(2)}) z^{(0,1,0)} + (\hat{q}_2^{(1)} \hat{q}_2^{(2)}) z^{(0,0,0)}$$

$$\hat{U}_1(z) = (\hat{U}_0(z) \underset{\leftarrow}{\otimes} (\hat{P}_1^{(3)} z^1 + \hat{q}_2^{(3)} z^0))$$
$$= (\hat{P}_1^{(1)} \hat{P}_1^{(2)} \hat{P}_1^{(3)}) z^{(1,1,1)} + (\hat{q}_2^{(1)} \hat{P}_1^{(2)} \hat{P}_1^{(3)}) z^{(0,1,1)} + (\hat{P}_1^{(1)} \hat{q}_2^{(2)} \hat{P}_1^{(3)}) z^{(1,0,1)} + (\hat{q}_2^{(1)} \hat{q}_2^{(2)} \hat{P}_1^{(3)}) z^{(0,0,1)}$$
$$+ (\hat{P}_1^{(1)} \hat{P}_1^{(2)} \hat{q}_2^{(3)}) z^{(1,1,0)} + (\hat{q}_{12}^{(1)} \hat{P}_1^{(2)} \hat{q}_2^{(3)}) z^{(0,1,0)} + (\hat{P}_1^{(1)} \hat{q}_2^{(2)} \hat{q}_2^{(3)}) z^{(1,0,0)} + (\hat{q}_2^{(1)} \hat{q}_2^{(2)} \hat{q}_2^{(3)}) z^{(0,0,0)}$$

Failure probability of group G_1 for r=2 obtained using of equation (8) as

$$\hat{F}_1 = (\hat{q}_2^{(1)} \hat{q}_2^{(2)} \hat{P}_1^{(3)}) + (\hat{q}_{12}^{(1)} \hat{P}_1^{(2)} \hat{q}_2^{(3)}) + (\hat{P}_1^{(1)} \hat{q}_2^{(2)} \hat{q}_2^{(3)}) + (\hat{q}_2^{(1)} \hat{q}_2^{(2)} \hat{q}_2^{(3)})$$

Remove the elements from FUGF of G_1 at which the system is failed then U_1 (z) written as

$$\hat{U}_1^*(z) = (\hat{P}_1^{(1)} \hat{P}_1^{(2)} \hat{P}_1^{(3)}) + (\hat{q}_2^{(1)} \hat{P}_1^{(2)} \hat{P}_1^{(3)}) + (\hat{P}_1^{(1)} \hat{q}_2^{(2)} \hat{P}_1^{(3)}) + (\hat{P}_1^{(1)} \hat{P}_1^{(2)} \hat{q}_2^{(3)})$$

FUGF of second group G_2 is determined from equation (11) as

$$\hat{U}_2(z) = \hat{U}_1^*(z) \underset{\leftarrow}{\otimes} \hat{u}_4(z)$$

$$\hat{U}_2(z) = (\hat{P}_1^{(1)} \hat{P}_1^{(2)} \hat{P}_1^{(3)} \hat{P}_1^{(4)}) z^{(1,1,1)} + (\hat{q}_2^{(1)} \hat{P}_1^{(2)} \hat{P}_1^{(3)} \hat{P}_1^{(4)}) z^{(1,1,1)} + (\hat{P}_1^{(1)} \hat{q}_2^{(2)} \hat{P}_1^{(3)} \hat{P}_1^{(4)}) z^{(0,1,1)} + (\hat{P}_1^{(1)} \hat{P}_1^{(2)} \hat{q}_2^{(3)} \hat{P}_1^{(4)}) z^{(1,0,1)}$$

$$+ (\hat{P}_1^{(1)} \hat{P}_1^{(2)} \hat{P}_1^{(3)} \hat{q}_1^{(4)}) z^{(1,1,0)} + (\hat{q}_2^{(1)} \hat{P}_1^{(2)} \hat{P}_1^{(3)} \hat{q}_1^{(4)}) z^{(1,1,0)} + (\hat{p}_1^{(1)} \hat{q}_2^{(2)} \hat{p}_1^{(3)} \hat{q}_1^{(4)}) z^{(0,1,0)} + (\hat{P}_1^{(1)} \hat{P}_1^{(2)} \hat{q}_2^{(3)} \hat{q}_1^{(4)}) z^{(1,0,0)}$$

Failure probability of group G_2 is given as

$$\hat{F}_2 = (\hat{P}_1^{(1)} \hat{q}_2^{(2)} \hat{P}_1^{(3)} \hat{q}_1^{(4)}) + (\hat{P}_1^{(1)} \hat{P}_1^{(2)} \hat{q}_2^{(3)} \hat{q}_1^{(4)})$$

Reliability of considered system is obtained using equation (10) as

$$\hat{R}_s = (\hat{P}_1^{(1)} \hat{P}_1^{(2)} \hat{P}_1^{(3)} \hat{P}_1^{(4)}) + (\hat{q}_2^{(1)} \hat{P}_1^{(2)} \hat{P}_1^{(3)} \hat{P}_1^{(4)}) + (\hat{P}_1^{(1)} \hat{q}_2^{(2)} \hat{P}_1^{(3)} \hat{P}_1^{(4)}) + (\hat{P}_1^{(1)} \hat{P}_1^{(2)} \hat{q}_2^{(3)} \hat{P}_1^{(4)})$$
$$+ (\hat{P}_1^{(1)} \hat{P}_1^{(2)} \hat{P}_1^{(3)} \hat{q}_1^{(4)}) + (\hat{q}_2^{(1)} \hat{P}_1^{(2)} \hat{P}_1^{(3)} \hat{q}_1^{(4)})$$

$\tilde{R}_s = (0.60858, 0.7722987, 0.952288)$

Failure probability of system is

$$\tilde{F}_s = (\widehat{q}_2^{(1)}\widehat{q}_2^{(2)}\widehat{P}_1^{(3)}) + (\widehat{q}_{12}^{(1)}\widehat{P}_1^{(2)}\widehat{q}_2^{(3)}) + (\widehat{P}_1^{(1)}\widehat{q}_2^{(2)}\widehat{q}_2^{(3)}) + (\widehat{q}_2^{(1)}\widehat{q}_2^{(2)}\widehat{q}_2^{(3)}) + (\widehat{p}_1^{(1)}\widehat{q}_2^{(2)}\widehat{p}_1^{(3)}\widehat{q}_1^{(4)}) + (\widehat{P}_1^{(1)}\widehat{P}_1^{(2)}\widehat{q}_2^{(3)}\widehat{q}_1^{(4)})$$

α cut bound of reliability of demonstrated system is determined using equation (3) shown in Table 1

Table 1. Reliability bounds of the considered system

α -cut	Reliability Bound $[\widehat{R}_s^L, \widehat{R}_s^U]$
0	[0.60858, 0.952288]
.1	[0.624952, 0.934289]
.2	[0.641324, 0.91629]
.3	[0.657696,0.898291]
.4	[0.674067, 0.880292]
.5	[0.690439, 0.862293]
.6	[0.7068110, 0.844294]
.7	[0.723183, 0.826295]
.8	[0.739555, 0.808297]
.9	[0.755927, 0.790298]
1.0	[0.772299, 0.772299]

Figure 2. Reliability of consecutive 2-out of-3-from 4: G system

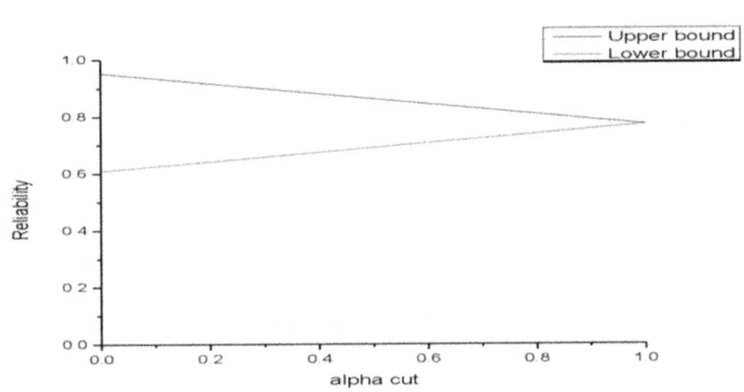

From the numerical example we get that as alpha cut is increases within their range the upper bound and lower bounds of reliability are getting closer the exact value of reliability of demonstrated system. Shown in Figure1.

CONCLUSION

In this chapter, we analyzed the formulae for fuzzy reliability of consecutive *r*-out of -*k*-from *n*: G system using FUGF. To obtain the reliability of the system in which the probabilities of system elements cannot be precisely calculate but can be approximated by fuzzy values. Fuzzy reliability discussed in Table 1 and Figure 1 in theform of upper and lower bound from membership function. Fuzzy reliability is determined in form of interval which is increasing 0 to 1 in the form of alpha cut.

REFERENCES

Baziuk, P. A., Rivera, S. S., & Núñez Mc Leod, J. (2016). Fuzzy human reliability analysis: Applications and contributions review. *Advances in Fuzzy Systems*, *2016*, 2016. doi:10.1155/2016/4612086

Cai, K. Y., Wen, C. Y., & Zhang, M. L. (1993). Fuzzy states as a basis for a theory of fuzzy reliability. *Microelectronics and Reliability*, *33*(15), 2253–2263. doi:10.1016/0026-2714(93)90065-7

Chongshan, G. (2009, November). Fuzzy availability analysis of a repairable consecutive-2-out-of-3: F System. In *2009 IEEE International Conference on Grey Systems and Intelligent Services (GSIS 2009)* (pp. 434-437). IEEE.

Ding, Y., & Lisnianski, A. (2008). Fuzzy universal generating functions for multi-state system reliability assessment. *Fuzzy Sets and Systems*, *159*(3), 307–324. doi:10.1016/j.fss.2007.06.004

Ding, Y., Zuo, M. J., Lisnianski, A., & Li, W. (2010). A framework for reliability approximation of multi-state weighted *k*-out-of-*n* systems. *IEEE Transactions on Reliability*, *59*(2), 297–308. doi:10.1109/TR.2010.2048659

Dubois, D., & Prade, H. (1988). *Possibility theory: An approach to computerized processing of uncertainty*. Plenum Press. doi:10.1007/978-1-4684-5287-7

Griffith, W. (1986). On consecutive k-out-of-n failure system and their generalizations. In B. A. North-Holland (Ed.), *Reliability and Quality Control* (pp. 157–165). Elsevier.

Habib, A. S., Al-Seedy, R. O., & Radwan, T. (2007). Reliability evaluation of multi-state consecutive *k*-out-of-*r*-from-*n*: G system. *Applied Mathematical Modelling*, *31*(11), 2412–2423. doi:10.1016/j.apm.2006.09.006

Ji, Z., Xia, Y., Sun, Q., & Cao, G. (2014). Interval-valued possibilistic fuzzy C-means clustering algorithm. *Fuzzy Sets and Systems*, *253*, 138–156. doi:10.1016/j.fss.2013.12.011

Kumar, A., & Ram, M. (2018). System reliability analysis based on weibull distribution and hesitant fuzzy set. *International Journal of Mathematical. Engineering and Management Sciences*, *3*(4), 513–521.

Kumar, A., Singh, S. B., & Ram, M. (2020). Systems reliability assessment using hesitant fuzzy set. *International Journal of Operational Research*, *38*(1), 1–18. doi:10.1504/IJOR.2020.106357

Kumar, J., & Goel, M. (2017). Fuzzy reliability analysis of a pulping system in paper industry with general distributions for all random variables. *Cogent Mathematics*, *4*(1), 1285467. doi:10.1080/23311835.2017.1285467

Levitin, G. (2005). *Universal Generating Functions in Reliability Analysis and Optimization*. Springer –Verlag.

Song, X., Zhai, Z., Zhu, P., Guo, Y., & Zhang, Y. (2017, July). *Reliability approximation evaluation of multi-stated weighted k-out-of-n systems. In 2017 Prognostics and System Health Management Conference*. PHM-Harbin.

Sun, H., & Liao, J. (1990). The reliability (*n; f; k*) system. *Journal of Electronics (China)*, *12*, 436–439.

Tong, Y. L. (1985). Rearrangement inequality for the longest run with an application to network reliability. *Journal of Applied Probability*, *22*(2), 286–393. doi:10.2307/3213781

Ushakov, I. (1986). Universal Generating Function. *Soviet Journal of Computer Systems*, *24*, 118–129.

Zadeh, L. A. (1965). Fuzzy set. *Information and Control*, *8*(3), 338–353. doi:10.1016/S0019-9958(65)90241-X

Chapter 10
Fuzzy System Reliability Using Fuzzy Lifetime Distribution Emphasizing Octagonal Intuitionistic Fuzzy Numbers

Pawan Kumar
Government Degree College, Bhikiyasain, India

ABSTRACT

The present study proposes to determine the fuzzy reliability of different systems in which the lifetime of components are following fuzzy exponential distribution where fuzzy reliability function and its α-cut set are presented. The fuzzy reliability of different systems is defined on the basis of octagonal intuitionistic fuzzy numbers. The fuzzy reliability functions of k-out-of-m system, series system, parallel systems, and their fuzzy mean time to failure are discussed respectively using the concept of α-cut of octagonal intuitionistic fuzzy numbers. Finally, some numerical examples are discussed to illustrate how to calculate the fuzzy system reliability and α-cut of fuzzy mean time to failure (FMTTF).

INTRODUCTION

The reliability is one of the most important engineering tasks in design of engineering systems. The reliability of an item is the probability that the item will perform its stated function satisfactorily for some specified condition under specific time period. The reliability function is the most frequently used function in lifetime data analysis. Exponential distribution is mostly used in used the studies of reliability

DOI: 10.4018/978-1-7998-7564-2.ch010

modelling. However, in real situation, collections of data or system parameters are often fuzzy/imprecise because of incomplete or vague information. To tide over this problem, fuzzy set theory had been used in the evaluation of the reliability of the systems (Onisawa and Kacprzyk 1995). The fundamental works on fuzzy set theory done by Zadeh (1978).

Zadeh (1965) defined concept of fuzzy set. Atanassov (1986) defined a new concept extension of fuzzy set named intuitionistic fuzzy set. Baloui Jamkhaneh (2011 and 2013) discussed reliability estimation using fuzzy lifetime distribution and fuzzy environments. Kumar & Singh (2015) evaluated fuzzy system reliability using intuitionistic fuzzy Weibull lifetime distribution when lifetime follows intuitionistic fuzzy numbers. Selva kumara & Lavanya (2017) defined octagonal intuitionistic fuzzy numbers and their α-cut set. Menaka (2017) defined a new concept of octagonal intuitionistic fuzzy number and its membership function. Baloui Jamkhaneh (2017) calculating system reliability when system follows generalized intuitionistic fuzzy exponential lifetime distribution. Kumar & Ram (2018) analyze system reliability of systems based on Weibull distribution and lifetime follows Hesitant Fuzzy Set. Negi et al. (2018) analyzed reliability of non-repairable k-out-of-n system using belief universal generating function. Sahaya Sudha & Vijayalakshmi (2018) discussed the concept of hexagonal intuitionistic fuzzy number and its arithmetic operations. Kumar & Kumari (2019) evaluating fuzzy system reliability using lifetime considered as hexagonal fuzzy numbers.

Therefore, in this chapter author analyzed the fuzzy reliability of different complex systems and α-cut set when lifetime follows exponential lifetime distribution. The failure rate of system is considered as octagonal intuitionistic fuzzy number. Section 2 introduces the intuitionistic fuzzy sets and intuitionistic fuzzy numbers and defines octagonal intuitionistic fuzzy numbers; section 3 describes the fuzzy probability theory. In section 4, the author evaluates fuzzy reliability function and FMTTF for series and parallel system and section 5 discuss the FMTTF and fuzzy reliability function of k-out-of-m system. Finally, in Section 6 discuss the conclusion of the study.

DEFINITIONS

[Atanassov (8)] Intuitionistic Fuzzy Set [IFS]:

An intuitionistic fuzzy set $\tilde{A}$ on universe of discourse X is given by $\tilde{A} = \{< x, \mu_{\tilde{A}}(x), \nu_{\tilde{A}}(x) >: x \in X\}$ with $\mu_{\tilde{A}}(x) : X \to [0, 1]$ and $\nu_{\tilde{A}}(x) : X \to [0, 1]$.

$\mu_{\tilde{A}}(x)$ and $\nu_{\tilde{A}}(x)$ define the degree of membership and degree of non-membership of the element $x \in X$ to the set $\tilde{A}$, for every $x \in X$, $0 \le \mu_{\tilde{A}}(x) + \nu_{\tilde{A}}(x) \le 1$.

[Atanassov (8)] Intuitionistic Fuzzy Numbers [IFN]

An intuitionistic fuzzy subset $\tilde{A}^I = \{< x, \mu_{\tilde{A}^I}(x), \upsilon_{\tilde{A}^I}(x) >| \, x \in X\}$ of the real line R is called an intuitionistic fuzzy number if it follows the given conditions:

- There exist $m \in R : \mu_{\tilde{A}^I}(m) = 1$ and $\upsilon_{\tilde{A}^I}(m) = 0$
- $\mu_{\tilde{A}^I}$ is a continuous function from $R \to [0,1]$ s.t.
- $0 \leq \mu_{\tilde{A}^I}(x) + \upsilon_{\tilde{A}^I}(x) \leq 1 \forall x \in X$

[Menaka (7)] Octagonal Intuitionistic Fuzzy Numbers

An octagonal IFN is specified by
$\tilde{A}^I_{oc} = \{(a_1, a_2, a_3, a_4, a_5, a_6, a_7, a_8); (a'_1, a'_2, a'_3, a_4, a_5, a'_6, a'_7, a'_8)\}$ where
$a_1 \leq a_2 \leq a_3 \leq a_4 \leq a_5 \leq a_6 \leq a_7 \leq a_8 \leq a'_1 \leq a'_2 \leq a'_3 \leq a'_6 \leq a'_7 \leq a'_8$
are real numbers and it's membership function and non-membership function are given below-

$$\mu_{\tilde{A}}(x) = \begin{cases} 0, & for\ x < a_1 \\ k\dfrac{(x - a_1)}{(a_2 - a_1)}, & for\ a_1 \leq x \leq a_2 \\ k, & for\ a_2 \leq x \leq a_3 \\ k + (1-k)\dfrac{(x - a_3)}{(a_4 - a_3)}, & for\ a_3 \leq x \leq a_4 \\ 1, & for\ a_4 \leq x \leq a_5 \\ k + (1-k)\dfrac{(a_6 - x)}{(a_6 - a_5)}, & for\ a_5 \leq x \leq a_6 \\ k, & for\ a_6 \leq x \leq a_7 \\ k\dfrac{(a_8 - x)}{(a_8 - a_7)}, & for\ a_7 \leq x \leq a_8 \\ 0, & for\ x \geq a_8 \end{cases}$$

and non-membership function where $0<k<1$.

$$\upsilon_{\tilde{A}}(x) = \begin{cases} 1, & for\ x < a'_1 \\ k + (1-k)\dfrac{(a'_1 - x)}{(a_2 - a_1)}, & for\, a'_1 \le x \le a'_2 \\ k, & for\ a'_2 \le x \le a'_3 \\ k\dfrac{(a'_3 - x)}{(a_4 - a'_3)}, & for\ a'_3 \le x \le a_4 \\ 0, & for\ a_4 \le x \le a_5 \\ k\dfrac{(x - a_5)}{(a'_6 - a_5)}, & for\ a_5 \le x \le a'_6 \\ k, & for\ a'_6 \le x \le a'_7 \\ k + (1-k)\dfrac{(x - a'_7)}{(a'_8 - a'_7)}, & for\ a'_7 \le x \le a'_8 \\ 1, & for\ x \ge a'_8 \end{cases}$$

Figure 1. Octagonal Intuitionistic Fuzzy Number for k = 0.5

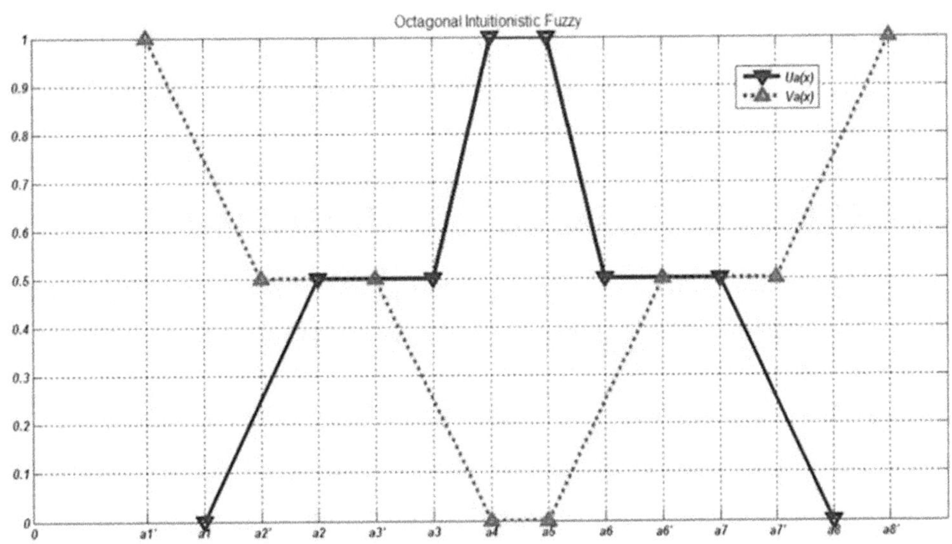

[Menaka (7)] Alpha Cut of Octagonal Intuitionistic Fuzzy Number

Alpha cut of membership function

$$\tilde{A}_{\alpha} = \begin{cases} l_1(\alpha), l_2(\alpha), & for \alpha \in [0, k] \\ s_1(\alpha), s_2(\alpha), & for \alpha \in [k, 1] \end{cases}$$

Where $l_1(\alpha) = a_1 + \frac{\alpha}{k}(a_2 - a_1)$, $\quad l_2(\alpha) = a_8 - \frac{\alpha}{k}(a_8 - a_7)$,

$s_1(\alpha) = a_3 + \frac{(\alpha - k)}{(1-k)}(a_4 - a_3)$, $\quad s_2(\alpha) = a_6 - \frac{(\alpha - k)}{(1-k)}(a_6 - a_5)$,

Alpha cut of non-membership function

$$\tilde{A}'_{\alpha} = \begin{cases} l'_1(\alpha), l'_2(\alpha), & for \alpha \in [0, k] \\ s'_1(\alpha), s'_2(\alpha), & for \alpha \in [1, k] \end{cases}$$

where $l'_1(\alpha) = a'_1 - \frac{(\alpha - k)}{(1-k)}(a'_2 - a'_1)$, $\quad l'_2(\alpha) = a'_7 + \frac{(\alpha - k)}{(1-k)}(a'_8 - a'_7)$,

$s'_1(\alpha) = a'_3 - \frac{\alpha}{k}(a_4 - a'_3)$, $\quad s'_2(\alpha) = a_5 + \frac{\alpha}{k}(a'_6 - a_5)$,

INTUITIONISTIC FUZZY EXPONENTIAL DISTRIBUTION

The exponential distribution is most important useful distribution for reliability analysis. Its probability density function (pdf) is defined as

$$f(x) = \lambda e^{-\lambda x}, \quad x > 0, \tag{1}$$

Here $\lambda > 0$ is the lifetime parameter of the distribution.

Due to the uncertainty and inaccuracy of data, the estimation of precise value of lifetime parameters becomes very difficult. To tackle this issue here the lifetime of exponential distribution λ is replaced by octagonal intuitionistic fuzzy number $\tilde{\lambda}$. In this case, the fuzzy probability of event Xϵ[c, d], c $\geq$ 0 $\tilde{P}(c \leq X \leq d)$ and its α-cut set is given by:

$$\tilde{P}(c \le X \le d)[\alpha] = \left\{ \int_c^d \lambda e^{-\lambda x} dx : \lambda \in \tilde{\lambda}[\alpha] \right\} = [P_L[\alpha], P_U[\alpha]]$$

$$\tilde{P}'(c \le X \le d)[\alpha] = \left\{ \int_c^d \lambda e^{-\lambda x} dx : \lambda \in \tilde{\lambda}'[\alpha] \right\} = [P'_L[\alpha], P'_U[\alpha]] \tag{2}$$

where $\tilde{\lambda}[\alpha]$ is the α-cut of membership function of octagonal intuitionistic fuzzy number and $\tilde{\lambda}'[\alpha]$ is the α-cut of non-membership function of octagonal intuitionistic fuzzy number.

$$P_L[\alpha] = \min \left\{ \int_c^d \lambda e^{-\lambda x} dx \mid \lambda \in \tilde{\lambda}[\alpha] \right\} \forall \alpha$$

$$P_U[\alpha] = \max \left\{ \int_c^d \lambda e^{-\lambda x} dx \mid \lambda \in \tilde{\lambda}[\alpha] \right\} \forall \alpha$$

and

$$P'_L[\alpha] = \min \left\{ \int_c^d \lambda e^{-\lambda x} dx \mid \lambda \in \tilde{\lambda}'[\alpha] \right\} \forall \alpha$$

$$P'_U[\alpha] = \max \left\{ \int_c^d \lambda e^{-\lambda x} dx \mid \lambda \in \tilde{\lambda}'[\alpha] \right\} \forall \alpha$$

FUZZY HAZARD FUNCTION

Hazard rate at any point in the life of an item is the incremental change in the number of failures per associated incremental change in time. If hazard rate is in the form of fuzzy then hazard rate is known as fuzzy hazard rate. Hazard function is also known as the failure rate. Mathematically, the hazard function is defined as

$$\tilde{H}(t) = \lim_{\Delta t \to 0} \frac{\tilde{P}(t < x < t + \Delta t \mid X > t)}{\Delta t} = \frac{f(t)}{S(t)} \quad (3)$$

If the lifetime follows intuitionistic fuzzy exponential distribution and failure rate assumed as octagonal intuitionistic fuzzy number, then the fuzzy hazard function is given by

$$\tilde{H}(t)[\alpha] = \{\lambda \mid \lambda \in \tilde{\lambda}[\alpha]\}$$

$$\tilde{H}'(t)[\alpha] = \{\lambda \mid \lambda \in \tilde{\lambda}'[\alpha]\}$$

Where $\tilde{H}(t)[\alpha]$ and $\tilde{H}'(t)[\alpha]$ are the alpha cut of membership function and non-membership function respectively. Now alpha cut of membership function is as

$$\tilde{H}(t)[\alpha] = \begin{cases} a_1 + \frac{\alpha}{k}(a_2 - a_1), \quad a_8 - \frac{\alpha}{k}(a_8 - a_7); \; for \; \alpha \in [0, k] \\ a_3 + \frac{(\alpha - k)}{(1-k)}(a_4 - a_3), \quad a_6 - \frac{(\alpha - k)}{(1-k)}(a_6 - a_5); \; for \; \alpha \in [1, k] \end{cases}$$

And alpha cut of non-membership function is given by

$$\tilde{H}'(t)[\alpha] = \begin{cases} a'_1 - \frac{(\alpha - k)}{(1-k)}(a'_2 - a'_1), \quad a'_7 + \frac{(\alpha - k)}{(1-k)}(a'_8 - a'_7); \; for \; \alpha \in [0, k] \\ a'_3 - \frac{\alpha}{k}(a_4 - a'_3), \quad a_5 + \frac{\alpha}{k}(a'_6 - a_5); \; for \; \alpha \in [1, k] \end{cases}$$

FUZZY RELIABILITY FUNCTION

Fuzzy reliability function $\tilde{S}(t)$ is the fuzzy probability in which a unit preserve beyond time *t*. Let the random variable *X* denote lifetime of components. If the cumulative distribution function is $F_X(t) = P(X \leq t)$ then the fuzzy reliability function at time *t* is defined as

$$\tilde{S}(t) = \tilde{P}(X > t) = 1 - F_X(t), \quad t > 0 \quad (4)$$

The fuzzy unreliability function $\tilde{Q}(t)$ is the probability of failure in the time interval [0, *t*], is given by

$$\tilde{Q}(t) = \tilde{P}(X \le t) = F_X(t), \quad t > 0 \tag{5}$$

Let the lifetime parameter/failure rate of a component has fuzzy exponential distribution and lifetime parameter/failure rate $\tilde{\lambda}$ follow octagonal intuitionistic fuzzy number.

So the fuzzy reliability of a component is defined as

$$\tilde{S}(t)[\alpha] = \{\int_t^{\infty} \lambda e^{-\lambda x} dx \mid \lambda \in \tilde{\lambda}[\alpha]\} = \{e^{-\lambda t} \mid \lambda \in \tilde{\lambda}[\alpha]\} \tag{6}$$

The α-cut of fuzzy reliability function when failure rate follow octagonal intuitionistic fuzzy numbers is given by

$$\tilde{S}(t)[\alpha] = \begin{cases} \exp[-\{a_1 + \frac{\alpha}{k}(a_2 - a_1)\}.t], \exp[-\{a_8 - \frac{\alpha}{k}(a_8 - a_7)\}.t]; & \textit{for } \alpha \in [0, k] \\ \exp[-\{a_3 + \frac{(\alpha - k)}{(1-k)}(a_4 - a_3)\}.t], \exp[-\{a_6 - \frac{(\alpha - k)}{(1-k)}(a_6 - a_5)\}.t]; & \textit{for } \alpha \in [k, 1] \end{cases} \tag{7a}$$

$$\tilde{S}'(t)[\alpha] = \begin{cases} \exp[-\{a'_1 - \frac{(\alpha - k)}{(1-k)}(a'_2 - a'_1)\}.t], \exp[-\{a'_7 + \frac{(\alpha - k)}{(1-k)}(a'_8 - a'_7)\}.t]; & \textit{for } \alpha \in [k, 0] \\ \exp[-\{a'_3 - \frac{\alpha}{k}(a_4 - a'_3)\}.t], \exp[-\{a_5 + \frac{\alpha}{k}(a'_6 - a_5)\}.t]; & \textit{for } \alpha \in [1, k] \end{cases} \tag{7b}$$

Here $\tilde{S}(t)[\alpha]$ and $\tilde{S}'(t)[\alpha]$ are the α-cut of membership and non-membership function of fuzzy reliability respectively.

FUZZY MEAN TIME TO FAILURE

Fuzzy mean time to failure (*FMTTF*) is the requisite time to failure. It is defined as

$$FMTTF[\alpha] = \int_0^{\infty} x f(x) dx \mid \lambda \in \tilde{\lambda}[\alpha]$$

$$= \int_0^\infty S(t)dt \mid \lambda \in \tilde{\lambda}[\alpha] = \{\frac{1}{\lambda} \mid \lambda \in \tilde{\lambda}[\alpha]\} \quad (8)$$

when the lifetime follows fuzzy exponential distribution and the parameters λ follow the octagonal intuitionistic fuzzy number then α-cut of FMTTF is as

$$FMTTF[\alpha] = \begin{cases} \dfrac{1}{\{a_1 + \frac{\alpha}{k}(a_2 - a_1)\}}, \dfrac{1}{\{a_8 - \frac{\alpha}{k}(a_8 - a_7)\}}; & for\ \alpha \in [0,k] \\ \dfrac{1}{\{a_3 + \frac{(\alpha-k)}{(1-k)}(a_4 - a_3)\}}, \dfrac{1}{\{a_6 - \frac{(\alpha-k)}{(1-k)}(a_6 - a_5)\}}; & for\ \alpha \in [k,1] \end{cases} \quad (9a)$$

$$FMTTF'[\alpha] = \begin{cases} \dfrac{1}{\{a'_1 - \frac{(\alpha-k)}{(1-k)}(a'_2 - a'_1)\}}, \dfrac{1}{\{(a'_7 + \frac{(\alpha-k)}{(1-k)}(a'_8 - a'_7)\}}; & for\ \alpha \in [k,0] \\ \dfrac{1}{\{a'_3 - \frac{\alpha}{k}(a_4 - a'_3)\}}, \dfrac{1}{\{a_5 + \frac{\alpha}{k}(a'_6 - a_5)\}}; & for\ \alpha \in [1,k] \end{cases} \quad (9b)$$

Here $FMTTF[\alpha]$ and $FMTTF'[\alpha]$ are the α-cut of membership and non-membership function of FMTTF respectively.

Example 1

Let lifetime of electronic component is modeled by an Exponential distribution with parameter λ follows octagonal intuitionistic fuzzy number.
$\tilde{\lambda} = \{(0.70, 0.72, 0.74, 0.76, 0.78, 0.80, 0.82, 0.84); (0.69, 0.71, 0.73, 0.76, 0.78, 0.81, 0.83, 0.85)\}$
and k=1/2. Find the α-cut of fuzzy reliability and FMTTF.

Solution: The alpha cut of octagonal intuitionistic fuzzy numbers is given by

$$\tilde{\lambda}[\alpha] = \begin{cases} 0.70 + \dfrac{0.02\alpha}{0.5}, \quad 0.84 - \dfrac{0.02\alpha}{0.5}; & for\,\alpha \in [0, 0.5] \\ 0.74 + \dfrac{0.02(\alpha - 0.5)}{0.5}, \quad 0.80 - \dfrac{0.02(\alpha - 0.5)}{0.5}; & for\,\alpha \in [0.5, 1] \end{cases}$$

$$\tilde{\lambda}'[\alpha] = \begin{cases} 0.69 + \dfrac{0.02(\alpha - 0.5)}{0.5}, \quad 0.83 - \dfrac{0.02(\alpha - 0.5)}{0.5}; \quad for\, \alpha \in [0.5, 0] \\ 0.73 - \dfrac{0.03\alpha}{0.5}, \quad 0.78 - \dfrac{0.03\alpha}{0.5}; \; for\, \alpha \in [0.5, 1] \end{cases}$$

Alpha cut of fuzzy reliability using equation (7a) and (7b) is as follows:

$$\tilde{S}(t)[\alpha] = \begin{cases} \exp[-\{0.70 + \dfrac{0.02\alpha}{0.5}\}.t], \exp[-\{0.84 - \dfrac{0.02\alpha}{0.5}\}.t]; \qquad for\ \ \alpha \in [0, 0.5] \\ \exp[-\{0.74 + \dfrac{0.02(\alpha - 0.5)}{0.5}\}.t], \exp[-\{0.80 - \dfrac{0.02(\alpha - 0.5)}{0.5}\}.t]; for\ \ \alpha \in [0.5, 1] \end{cases}$$

and

$$\tilde{S}'(t)[\alpha] = \begin{cases} \exp[-\{0.69 + \dfrac{0.02(\alpha - 0.5)}{0.5}\}.t], \exp[-\{0.83 - \dfrac{0.02(\alpha - 0.5)}{0.5}\}.t]; \quad for\ \ \alpha \in [0.5, 0] \\ \exp[-\{0.73 - \dfrac{0.03\alpha}{0.5}\}.t], \exp[-\{0.78 - \dfrac{0.03\alpha}{0.5}\}.t]; \qquad for\ \ \alpha \in [1, 0.5] \end{cases}$$

Now if t=0.5 then alpha cut of fuzzy reliability is as:

$$\tilde{S}(t)[\alpha] = \begin{cases} \exp[-\{0.35 + 0.02\alpha\}], \quad \exp[-\{0.42 - 0.02\alpha\}]; \qquad for\ \ \alpha \in [0, 0.5] \\ \exp[-\{0.37 + 0.02(\alpha - 0.5)\}], \quad \exp[-\{0.40 - 0.02(\alpha - 0.5)\}]; for\ \ \alpha \in [0.5, 1] \end{cases}$$

and

$$\tilde{S}'(t)[\alpha] = \begin{cases} \exp[-\{0.345 + 0.02(\alpha - 0.5)\}], \quad \exp[-\{0.415 - 0.02(\alpha - 0.5)\}]; \quad for\ \ \alpha \in [0.5, 0] \\ \exp[-\{0.365 - 0.03\alpha\}], \quad \exp[-\{0.39 - 0.03\alpha\}]; \qquad for\ \ \alpha \in [1, 0.5] \end{cases}$$

The alpha cut of FMTTF using equation (9a) and (9b) is as

$$FMTTF[\alpha]=\begin{cases}\dfrac{1}{\{0.70+\dfrac{0.02\alpha}{0.5}\}},\ \dfrac{1}{\{0.84-\dfrac{0.02\alpha}{0.5}\}};\quad for\ \alpha\in[0,0.5]\\ \dfrac{1}{\{0.74+\dfrac{0.02(\alpha-0.5)}{0.5}\}},\ \dfrac{1}{\{0.80-\dfrac{0.02(\alpha-0.5)}{0.5}\}};\quad for\ \alpha\in[0.5,1]\end{cases}$$

and

$$FMTTF'[\alpha]=\begin{cases}\dfrac{1}{\{0.69+\dfrac{0.02(\alpha-0.5)}{0.5}\}},\ \dfrac{1}{\{0.83-\dfrac{0.02(\alpha-0.5)}{0.5}\}};\quad for\ \alpha\in[0.5,0]\\ \dfrac{1}{\{0.73-\dfrac{0.03\alpha}{0.5}\}},\ \dfrac{1}{\{0.78-\dfrac{0.03\alpha}{0.5}\}};\quad for\ \alpha\in[1,0.5]\end{cases}$$

k-out-of-*m* System

A m-component system that *works* if and only if at least k of the m components work is called a *k-out-of-m*. Each component can be represented by a Bernoulli random variable y_i with fuzzy reliability (fuzzy survival probability) $\tilde{S}(t)$ and unreliability (fuzzy failure probability) $\tilde{Q}(t)$. The number of working components follows the binomial distribution and a k-out-of-m system consisting of m independent and identical components, the fuzzy system reliability function is given by

$$\tilde{R}(t)[\alpha]=\tilde{P}(z\geq k)=\left\{\sum_{j=k}^{m}{}^{m}C_{j}\,\{\tilde{S}(t)\}^{j}\{\tilde{Q}(t)\}^{m-j}\mid \tilde{S}(t)\in\tilde{S}(t)[\alpha]\,\&\,\tilde{Q}(t)\in\tilde{Q}(t)[\alpha]\,and\,\tilde{S}(t)+\tilde{Q}(t)=1\right.$$

Both parallel and series systems are special cases of the *k-out-of-m* system. A series system is equivalent to a 1-*out-of-m* system. In the first case, *k-out-of-m* system is reducing at series systems and it's fuzzy reliability function with modeled fuzzy exponential is as:

$$\tilde{R}(t)[\alpha]=\left\{S(t)^{m}\mid S_{\alpha}\right\}=\tilde{S}(t)[\alpha]$$

$$\tilde{S}(t)[\alpha]=\begin{cases}\exp[-\{a_1+\frac{\alpha}{k}(a_2-a_1)\}.mt],\exp[-\{a_8-\frac{\alpha}{k}(a_8-a_7)\}.mt]; & for\ \alpha\in[0,k]\\ \exp[-\{a_3+\frac{(\alpha-k)}{(1-k)}(a_4-a_3)\}.mt],\exp[-\{a_6-\frac{(\alpha-k)}{(1-k)}(a_6-a_5)\}.mt]; & for\ \alpha\in[k,1]\end{cases}$$

$$\tilde{S}'(t)[\alpha]=\begin{cases}\exp[-\{a'_1-\frac{(\alpha-k)}{(1-k)}(a'_2-a'_1)\}.mt],\exp[-\{a'_7+\frac{(\alpha-k)}{(1-k)}(a'_8-a'_7)\}.mt]; & for\ \alpha\in[k,0]\\ \exp[-\{a'_3-\frac{\alpha}{k}(a_4-a'_3)\}.mt],\exp[-\{a_5+\frac{\alpha}{k}(a'_6-a_5)\}.mt]; & for\ \alpha\in[1,k]\end{cases}$$

In the second case, a parallel system is equivalent to *m-out-of-m* system and its α-cut of fuzzy reliability function with fuzzy exponential and failure rate is considered as octagonal intuitionistic fuzzy numbers is as:

$$\tilde{R}(t)[\alpha]=\left\{1-\tilde{Q}(t)^m \mid S_\alpha\right\}$$

$$=\left\{1-(1-\tilde{S}(t))^m \mid S_\alpha\right\}$$

$$\tilde{S}(t)[\alpha]=\begin{cases}1-\{1-\exp[-\{a_1+\frac{\alpha}{k}(a_2-a_1)\}.t]\}^m,1-\{1-\exp[-\{a_8-\frac{\alpha}{k}(a_8-a_7)\}.t]\}^m; & for\ \alpha\in[0,k]\\ 1-\{1-\exp[-\{a_3+\frac{(\alpha-k)}{(1-k)}(a_4-a_3)\}.t]\}^m,1-\{1-\exp[-\{a_6-\frac{(\alpha-k)}{(1-k)}(a_6-a_5)\}.t]\}^m; & for\ \alpha\in[k,1]\end{cases}$$

$$\tilde{S}'(t)[\alpha]=\begin{cases}1-\{1-\exp[-\{a'_1-\frac{(\alpha-k)}{(1-k)}(a'_2-a'_1)\}.t]\}^m,1-\{1-\exp[-\{a'_7+\frac{(\alpha-k)}{(1-k)}(a'_8-a'_7)\}.t]\}^m; & for\ \alpha\in[k,0]\\ 1-\{1-\exp[-\{a'_3-\frac{\alpha}{k}(a_4-a'_3)\}.t]\}^m,1-\{1-\exp[-\{a_5+\frac{\alpha}{k}(a'_6-a_5)\}.t]^m; & for\ \alpha\in[1,k]\end{cases}$$

MTTF of *k*-out-of-*m* system is calculated as follows:

$$FMTTF[\alpha]=\int_0^\infty\{\sum_{j=k}^{m}{}^mC_j(S(t))^jQ(t)^{m-j}\mid S_\alpha\}$$

$$=\left\{\sum_{j=k}^{m}\frac{1}{\lambda.j}\mid\lambda\in\tilde{\lambda}[\alpha]\right]$$

$$FMTTF[\alpha]=\begin{cases}\sum_{j=k}^{m}\frac{1}{\{a_1+\frac{\alpha}{k}(a_2-a_1)\}.j},\ \sum_{j=k}^{m}\frac{1}{\{a_8-\frac{\alpha}{k}(a_8-a_7)\}.j}; & for\ \alpha\in[0,k]\\ \sum_{j=k}^{m}\frac{1}{\{a_3+\frac{(\alpha-k)}{(1-k)}(a_4-a_3)\}.j},\ \sum_{j=k}^{m}\frac{1}{\{a_6-\frac{(\alpha-k)}{(1-k)}(a_6-a_5)\}.j}; & for\ \alpha\in[k,1]\end{cases} \tag{10a}$$

$$FMTTF'[\alpha]=\begin{cases}\sum_{j=k}^{m}\frac{1}{\{a'_1-\frac{(\alpha-k)}{(1-k)}(a'_2-a'_1)\}.j},\ \sum_{j=k}^{m}\frac{1}{\{(a'_7+\frac{(\alpha-k)}{(1-k)}(a'_8-a'_7)\}.j}; & for\ \alpha\in[k,0]\\ \sum_{j=k}^{m}\frac{1}{\{a'_3-\frac{\alpha}{k}(a_4-a'_3)\}.j},\ \sum_{j=k}^{m}\frac{1}{\{a_5+\frac{\alpha}{k}(a'_6-a_5)\}.j}; & for\ \alpha\in[1,k]\end{cases} \tag{10b}$$

Example 2: let us consider a 2-out-of-3 system consisting 2 independent and identical components. Failure rate is considered as ex.1then the fuzzy reliability is given as:

$$\tilde{R}(t)[\alpha]=\tilde{P}(z\geq 2)=\left\{\sum_{j=2}^{3}{}^{m}C_j\,\{\tilde{S}(t)\}^{j}\{1-\tilde{S}(t)\}^{m-j}\mid \tilde{S}\in\tilde{S}[\alpha]\right.$$

$$=\left(\tilde{S}(t)^2[3-2\tilde{S}(t)]\mid \tilde{S}(t)\in\tilde{S}(t)[\alpha]\right)$$

α-cut of fuzzy reliability is as:

$$\tilde{R}(t)[\alpha]=\begin{cases}\begin{bmatrix}\exp[-\{1.4+\frac{0.04\alpha}{0.5}\}.t](3-2\exp[-\{0.70+\frac{0.02\alpha}{0.5}\}.t])\\ \exp[-\{1.68-\frac{0.04\alpha}{0.5}\}.t]\left[3-2\exp[-\{0.84-\frac{0.02\alpha}{0.5}\}.t]\right]\end{bmatrix}; & for\ \alpha\in[0,0.5]\\ \begin{bmatrix}\exp[-\{1.48+\frac{0.04(\alpha-0.5)}{0.5}\}.t]\left(3-2\exp[-\{0.74+\frac{0.02(\alpha-0.5)}{0.5}\}.t]\right)\\ \exp[-\{1.6-\frac{0.04(\alpha-0.5)}{0.5}\}.t]\left\{3-2\exp[-\{0.80-\frac{0.02(\alpha-0.5)}{0.5}\}.t]\right\}\end{bmatrix}; & for\ \alpha\in[0.5,1]\end{cases}$$

$$\tilde{R}'(t)[\alpha] = \begin{cases} \begin{bmatrix} \exp[-\{1.38 + \frac{0.04(\alpha-0.5)}{0.5}\}.t]\{3 - 2\exp[-\{0.69 + \frac{0.02(\alpha-0.5)}{0.5}\}.t]\}, \\ \exp[-\{1.66 - \frac{0.04(\alpha-0.5)}{0.5}\}.t]\{3 - 2\exp[-\{0.83 - \frac{0.02(\alpha-0.5)}{0.5}\}.t]\} \end{bmatrix}; for\ \alpha \in [0.5, 0] \\ \begin{bmatrix} \exp[-\{1.46 - \frac{0.06\alpha}{0.5}\}.t]\{3 - 2\exp[-\{0.73 - \frac{0.03\alpha}{0.5}\}.t]\}, \\ \exp[-\{1.56 - \frac{0.03\alpha}{0.5}\}.t]\{3 - 2\exp[-\{0.78 - \frac{0.03\alpha}{0.5}\}.t]\} \end{bmatrix}; for\ \alpha \in [1, 0.5] \end{cases}$$

For t=0.5

$$\tilde{R}(t)[\alpha] = \begin{cases} \begin{bmatrix} \exp[-\{0.7 + 0.04\alpha\}]\{3 - 2\exp[-\{0.35 + 0.02\alpha\}]\}, \\ \exp[-\{0.84 - 0.04\alpha\}]\{3 - 2\exp[-\{0.42 - 0.02\alpha\}]\} \end{bmatrix}; for\ \alpha \in [0, 0.5] \\ \begin{bmatrix} \exp[-\{0.74 + 0.04(\alpha-0.5)\}]\{3 - 2\exp[-\{0.37 + 0.02(\alpha-0.5)\}]\}, \\ \exp[-\{0.8 - 0.04(\alpha-0.5)\}]\{3 - 2\exp[-\{0.4 - 0.02(\alpha-0.5)\}]\} \end{bmatrix}; for\ \alpha \in [0.5, 1] \end{cases}$$

$$\tilde{R}'(t)[\alpha] = \begin{cases} \begin{bmatrix} \exp[-\{0.69 + 0.04(\alpha-0.5)\}]\{3 - 2\exp[-\{0.345 + 0.02(\alpha-0.5)\}]\}, \\ \exp[-\{0.83 - 0.04(\alpha-0.5)\}]\{3 - 2\exp[-\{0.415 - 0.02(\alpha-0.5)\}]\} \end{bmatrix}; for\ \alpha \in [0.5, 0] \\ \begin{bmatrix} \exp[-\{0.73 - 0.06\alpha\}]\{3 - 2\exp[-\{0.365 - 0.03\alpha\}]\}, \\ \exp[-\{0.78 - 0.03\alpha\}]\{3 - 2\exp[-\{0.39 - 0.03\alpha\}]\} \end{bmatrix}; for\ \alpha \in [1, 0.5] \end{cases}$$

when α=0

$$\tilde{R}(t)[\alpha] = \begin{cases} e^{-1.4.t}\{3 - 2e^{-0.70.t}\}, e^{-1.68.t}\{3 - 2e^{-0.84.t}\};\ for\ \alpha \in [0, 0.5] \\ e^{-1.48.t}\{3 - 2e^{-0.74.t}\}, e^{-1.6.t}\{3 - 2e^{-0.80.t}\};\ for\ \alpha \in [0.5, 1] \end{cases}$$

$$\tilde{R}'(t)[\alpha] = \begin{cases} e^{-1.38.t}\{3 - 2e^{-0.69.t}\}, e^{-1.66.t}\{3 - 2e^{-0.83.t}\};\ for\ \alpha \in [0.5, 0] \\ e^{-1.46.t}\{3 - 2e^{-0.73.t}\}, e^{-1.56.t}\{3 - 2e^{-0.78.t}\};\ for\ \alpha \in [1, 0.5] \end{cases}$$

The alpha cut of FMTTF using equation (10a) & (10b) is as

$$FMTTF[\alpha] = \begin{cases} \dfrac{5}{\{4.2 + \dfrac{0.12\alpha}{0.5}\}}, & \dfrac{5}{\{5.04 - \dfrac{0.12\alpha}{0.5}\}}; \quad for \ \alpha \in [0, 0.5] \\ \dfrac{5}{\{4.44 + \dfrac{0.12(\alpha - 0.5)}{0.5}\}}, & \dfrac{5}{\{4.8 - \dfrac{0.12(\alpha - 0.5)}{0.5}\}}; \quad for \ \alpha \in [0.5, 1] \end{cases}$$

and

$$FMTTF\,'[\alpha] = \begin{cases} \dfrac{5}{\{4.14 + \dfrac{0.12(\alpha - 0.5)}{0.5}\}}, & \dfrac{5}{\{4.98 - \dfrac{0.12(\alpha - 0.5)}{0.5}\}}; \quad for \ \alpha \in [0.5, 0] \\ \dfrac{5}{\{4.38 - \dfrac{0.18\alpha}{0.5}\}}, & \dfrac{5}{\{4.68 - \dfrac{0.18\alpha}{0.5}\}}; \quad for \ \alpha \in [1, 0.5] \end{cases}$$

when α=0

$$FMTTF[\alpha] = \begin{cases} 1.19047, & 0.99206; \quad for \ \alpha \in [0, 0.5] \\ 1.06837, & 1.09649; \quad for \ \alpha \in [0.5, 1] \end{cases}$$

$$FMTTF\,'[\alpha] = \begin{cases} 1.14155, & 1.05485; \quad for \ \alpha \in [0.5, 0] \\ 1.14155, & 1.06837; \quad for \ \alpha \in [1, 0.5] \end{cases}$$

CONCLUSION

In this chapter author defined Octagonal intuitionistic fuzzy number for evaluating fuzzy reliability and FMTTF of systems. The fuzzy probability theory has been successfully applied for evaluating the fuzzy reliability of systems whenever the lifetime rate of components contains fuzziness and randomness respectively, conventional reliability system is not feasible. Thus author applied successfully the fuzzy distribution to overcome this difficulty using α-cut of octagonal intuitionistic fuzzy number. FMTTF, fuzzy hazard function and their α-cut set have been successfully investigated using lifetimes follows octagonal intuitionistic fuzzy number. Numerical examples have also been taken to illustrate the methodology. It also helps us to solve many problems in future which have eight parameters.

REFERENCES

Atanassov, K. (1986). Intuitionistic fuzzy sets. *Fuzzy Sets and Systems*, *20*(1), 87–96. doi:10.1016/S0165-0114(86)80034-3

Baloui Jamkhaneh, E. (2011). An evaluation of the systems reliability using fuzzy lifetime distribution, *Journal of Applied Mathematics. Islamic Azad University of Lahijan*, *7*, 73–80.

Baloui Jamkhaneh, E. (2012). Reliability estimation under the fuzzy environments,140804. *The Journal of Mathematics and Computer Science*, *5*(01), 28–39. doi:10.22436/jmcs.05.01.04

Baloui Jamkhaneh, E. (2017). System Reliability using Generalized Intuitionistic Fuzzy Exponential Lifetime Distribution. *International Journal of Soft Computing and Engineering*, *7*(4), 1–7.

Kumar, A., & Ram, M. (2018). System Reliability Analysis Based On Weibull Distribution and Hesitant Fuzzy Set, *International Journal of Mathematical. Engineering and Management Sciences*, *3*(4), 513–521.

Kumar, D., & Kumari, A. (2019). An Evaluation of the System Reliability Using Fuzzy Lifetime Distribution Emphasizing Hexagonal Fuzzy. *IOSR Journal of Engineering*, *9*(8), 17–25.

Kumar, P., & Singh, S. B. (2015). Fuzzy system reliability using intuitionistic fuzzy Weibull lifetime distribution. *International Journal of Reliability and Applications*, *16*(1), 15–26.

Menaka, G. (2017). Ranking of Octagonal intuitionistic fuzzy number. *IOSR Journal of Mathematics*, *13*(3), 63–71. doi:10.9790/5728-1303026371

Negi, S., Jaiswal, N., & Singh, S. B. (2018). Reliability Assessment of Non-Repairable *k*-Out-Of-*n* System using Belief Universal Generating Function. *International Journal of Performability Engineering*, *14*(5), 831–840. doi:10.23940/ijpe.18.05.p1.831840

Sahaya Sudha, A. & Vijayalakshmi, K. R. (2018). Arithmetic Operations of Hexagonal Intuitionistic Fuzzy Number Using Extension Principle. *International Journal of Mathematics and its Applications*, *6*(2A), 219-225.

Selva Kumara, K., & Lavanya, S. (2017). Solving Fuzzy Game Problem Using Octagonal Intuitionistic Fuzzy Numbers. *International Journal of Pure and Applied Mathematics*, *113*(4), 188-198.

Zadeh, L. A. (1965). Fuzzy sets. *Information and Control*, *8*(3), 338–353. doi:10.1016/S0019-9958(65)90241-X

Chapter 11
Fuzzy System Reliability Using Hesitant Fuzzy Element and Score Function

Deepak Kumar
Kumaun University, DSB Campus, India

S. B. Singh
G. B. Pant University of Agriculture and Technology, India

Anita Kumari
Kumaun University, India

ABSTRACT

The present work introduces a new technique for analyzing the fuzzy reliability of a system stem under hesitant fuzzy environment where the reliability of a component/ unit of a system is represented by hesitant fuzzy element (HFE). In this study, the authors evaluate the fuzzy reliability for some complex systems (series configuration, parallel configuration, and bridge configuration) using score function which is very useful in reliability modelling (especially in decision making, risk analysis, and optimization problems). Reliability modelling under hesitant situations plays a key role in the reliability engineering field. The score function used in this study is helpful for a simple comparison between reliabilities of any two components as form of HFEs. Hesitancy is the most common problem in human behaviour, for which hesitant fuzzy set can be considered as a useful tool allowing several possible degrees of membership of an element to a set. Additionally, a numerical example is taken for demonstration of the present technique.

DOI: 10.4018/978-1-7998-7564-2.ch011

INTRODUCTION

Earlier the reliability theory has been successfully used for solving various engineering problems related to decision making. But in fuzzy environment decision-maker feel difficulties to make his decision more efficiently due to different types of situations like hesitancy, incomplete and vague information etc. Hesitant fuzzy set (HFS) (Torra, 2010) is a useful mathematical tool to handle such type of situations because it allows several values (between 0 and 1) in its membership degree in decision making process. Therefore, the concept of HFS is developed in this paper for analyzing fuzzy system reliability of different types of complex systems.

HFS (Torra and Narukawa, 2009) can reflect human's hesitance more objectively than the other classical extensions of fuzzy set (intuitionistic fuzzy set (Atanassov, 1986 and Atanassov, 1999) type-2 fuzzy set (Karnik & Mendel, 2001) (T-2FS), fuzzy multi-sets etc). (Torra, 2010) Explained the concept of hesitant fuzzy sets with some set theoretic operations such as union, intersection and complement and also discussed relationships among hesitant fuzzy sets and other generalizations of fuzzy sets.

In recent decades, fuzzy reliability theory of various type of systems in different conditions was studied by several authors (Chen, 1994, 1996), (Kumar and Singh, 2013), (Kumar and Singh, 2014) (Kumar and Singh, 2018) (Kumar and Singh, 2018) and Ram et al., 2013). As an addition of these efforts, here we investigate the implementations of HFSs in fuzzy reliability evaluation of different types of complex systems by introducing a new method.

For describing the hesitant fuzzy information in the reliability evaluation more realistically, we introduce the concept of hesitant fuzzy set (HFS) in the system reliability where the reliability of the component/unit of the system is represented by hesitant fuzzy element (HFE). Some definitions as well as results about hesitant fuzzy set and notions used in the current study are discussed in Section 2 and Section 3 respectively.

In the fuzzy environment, the reliability of any component/unit can be expressed in form of HFE. For a HFS, the elements in membership hesitancy function are almost in the partial order. Hence, we propose to use the score function and deviation degree for getting reliable access to the comparison of reliabilities of the given components/units as form of HFEs. Few methods have been generated to evaluate the fuzzy reliability of different types of systems in Section 4. Section 5 demonstrates some examples. Finally, the brief conclusion of the study is given in Section 6.

PRELIMINARIES

Here we have a introduction part about dual hesitant fuzzy set theory and some different complex systems.

Definition 2.1 (Torra, 2010): Let X be a reference set. A hesitant fuzzy set (HFS) E on X is expressed by

$$E = \left\{x, h_E\left(x\right)/x \in X\right\}, \text{ where } h_E\left(x\right) \in \left[0,1\right],$$

denoting the possible membership degree of the element $x \in X$ to the set E with the condition $\gamma \in h_E(x),\ 0 \le \gamma \le 1$.

Xia and Xu (2013) called $h = h_E\left(x\right)$ a hesitant fuzzy element (HFE). If we take three HFEs denoted by h, h_1 and h_2. Torra (2010) and Xia and Xu (2013) defined some operations on the HFEs h, h_1 and h_2

(i) $h^c = \bigcup_{\gamma \in h}(1-\gamma),\ h \neq \varphi$

(ii) $h_1 \oplus h_2 = \bigcup_{\gamma_1 \in h_1, \gamma_2 \in h_2} \left\{\gamma_1 + \gamma_2 - \gamma_1 \gamma_2\right\}$

(iii) $h_1 \otimes h_2 = \bigcup_{\gamma_1 \in h_1, \gamma_2 \in h_2} \left\{\gamma_1 \gamma_2\right\}$

Definition 2.2: Let h be a HFE, the score function of h is defined as follows:

$$s\left(h\right) = \frac{1}{\# h}\sum_{\gamma \in h} \gamma$$

$\# h$ being the total number of elements in h.

Let h_1 and h_2 be two HFEs, then

if $s\left(h_1\right) > s\left(h_2\right)$, then $h_1 > h_2$;

if $s\left(h_1\right) = s\left(h_2\right)$, then $h_1 = h_2$

Example: Let $h_1 = \left\{0.1, 0.4, 0.5, 0.8\right\}$ and $h_2 = \left\{0.3, 0.5, 0.7, 0.9\right\}$ be two HFEs, the results obtained using the score function are

$$s\left(h_1\right) = 0.45$$

$s(h_2) = 0.6$

here if $s(h_2) > s(h_1)$ then $h_2 > h_1$.

Definition 2.3: A system is said to be in series configuration, if its components/ units are arranged in a series manner.

Figure 1. Series System

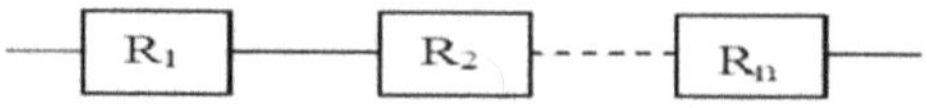

Definition 2.4: A system is said to be in parallel configuration if its units or components are connected in parallel.

Figure 2. Parallel System

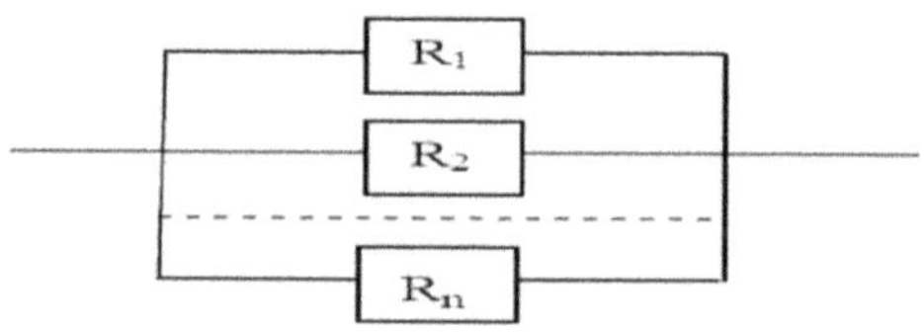

Definition 2.5: If series networks are arranged in parallel manner then this combination is said to be a parallel-series combination (Fig. 3). On other hand, if parallel networks are arranged in series manner then this combination is said to be a series-parallel combination (Fig. 4). Collectively above combinations is called bridge configuration.

Figure 3. Parallel-Series System

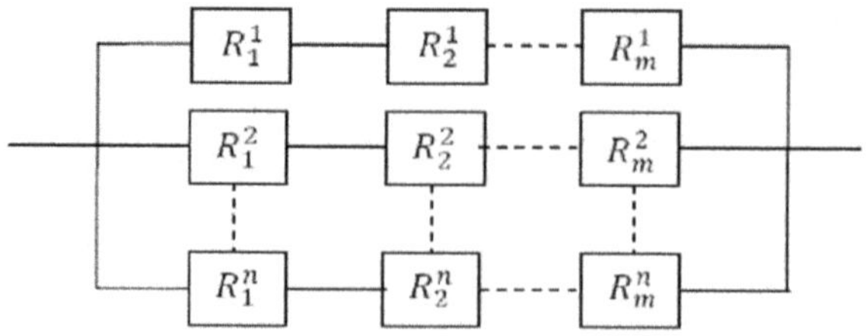

Figure 4. Series-Parallel System

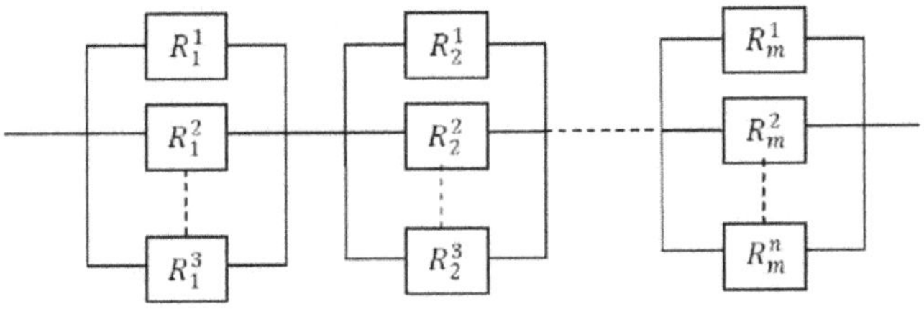

NOTIONS

The following notions are associated with the present study:

R_s Fuzzy reliability of series system
R_P Fuzzy reliability of parallel system
R_{PS} Fuzzy reliability of parallel-series system
R_{SP} Fuzzy reliability of series-parallel system

FUZZY SYSTEM RELIABILITY APPRAISAL USING HESITANT FUZZY APPROACH

In this method, mainly reliability of the components/units of a given system are taken as HFEs.

Series System: Here we take n components or units arranged as Fig. 1.

If the reliability of any i^{th} component or unit $R_i = h_i$ .denote hesitant fuzzy element .where i=1 to n), then the fuzzy reliability

$$R_s = \underset{i=1}{\overset{n}{\otimes}} R_i$$
$$R_s = \underset{i=1}{\overset{n}{\otimes}} h_i$$
$$R_s = h_1 \otimes h_2 \otimes h_3 \otimes \ldots \otimes h_n$$
$$R_s = \bigcup_{\substack{\gamma_i \in h_i \\ i=1,2,\ldots,n}} \left\{ \gamma_1 . \gamma_2 . \gamma_3 . \ldots . \gamma_n \right\}$$

Parallel System: In this case, we assume n components/units which are connected in parallel (Fig. 2).

If the reliability of any i[th] component or unit .denote hesitant fuzzy element (where i=1 to n), then the fuzzy reliability is given as follows

$$R_p = \left(\underset{i=1}{\overset{n}{\otimes}} R_i^c \right)^c$$

$$\left(R_p\right)^c = \underset{i=1}{\overset{n}{\otimes}} R_i^c$$

$$\left(R_p\right)^c = \underset{i=1}{\overset{n}{\otimes}} h_i^c$$

$$\left(R_p\right)^c = h_1^c \otimes h_2^c \otimes h_3^c \otimes \ldots \otimes h_n^c$$

$$\left(R_p\right)^c = \bigcup_{\substack{\gamma_i \in h_i \\ i=1,2,\ldots,n}} \left\{ p_1 p_2 \ldots p_n / p_i = 1 - \gamma_i \right\}$$

Parallel-Series System: In Fig. 3, let R_i^j be the reliability of the jth component (where j=1 to m) of ith branch (where i=1 to n) then the fuzzy reliability .

$$R_{ps}^c = \underset{i=1}{\overset{n}{\otimes}} \left[\underset{j=1}{\overset{m}{\otimes}} R_i^j \right]^c$$

$$R_{ps}^c = \underset{i=1}{\overset{n}{\otimes}} \left[\underset{j=1}{\overset{m}{\otimes}} h_i^j \right]^c$$

$$R_{ps}^c = \underset{i=1}{\overset{n}{\otimes}} \left[h_i^1 \otimes h_i^2 \otimes \ldots \otimes h_i^m \right]^c$$

$$R_{ps}^c = \underset{i=1}{\overset{n}{\otimes}} \left[\bigcup_{\substack{\gamma_i^j \in h_i^j \\ i=1,2,\ldots,n \\ j=1,2,\ldots,m}} \left\{ \left(s_i^1 . s_i^2 \ldots . s_i^m \right) / s_i^j = 1 - h_i^j \right\} \right]^c$$

$$R_{ps}^c = \underset{i=1}{\overset{n}{\otimes}} \left[\bigcup_{\substack{\gamma_i^j \in h_i^j \\ i=1,2,\ldots,n \\ j=1,2,\ldots,m}} \left\{ 1 - \left(s_i^1 . s_i^2 \ldots . s_i^m \right) / s_i^j = 1 - h_i^j \right\} \right]$$

$$R_{ps}^c = \left[\bigcup_{\substack{\gamma_i^j \in h_i^j \\ i=1,2,\ldots,n \\ j=1,2,\ldots,m}} \left\{ \left(q_1 . q_2 \ldots q_n \right) / 1 - \left(s_i^1 . s_i^2 \ldots . s_i^m \right) = q_i \, , \; s_i^j = 1 - h_i^j \right\} \right]$$

Series-parallel System: In Fig. 4, let R_i^j denote the reliability of the ith component (where i=1 to n) of jth subsystem (where j=1 to m) then the fuzzy reliability

$$R_{sp} = \bigotimes_{j=1}^{m}\left[\bigotimes_{i=1}^{n} R_i^j\right]^C$$

$$R_{sp} = \bigotimes_{j=1}^{m}\left[\bigotimes_{i=1}^{n} h_i^j\right]^C$$

$$R_{sp} = \bigotimes_{j=1}^{m}\left[h_1^j \otimes h_2^j \otimes \ldots \otimes h_n^j\right]^C$$

$$R_{sp} = \bigotimes_{j=1}^{m}\left[\bigcup_{\substack{\gamma_i^j \in h_i^j,\\ i=1,2,\ldots,n\\ j=1,2,\ldots,m}} \left\{\gamma_1^j \gamma_2^j \ldots\ldots \gamma_n^j\right\}\right]^C$$

$$R_{sp} = \bigotimes_{j=1}^{m}\left[\bigcup_{\substack{\gamma_i^j \in h_i^j,\\ i=1,2,\ldots,n\\ j=1,2,\ldots,m}} \left\{1-\left(\gamma_1^j \gamma_2^j \ldots\ldots \gamma_n^j\right)\right\}\right]$$

$$R_{sp} = \left[\bigcup_{\substack{\gamma_i^j \in h_i^j,\\ i=1,2,\ldots,n\\ j=1,2,\ldots,m}} \left\{r^1.r^2\ldots r^m / r^j = 1-\left(\gamma_1^j \gamma_2^j \ldots\ldots \gamma_n^j\right)\right\}\right]$$

EXAMPLES

Here some examples are illustrated to demonstrate the methods discussed in this study. Consider a system consists of three components. Let we have given reliabilities of three components/units (as form of HFEs)

$$R_1 = h_1 = \left\{0.3, 0.8\right\},$$
$$R_2 = h_2 = \left\{0.6\right\},$$
$$R_3 = h_3 = \left\{0.1,\ 0.5\right\}$$

Now fuzzy reliabilities of different configurations calculated as follows

Series System

Let these three components/units be in series configuration (Fig. 5).

Figure 5. Series system of three components/units

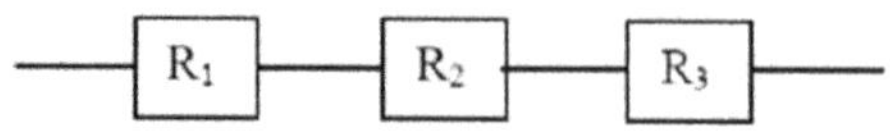

$$R_s = \overset{3}{\underset{i=1}{\otimes}} R_i = \overset{3}{\underset{i=1}{\otimes}} h_i$$
$$R_s = h_1 \otimes h_2 \otimes h_3$$
$$R_s = \{0.018, 0.09, 0.048, 0.24\}$$

So by using definition 2.2, score function of R_s

$$S(R_s) = \{0.099\}$$

Parallel System

Now let given three components/units be in parallel configuration (Fig. 6).

Figure 6. Parallel system of three components/units

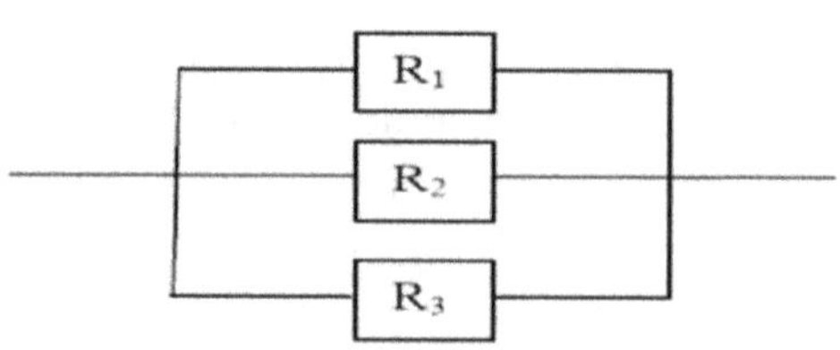

Fuzzy reliability

$$R_p = \left(\bigotimes_{i=1}^{3} R_i^c \right)^c$$

$$R_p = \left(\bigotimes_{i=1}^{3} h_i^c \right)^c$$

$$R_p = \{ 0.748, 0.86, 0.928, 0.96 \}$$

So by using definition 2.2, score function of R_p

$$S(R_p) = \{ 0.874 \}$$

Parallel-Series System

Let 9 components/units are connected as figure 7 such that

$$R_1^1 = R_1^2 = R_1^3 = \{ 0.3, 0.8 \}$$

$$R_2^1 = R_2^2 = R_2^3 = \{ 0.6 \}$$

$$R_3^1 = R_3^2 = R_3^3 = \{ 0.1, 0.5 \}$$

Figure 7. Parallel-series system of nine components/units

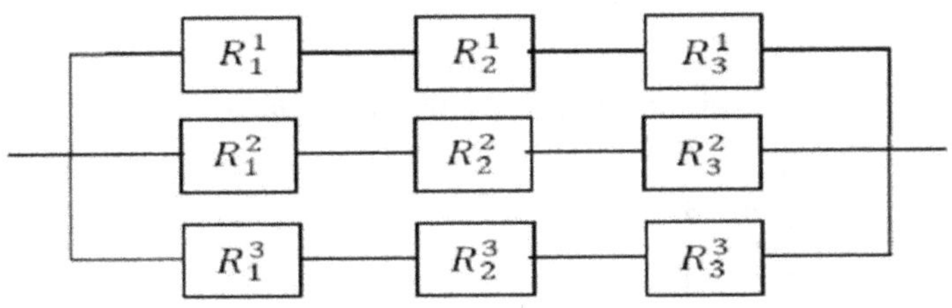

Then fuzzy reliability

$$\begin{aligned} R_{ps} = \{&0.053033832, 0.12246516, 0.081963552, 0.26711376, 0.12246516, 0.1868058, \\ &0.14927376, 0.3208488, 0.081963552, 0.14927376, 0.110009472, 0.28950336 \\ &0.26711376, 0.3208488, 0.28950336, 0.4327968, 0.12246516, 0.1868058, \\ &0.14927376, 0.3208488, 0.1868058, 0.246429, 0.2116488, 0.370644, \\ &0.14927376, 0.2116488, 0.17526336, 0.3415868, 0.3208488, 0.370644, \\ &0.3415968, 0.474384, 0.081963552, 0.14927376, 0.110009472, 0.28950336, \\ &0.14927376, 0.2116488, 0.17526336, 0.3415968, 0.1100094720.17526336, \\ &0.137198592, 0.31120896, 0.28950336, 0.3415968, 0.31120896, 0.28950336, \\ &0.3415968, 0.31120896, 0.4501248, 0.26711376, 0.3208488, 0.28950336, \\ &0.4327968, 0.3208488, 0.370644, 0.3415968, 0.474384, 0.28950336, 0.3415968, \\ &0.31120896, 0.4501248, 0.4327968, 0.474384, 0.4501248, 0.561024\} \end{aligned}$$

So by using definition 2.2, score function of R_{ps}

$$S\left(R_{ps}\right) = \left\{ 0.268567298 \right\}$$

Series -Parallel System

Let 9 components/units are connected as figure 8 such that

$$R_1^1 = R_2^1 = R_3^1 = \left\{0.3, 0.8\right\}$$

$$R_1^2 = R_2^2 = R_3^2 = \left\{ 0.6 \right\}$$

Figure 8. Series-parallel system of nine components/units

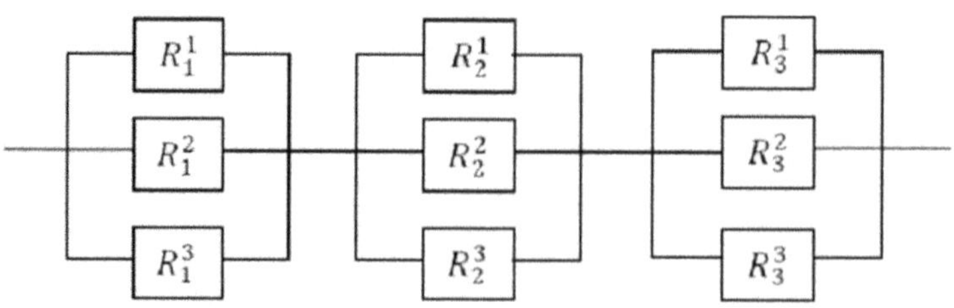

Then fuzzy reliability

$$R_{sp} = (0.418508992, 0.48117344, 0.519219712, 0.53712384\ 0.48117344, 0.5532208, 0.59696384, 0.6175488, 0.519219712, 0.59696384, 0.6441656320.66637824, 0.53712384, 0.6175488, 0.66637824, .68935680.48117344, 0.5532208, 0.59696384, 0.6175488, 0.5532208, 0.636056, 0.6863488, 0.710016, 0.7661568, 0.7925760.519219712.0.59696384, 0.644165632, 0.66637824, 0.59696384, 0.6863488, 0.74061824, 0.7661568, 0.644165632, 0.74061824, 0.799178752, 0.82673664, 0.66637824, 0.7661568, 0.82673664, 0.8552448\ 0.53712384, 0.6175488, 0.66637824, 0.6893568, 0.6175488, 0.710016, 0.7661568, 0.792576, 0.66637824, 0.7661568, 0.82673664, 0.85524480.6893568, 0.792576, 0.8552448, 0.884736$$

So by using definition 2.2, score function of R_{sp}

$$S(R_{sp}) = \{ 0.667627625 \}$$

Now it is clear that

$$S(R_p) > S(R_{sp}) > S(R_{ps}) > S(R_s) \Rightarrow R_p \succ R_{sp} \succ R_{ps} \succ R_s$$

Hence among all four cases, fuzzy reliability is superior in parallel system while inferior in series system.

CONCLUSION

Here we have developed a new technique for analyzing fuzzy reliability of a system in hesitant fuzzy environment. This technique is very useful where people's hesitancy and uncertainty present in decision making process. By introducing this technique, here we have seen application of hesitant fuzzy theory in reliability evaluation of different complex systems like series system, parallel system, series-parallel system and parallel-series system.It is clear from illustrations that fuzzy reliability is minimum in series configuration and maximum in parallel configuration. The present method uses the score function to make a important comparison between any two data of reliability of a component/unit as form of hesitant fuzzy element.

Hence we can conclude that this new technique has no restriction and it is applicable in many different fuzzy environments like subjectivity, ambiguity, uncertainty and hesitancy.

REFERENCES

Atanassov, K. T. (1986). Intuitionistic fuzzy sets. *Fuzzy Sets and Systems*, *20*(1), 87–96. doi:10.1016/S0165-0114(86)80034-3

Atanassov, K. T. (1999). *Intuitionistic fuzzy sets*. Springer Physica-Verlag. doi:10.1007/978-3-7908-1870-3

Chen, S. M. (1994). Fuzzy system reliability analysis using fuzzy number arithmetic operations. *Fuzzy Sets and Systems*, *64*(1), 31–38. doi:10.1016/0165-0114(94)90004-3

Chen, S. M. (1996). New method for fuzzy system reliability analysis. *Cybernetics and Systems: An International Journal*, *27*(4), 385–401. doi:10.1080/019697296126480

Karnik, N. N., & Mendel, J. M. (2001). Operations on type-2 fuzzy sets. *Fuzzy Sets and Systems*, *122*(2), 327–348. doi:10.1016/S0165-0114(00)00079-8

Kumar, A., & Singh, S. B. (2018). Signature reliability of linear multi-state sliding window system. *International Journal of Quality & Reliability Management*, *35*(10), 2403–2413. doi:10.1108/IJQRM-04-2017-0083

Kumar, D., & Singh, S. B. (2013). Reliability analysis of complex repairable system with reboot delay. *International Journal of Information and Computation Technology*, *3*, 341–344.

Kumar, D., & Singh, S. B. (2014, November). Evaluating fuzzy reliability using rough intuitionistic fuzzy set. In 2014 Innovative Applications of Computational Intelligence on Power, Energy and Controls with their impact on Humanity (CIPECH) (pp. 138-142). IEEE.

Kumar, P., & Singh, S. B. (2018). *Fuzzy reliability of systems using different types of level (λ, 1) interval-valued fuzzy numbers*. Reliability Engineering Theory and Applications, CRC Press Taylor & Francis Group. doi:10.1201/9781351130363-10

Ram, M., Singh, S. B., & Singh, V. V. (2013). Stochastic Analysis of a Standby System with Waiting Repair Strategy. *IEEE Transactions on Systems, Man, and Cybernetics. Systems*, *43*(3), 698–707. doi:10.1109/TSMCA.2012.2217320

Torra, V. (2010). Hesitant fuzzy sets. *International Journal of Intelligent Systems*, *25*, 529–539.

Torra, V., & Narukawa, Y. (2009). On hesitant fuzzy sets and decision. *The 18th IEEE international Conference on Fuzzy Systems*, 1378-1382.

Xia, M., & Xu, Z. S. (2011). Hesitant fuzzy information aggregation in decision making. *International Journal of Approximate Reasoning*, *52*(3), 395–407. doi:10.1016/j.ijar.2010.09.002

Zadeh, L. A. (1965). Fuzzy sets. *Information and Control*, *8*(3), 338–353. doi:10.1016/S0019-9958(65)90241-X

Chapter 12
Fuzzy Reliability Theory:
A Bibliometric-Based Review

Ingrid N. Pinto-López
https://orcid.org/0000-0002-1580-1375
UPAEP Universidad, Mexico

Cynthia M. Montaudon-Tomas
https://orcid.org/0000-0002-2595-6960
UPAEP Universidad, Mexico

ABSTRACT

This chapter analyzes fuzzy reliability theory using bibliometric analysis. Different aspects of fuzzy have already been analyzed using bibliometric analysis, and a series of bibliometric tools have also been used. VOSviewer software was used to identify maps showing the most relevant trends. The analysis includes scientific articles, citations, journals, authors, universities, keywords, and countries. Results show that countries belonging mainly to Asia are at the avant-garde in terms of research in the field, China and India being the most productive countries in terms of the number of articles published, citations, and universities invested in the topic. Other countries in North America, such as Canada and the United States, and in Europe, the UK, Poland, Italy, and France, also show a great interest in this area of science. Research on the topic is relatively recent. The first articles were published in 1991; therefore, it presents excellent opportunities that will quite possibly attract researchers and universities from different regions of the world.

DOI: 10.4018/978-1-7998-7564-2.ch012

INTRODUCTION

The origins of fuzzy reliability theory come from the consideration of reliability aspects of computer systems, in which system states could not be simply classified as failed or functioning (Huang & He, 2008; Zadeh, 1965). Various forms of fuzzy reliability theory can be identified, including profust reliability theory (Cai, Wen, & Zhang, 1993) posbist reliability theory (Cai, Wen & Zhang, 1991; Cappelle & Kerre, 1993), and posfust reliability theory (Kerre, Onisawa, Cappelle & Gazdik, 1998).

The works on the fuzzy reliability theory can be traced to the 1990s (Kacprzyk & Onisawa 1995), being the most relevant works those of Cai (1996) in the United States and Rotshtein & Shtovba (1997) in Ukraine. Theories were also developed by Capelle & Kerre, (1993) and Nikolaidis, Chen, Cudney, Haftka, & Rosca (2004). The applications of fuzzy ideas in reliability theory dealing with lacking inaccuracy or fluctuation data can be seen in many areas (Bamrungsetthapong & Pongpullponsak, 2014). For this study, fuzzy reliability theory was differentiated from the conventional reliability theory (Cai et al., 1991), which has been used in different contexts and settings (Chaube & Singh, 2016; Xiaoning, 2008).

Bibliometrics has been named the "Science of evaluation and analysis of Sciences" (Martínez, Díaz, Lima, Herrera & Herrera-Viedma, 2014). It applies mathematical processes and statistical methods (Garfield, 1979) to various sources, mainly publications in journals, to analyze and evaluate the quality of the scientific activity carried out by individual researchers, by research groups, by universities, and by countries (Martínez et al., 2014). Bibliometric analysis can be carried out through studies of performance, based on quantitative indicators such as the number of publications and/or impact indicators (Garfield, 1972; Martínez et al., 2014); and studies of content, based on science maps constructed through co-occurrences of terms that allow identifying the underlying conceptual structure of a scientific discipline (Cobo, López-Herrera, Herrera-Viedma & Herrera, 2011; Martínez et al., 2014).

This methodology helps evaluate the impact and performance of scientific publications through bibliometric indicators such as articles, citations, authors, journals, institutions, languages, h-index, and countries, among others; as well as various content indicators such as co-occurrences, bibliographic coupling, co-citations, etc., within a specific line of research.

Research using this methodology has increased, particularly due to the development and growth of information and communication technology tools (Pinto-López, Montaudon-Tomas & Yañez-Moneda, 2020), which has allowed access for researchers, universities, governments, and countries around the world to various scientific databases such as the Web of Science, Scopus, Google Scholar, Ebsco, Scielo, ScienceDirect, Springer, and JStor. Similarly, software tools such as VOSviewer have been developed to analyze content indicators (Van & Waltman, 2010).

Bibliometric analysis is a distinctive approach to providing a complementary perspective of research in a field closely related to scientometrics, informetrics, and webometrics. The field has substantially benefited from computerized data treatment from the last few years, and more analytical methods have become available, including mapping tools. It helps to recognize research trends by measuring scientific progress in specific disciplines.

This analysis will create an introductory perspective to the field of fuzzy reliability theory and a general overview of the state of the field, explaining the moment in which research started to be produced and its evolution. It also analyses document type, the language of publication, publication output, authorship, publication patterns, distribution of subject category, distribution of author keywords, country of publication, most-frequently cited article, document distribution, and connections between different variables. As a result, understanding of the potential and interest in the field, and the topics most frequently associated with fuzzy reliability theory will be obtained. It will also provide new insights on the topic, identifying the most influential works, the evolution of publications over time, and identifying areas of current research interest and potential directions in the future (Fahimnia, Sarkis, & Davarzani, 2015).

MAIN FOCUS OF THE CHAPTER

The primary focus of the chapter is to develop a bibliometric analysis of research into fuzzy reliability theory using information systems for the period between 1991 and 2019. The analysis was performed considering 1991 as the initial date because that is the year in which the first article registered in both the Web of Science and Scopus appeared. Bibliometric analysis was performed to identify the evolution and current state of research in fuzzy reliability theory, while mapping was used to shed light on existing connections with other related topics. The chapter helps provide an understanding of the research area.

METHODOLOGY

The number of publications based on bibliometric analysis has significantly increased over the past decade due to increased storage potential and search possibilities of IT technologies. Scientific repositories such as the Web of Science, Scopus, Google Academic, Scielo, among others, store information of millions of articles that have been published in the most relevant scientific journals in all disciplines and areas of knowledge. Additionally, visual tools such as VOSviewer allow connecting information from different data repositories to create bibliometric maps that provide better data visualization (Pinto-López et al., 2020).

Bibliometrics is a part of scientometrics, and as a discipline, it considers the quantitative analysis of products generated in the context of scientific research. Bibliometrics has mainly been used to evaluate scientific journals to analyze its production in terms of volume, trends, structure, visibility, and impact (Reyes et al., 2019). The analysis also considers institutions, authors, countries, key words, citations, language, and others (Blanco-Mesa et al., 2017), allowing the identification of the nuclear journals in each discipline, to establish acquisition policies adjusted to a specific budget, adopt publication discarding policies, study the dispersion and obsolescence of scientific literature, design standardization norms, design processes for indexation, classification, and confection of automatic summaries, predict the productivity of authors, individuals, organizations, countries and more (Sengupta, 1992; Spinak, 1998; Pinto-López et al., 2020).

The information used to perform bibliometric analysis comes from specific databases such as ISI Web of Science (WoS) and Scopus. Both scientific repositories comprise valuable information about high-quality multidisciplinary research that has been published in leading journals across the globe in sciences, social sciences, arts, humanities, technology, medicine, and others. Identifying and creating indexes of the most relevant publications and providing bibliographic information allows for the evaluation and analysis of the productivity and the scientific quality of research (Chen et al., 2012; Elsevier, 2020).

WoS incorporates other databases into the WoS's main collection. It includes scientific articles from all areas of science in more than 15,000 specialized journals and 50,000,000 classified articles in over 250 categories and 151 areas of research (Blanco-Mesa et al., 2017).

The content in Scopus comes from over 5,000 publishers that must be reviewed and selected by an independent Content Selection and Advisory Board (CSAB) to be, and continue to be, indexed in this database. Over 70,000,000 articles, 70,000 institutional profiles and 16,000,000 researcher's profiles have been included (Elsevier, 2020).

The process to identify and select the articles to be considered in the study is divided into three stages (see Diagram 1). In the first, the search criteria are identified and defined. In this case, the search is carried out with four similar keywords selected based on a literature review. Stage 2 consists of a series of searches in both Scopus and the Web of Science and identifying whether the results meet the established criteria. Otherwise, the publication is discarded. In the third and final stage, the analysis focuses on finding possible duplications in the databases. This process is carried out manually.

Table 1. Process to identify and select items

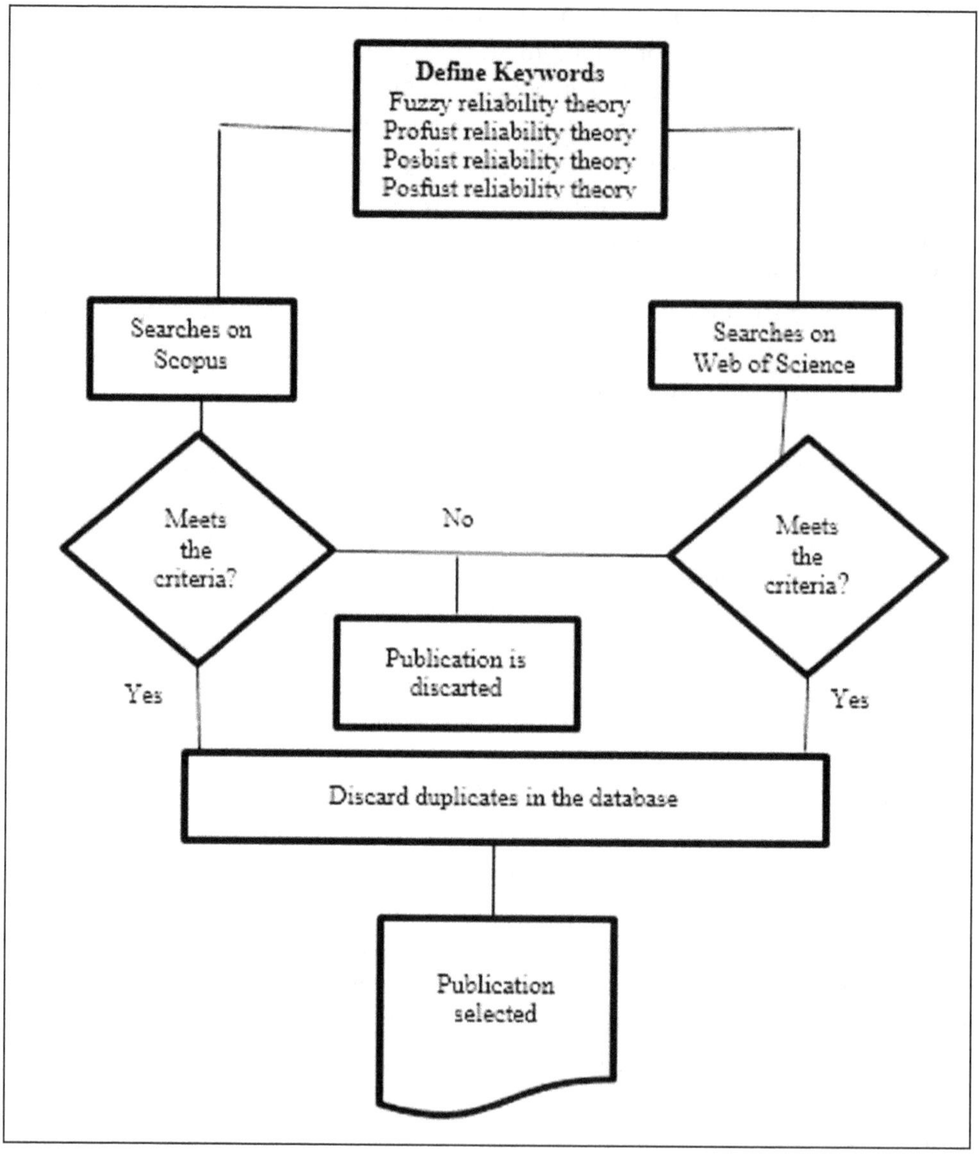

RESULTS AND DISCUSSION

The data analyzed was obtained from the WoS and Scopus in 2020, specifically on June 29th-30th and July 1st-7th. The keywords for the search strategy were "fuzzy reliability theory," "profust reliability theory," "posbist reliability theory," and "posfust reliability theory." The selected time frame considered all publications from

1991 to 2019. The citation indexes that were used are SCI-EXPANDED (Science Citation Index Expanded), SSCI (Social Sciences Citation Index), A&HCI (Arts & Humanities Citation Index), and ESCI (Emerging Sources Citation Index). The types of documents considered for the analysis were articles, proceeding papers, book reviews, meeting abstracts, and reviews. The analysis of the selected bibliometric indicators is shown in the following sections.

Number of Published Articles

Several coincidences were found for the topics analyzed. Sixteen results were identified on the WoS, while 117 were identified in Scopus. The 16 articles found on the WoS were also found among the results generated by Scopus.

From the 117 articles that were retrieved, 11.11% were identified as open access. The frequency related to the selected articles is shown in Figure 1. As can be observed, this is an area that has been seldomly analyzed, which is fairly recent and has numerous areas of opportunity. As has been explained earlier, the first article identified was published in 1991. The frequency of publications has not been constant over the years. Nevertheless, in the last decade (2010-2019), 65.81% of all publications were developed, being 2011 the year with the most published works (11.11%).

Figure 1. Number of annual publications in the WoS & Scopus since 1991

General Structure of Citations

One of the markers that show the relevance of published works is the number of citations. This helps identify the influence and popularity of the articles in a field (Blanco-Mesa et al., 2017) when analyzing the distribution of the number of citations per year. From 1991 to 2019, research on fuzzy reliability theory had an average of 16.13 citations per article. Table 2 presents the detail of published articles (TP) and total citations (TC). The number of articles that obtained citations each year is also presented (Figure 2).

As can be observed, the number of citations per year shows a rather rugged and unstable frequency, which can be explained by the fact that the topics being analyzed are both novel and recent, and the area of study has not yet been fully explored. The articles that have received more citations are those published in 1991. These articles explain the theoretical basis of fuzzy logic and are followed, in number of citations, by those published in 1996 and 2004, all of which have over 200 citations.

With regards to the number of citations, 47% of the articles analyzed have 1 to 9 citations; 9.40% have 10 to 24; 9.40% have 25 to 49; 2.56% have 50 to 74; 2.56% have 75 to 99; 1.71% have to 100 to 124, 1.71% have to 125 to 149, while the remaining 0.85% have 150 citations or more. Still, 24.78% of articles have not been cited yet.

In terms of the distribution of citations by decade, between 1991 and 1999, 37.14% of all citations were recorded, while between 2000 and 2009, it was 32.85%, and from 2010 to 2019, the percentage was 29.99%.

Figure 2. Number of annual citations in the WoS & Scopus since 1991

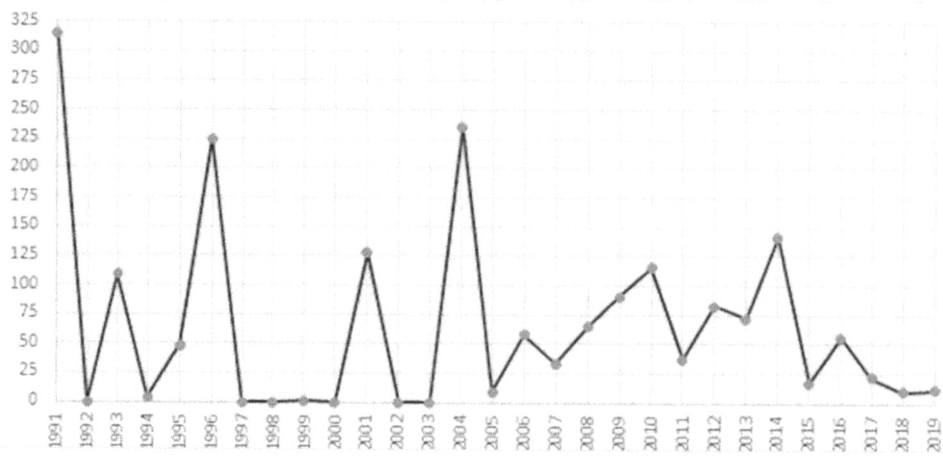

Table 2. Annual citation structure

Year	TP	TC	≥150	≥125	≥100	≥75	≥50	≥25	≥10	≥1
1991	3	314	1	-	-	1	-	-	1	-
1992	-	-	-	-	-	-	-	-	-	-
1993	1	109	-	-	1	-	-	-	-	-
1994	2	4	-	-	-	-	-	-	-	2
1995	2	48	-	-	-	-	-	1	-	1
1996	3	225	-	1	-	-	1	-	1	-
1997	-	-	-	-	-	-	-	-	-	-
1998	-	-	-	-	-	-	-	-	-	-
1999	1	1	-	-	-	-	-	-	-	1
2000	-	-	-	-	-	-	-	-	-	-
2001	3	128	-	-	1	-	-	-	1	-
2002	-	-	-	-	-	-	-	-	-	-
2003	-	-	-	-	-	-	-	-	-	-
2004	5	236	-	1	-	-	1	1	-	1
2005	3	9	-	-	-	-	-	-	-	2
2006	5	58	-	-	-	-	-	2	-	2
2007	4	33	-	-	-	-	-	-	2	2
2008	4	65	-	-	-	-	1	-	-	2
2009	5	91	-	-	-	1	-	-	-	2
2010	10	116	-	-	-	1	-	-	1	5
2011	13	37	-	-	-	-	-	-	1	9
2012	8	83	-	-	-	-	-	2	-	3
2013	9	72	-	-	-	-	-	-	3	4
2014	8	142	-	-	-	-	-	4	-	3
2015	3	17	-	-	-	-	-	-	-	3
2016	5	56	-	-	-	-	-	1	-	2
2017	7	22	-	-	-	-	-	-	1	3
2018	6	10	-	-	-	-	-	-	-	3
2019	7	11	-	-	-	-	-	-	-	5
Total	**117**	**1887**	**1**	**2**	**2**	**3**	**3**	**11**	**11**	**55**

Source: Own elaboration based on WoS & Scopus 2020. Abbreviations: TP: Total papers, TC: Total citations; ≥150, ≥125, ≥100, ≥75, ≥50, ≥25, ≥10, ≥1 = Number of papers with equal or more than 150, 125, 100, 75, 50, 25, 10 and 1 citations.

The 50 Most Cited Papers

The analysis of scientific research is essential to identify the impact of research in a specific field, notion, or topic (Merigó et al., 2017). Table 3 shows the 50 most cited articles regarding the topics that were analyzed. The table also considers authors, journals, year of publication and total citations (TC). This section identifies the total number of citations that each article has in each of the databases that were used (WoS and Scopus), as well as citations per year (C/Y). Additionally, the average citations per year are also obtained base on the citations of each article in the WoS and Scopus.

Another relevant point to analyze the structure of published documents about fuzzy reliability theory is co-citations. Figure 3 presents the influences of existing work-connections based on co-citations, which is used to identify when a publication makes references to two additional documents, and there is the possibility that both sources become are connected in terms of their content (Small, 1973). Figure 3a shows the influence of the existing connections, which were analyzed considering co-citations. The largest circles identify the main drivers in the topic. These connections are analyzed in more detail in figure 3b. The main clusters can be identified, and their interrelations with other clusters are visible. Four clusters have been identified. Among the most influential authors in the field are Singer, D., Chen, SM., Zadeh, LA., Cai, KY., Wen, CY., Zhang, M. Chaube, S., Singh, SB.

Keywords

Figure 4 shows the bibliometric map with the proximity relations of the keywords. Twelve clusters were identified, showing the connections between "fuzzy reliability theory" and other keywords. The most relevant notions include fuzzy sets, reliability, failure analysis, uncertainty analysis, system reliability, reliability theory, probability, membership functions, and numerical methods

The Most Productive Authors

Table 4 shows the most productive and influential authors in research about fuzzy reliability theory. This table is organized considering the authors reported in the WoS and Scopus. The WoS identified 28 authors. The ones that were considered had published two or more articles. Scopus reports 160 authors, and the ones that were considered had published three or more articles. With regards to authors, the names are indicated along with the total number of publications (TP), the number of citations per author (TC), the average citations per paper (TC/TP), and the number of articles per author, which are grouped in four intervals based on the number of

Table 3. The 50 most cited papers

R	Title	Author/s	Journal	Year	TC WoS	TC Scopus	C/Y WoS	C/Y Scopus
1	Fuzzy variables as a basis for a Theory of Fuzzy Reliability in the possibility context	Cai, KY., Wen, CY., Zhang, ML	FSS	1991	160	210	5.71	7.50
2	Posbist fault tree analysis of coherent systems	Huang, HZ., Tong, X., Zuo, MJ.	RESS	2004	108	140	7.20	9.33
3	System failure engineering and fuzzy methodology an introductory overview	Cai, K.Y.	FSS	1996		137		5.95
4	Fuzzy States as a basis for a Theory of Fuzzy Reliability	Cai, KY., Wen, CY., Zhang, ML	MR	1993	85	109	3.26	4.19
5	Reliability modeling of repairable systems using Petri nets and fuzzy Lambda-Tau methodology	Knezevic, J., Odoom, E.R.	RESS	2001		106		5.88
6	Posbist reliability behavior of typical systems with two types of failure	Cai, KY., Wen, CY., Zhang, ML	FSS	1991	70	91	2.50	3.25
7	Reliability assessment for fuzzy multi-state systems	Liu, Y., Huang, H.Z.	IJSS	2010		84		9.33
8	Evidential networks for reliability analysis and performance evaluation of systems with imprecise knowledge	Simon, C., Weber, P.	IEEE TR	2009		83		8.3
9	A general formal approach for fuzzy reliability analysis in the possibility context	Utkin, LV., Gurov, SV.	FSS	1996	62	68	2.69	2.95
10	Reliability and performance assessment for fuzzy multi-state elements	Liu, Y., Huang, H.Z., Levitin, G.	JRR	2008		63		5.72
11	Fuzzy reliability estimation using Bayesian approach	Wu, H.C.	CIE	2004		53		3.53
12	Developments and thoughts on operation reliability assessment of mechanical equipment	He, Z., Cao, H., Zi, Y., Li, B.	JME	2014		47		9.4
13	Measuring reliability under epistemic uncertainty: Review on non-probabilistic reliability metric	Kang, R., Zhang, Q., Zeng, Z., Zio, E., Li, X.	CJA	2016		45		15.00
14	Uncertain probabilities III: the continuous case	Buckley, JJ	SC	2004	42	42	2.80	2.80
15	Posbist reliability behavior of Fault-Tolerant systems	Cai, KY., Wen, CY., Zhang, ML	MR	1995	39	41	1.62	1.70
16	Behavior analysis of synthesis unit in fertilizer plant	Garg, H., Sharma, S.P.	IJQRM	2012		34		4.85
17	A Profust Reliability-Based Approach to Prognostics and Health Management	Zhao, Zhiyao., Quan, Quan., Cai, Kai-Yuan	IEEE TR	2014	24	31	4.8	6.2
18	Reliability, Availability and Maintainability Analysis of Industrial Systems Using PSO and Fuzzy Methodology	Garg, H.	JMSI	2014		30		6.00

Table 3. Continued

R	Title	Author/s	Journal	Year	TC WoS	TC Scopus	C/Y WoS	C/Y Scopus
19	A novel approach for analyzing fuzzy system reliability using different types of intuitionistic fuzzy failure rates of components	Kumar, M., Yadav, S.P.	ISAT	2012		29		4.14
20	Fuzzy Bayesian system reliability assessment based on exponential distribution	Wu, H.C.	AMM	2006		27		2.07
21	Repairable consecutive-k-out-of-n: F system with fuzzy states	Guan, JC., Wu, YQ.	FSS	2006	23	27	1.76	2.07
22	Applying weakest t-norm based approximate intuitionistic fuzzy arithmetic operations on different types of intuitionistic fuzzy numbers to evaluate reliability of PCBA fault	Kumar, M.	ASCJ	2014		25		5.00
23	Fuzzy reliability analysis in the case of random stress and fuzzy strength	Huang, H., Sun, Z., Guo, D., Zhou, F., Tian, Z.	JMS	2001		22		1.22
24	Reliability prediction for evolutionary product in the conceptual design phase using neural network-based fuzzy synthetic assessment	Liu, Y., Huang, H.Z., Ling, D.	IJSS	2013		20		3.33
25	Fuzzy system reliability evaluation using time-dependent intuitionistic fuzzy set	Kumar, M., Prasad Yadav, S., Kumar, S.	IJSS	2013		20		3.33
26	Evolution of reliability engineering discipline over the last six decades: a comprehensive review	Bhamare, S.S., Yadov, O.P., Rathore, A.	IJRS	2007		19		1.58
27	Structure functions with fuzzy states	Cutello, V., Montero, J., Yáñez, J.	FSS	1996		19		0.82
28	Fuzzy reliability theories	Kai-Yuan, C.	FSS	1991		13		0.46
29	A new structural reliability index based on uncertainty theory	Wang, P., Zhang, J., Zhai, H., Qiu, J.	CJA	2017		10		5.00
30	Optimal replacement policy for fuzzy multi-state element	Liu, Y., Huang, H.Z.	JMVLSC	2011		10		1.25
31	Reliability analysis of rangeland systems: the application of Profust theory	Mirakbari, M., Ganji, A.	SERRA	2010	10	10	1.11	1.11
32	Fuzzy sets in the evaluation of reliability	Hryniewicz, O.	SCI	2007		10		0.83
33	Reliability analysis of structure for fuzzy safety state	Wang, Z., Li, Y., Huang, H.Z., Liu, Y.	IASC	2012		9		1.28
34	System reliability analysis based on Weibull distribution and hesitant fuzzy set	Kumar, A., Ram, M.	IJMEMS	2018		8		8.00
35	Vibration reliability analysis on attractive transmission system of shearer	Zhou, D., Zhang, X.F., Yang, Z., Zhang, Y.M.	JCCS	2015		8		2.00

Table 3. Continued

R	Title	Author/s	Journal	Year	TC WoS	TC Scopus	C/Y WoS	C/Y Scopus
36	Optimization and mechanical accuracy reliability of a new type of forging manipulator	Chen, K., Ma, C., Zheng, M., Gao, F.	CJME	2015		8		2.00
37	Uncertain evaluation measure system and model for power grid catastrophic accident	Ma, C., Xiao, X., Yang, H., Li, H.	CSEE	2012		8		1.14
38	Fuzzy availability modeling and analysis of biscuit manufacturing plant: A case study	Kumar, K., Kumar, P.	IJSAEM	2011		8		1.00
39	Mixture models in profust reliability theory	Kai-Yuan, C., Chuan-Yuan, W., Ming-Lian, Z.	MR	1995		7		0.29
40	Optimization and Mechanical Accuracy Reliability of a New Type of Forging Manipulator	Chen, Kang., Ma Chunxiang., Zheng, Maoqui., Gao, Feng	CJME	2015	6	8	1.50	2.00
41	A modified profust-performance-reliability algorithm and its application to dynamic system	Zhao, Z., Quan, Q., Cai, K.Y.	JIFS	2017		6		3.00
42	Fuzzy reliability theory based on membership function	Chaube, S., Singh, S.B.	IJMEMS	2016		6		2.00
43	A new structural reliability index based on uncertainty theory	Wang, Pidong., Zhang, Jianguo., Zhai, Hao., Qiu, Jiwei	CJA	2017	5	10	2.50	5.00
44	A modified profust – performance – reliability algorithm and its application to dynamic systems	Zhao, Zhiyao., Quan, Quan., Cai, Kai-Yuan	JIFS	2017	5	6	2.50	3.00
45	Posbist Reliability Theory of k-out-of-n: G System	Huang, Hong-Zhong., Li, Yanfeng., Liu, Yu	JMVLSC	2010	5	5	0.55	0.55
46	Efficient numerical simulation methods for estimating fuzzy failure probability-based importance measure indices	Ling, C., Lu, Z., Feng, K., Sun, B.	SMO	2019		5		5.00
47	Research and application pf transformer condition evaluation model-based on multi-source information	Zhang, H., Chang, A., Wang, Q., Zhou, Q., Liang, Y.	HVA	2016		5		1.66
48	Dynamic, progressive failure evolution research on landslide under precipitation	Wang, Y., Li, X., Zhang, B., Li, M.	JHE	2013		5		0.83
49	Application of fuzzy mathematics to slope stability analysis	Wang, Y.X.	RSM	2010		5		0.55
50	General approach for fuzzy reliability analysis based on the fuzzy probability theory	Dong, Y.G., Chen, X.Z., Zhao, X.D., Quan, Z.W.	CJCM	2005		5		0.35

Source: Own elaboration based on the WoS & Scopus 2020. Abbreviations: R: Ranking; TC: Total citations; C/Y: Citations per year; FSS: Fuzzy Sets and Systems; RESS: Reliability Engineering & System Safety; MR: Microelectronics and Reliability; SC: Soft Computing; IEEE TR: IEEE Transactions on Reliability; SERRA: Stochastic Environmental Research and Risk Assessment; CJME: Chinese Journal of Mechanical Engineering; CJA: Chinese Journal of Aeronautics; JIFS: Journal of Intelligent & Fuzzy Systems; JMVLSC: Journal of Multiple-Valued Logic and Soft Computing; IJSS: International Journal of Systems Science; JRR: Journal of Risk and Reliability; JME: Journal of Mechanical Engineering; CIE: Computers and Industrial Engineering; CJA: Chinese Journal of Aeronautics; IJQRM: International Journal of Quality and Reliability Management; JMSI: Journal of Metrology Society of India; ISAT: ISA Transactions; AMM: Applied Mathematical Modelling; ASCJ: Applied Soft Computing Journal; JMS: Journal of Mechanical Strength; IJRS: International Journal of Reliability and Safety; JMVLSC: Journal of Multiple-Valued Logic and Soft Computing; SCI: Studies in Computational Intelligence; IASC: Intelligent Automation and Soft Computing; IJMEMS: International Journal of Mathematical, Engineering and Management Sciences; JCCS: Journal of the China Coal Society; CJME: Chinese Journal of Mechanical Engineering; CSEE: Chinese Society of Electrical Engineering; IJSAEM: International Journal of Systems Assurance Engineering and Management; SMO: Structural and Multidisciplinary Optimization; HVA: High Voltage Apparatus; JHE: Journal of Hydraulic Engineering; RSM: Rock and Soil Mechanics; CJCM: Chinese Journal of Computational Mechanics.

Figure 3a. Bibliometric mapping of articles co-citations with a threshold of 40 citations and the 100 most representative connections.

Figure 3b. Detailed bibliometric mapping of articles co-citations with a threshold of 40 citations and the 100 most representative connections.

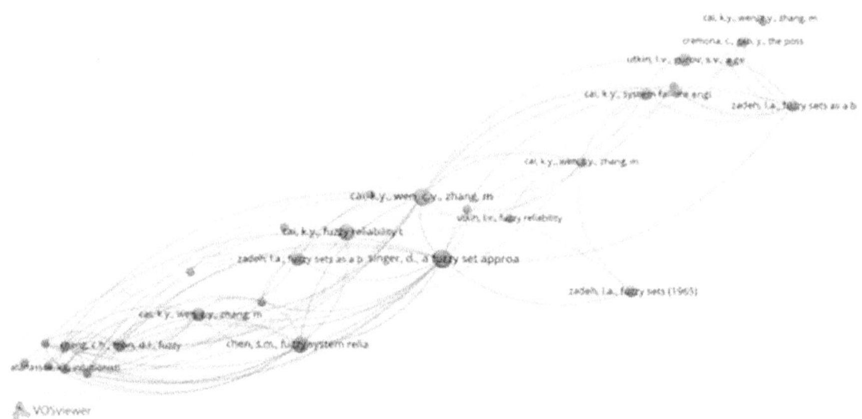

citations. These intervals are based on articles that have between 1 and 9 citations, articles between 10 and 24 citations, articles between 25 and 49 citations, and those with 50 citations and over. The university and country of origin of the authors are also identified.

It is important to acknowledge that 78.94% of the most relevant authors are from China and 26.66% from India, which shows the relevance of this topic in Asian countries.

Figure 5 shows the connection between productivity and author impact. Productivity can be understood as the number of publications and the impact as an average of citations per article. In Figure 5, the three most productive authors are

Figure 4. Bibliometric mapping of Co-occurrence of author and keyword of documents published with a threshold of 181 and the 100 most representative connections.

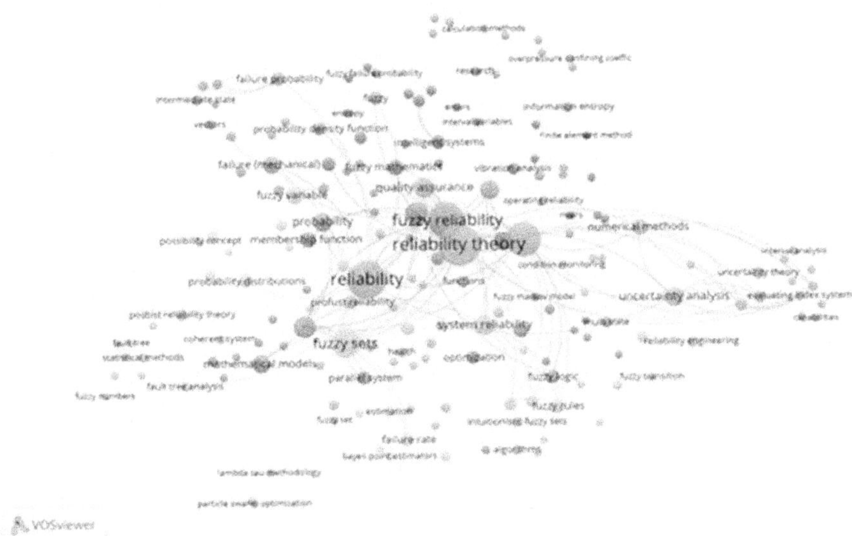

Cai, Chuan & Ming, while the three authors with the highest impact are Chuan-Yuan, Ming-Lian & Cai.

The bibliometrics maps presented in Figure 6a and Figure 6b depict the connections between authors. Thirteen clusters are identified. As can be observed in Figure 6a, the most prominent authors are Cai, Huang, Liu, Kumar, Ling, Wen, Cao, Kang, Wang, Zhang, among others. Figure 6b shows the existing connections between the authors in the core, which are not visible in Figure 6a.

The Most Influential Journals

Research about fuzzy reliability theory is published in a large number of journals, some of which are discipline-specific while others are multidisciplinary. The WoS database reported 12 journals, while Scopus reported 87. The journals that were considered were those which had at least two publications and one citation. In Table 5, the ranking of the 25 journals with more publications on the topic can be observed. This information is organized based on the total number of publications (TP). It also shows the total citations of each journal considering the number of published articles that were found in the WoS and Scopus (see Figure 7). The articles published by each journal were analyzed to identify those which were published in both databases. The analysis includes the average citations per year, the number of articles published by journal classified according to the number of citations in

Table 4. The most-cited authors

R	Name	TP	TC	C/P	≥ 50	≥ 25	≥ 10	≥ 1	U	C
					WoS					
1	Cai,-Kai-Yuan	7	388	55.42	3	2	-	2	UAA	China
2	Chuan-Yuan, Wen	5	358	71.60	3	1	-	1	UAA	China
3	Ming Liang, Zhang	4	198	49.50	2	1	-	1	UAA	China
4	Huang, Hong-Zhong	2	115	57.50	1	-	-	1	UEST	China
5	Quan, Quan	2	30	15.00	-	-	1	1	BU	China
6	Zhao, Zhiyao	2	30	15.00	-	-	1	1	BU	China
					SCOPUS					
1	Cai, Kai-Yuan	11	651	59.18	4	2	1	4	UAA	China
2	Huang, Hong-Zhong	9	356	39.55	3	-	3	4	UEST	China
3	Liu, Yu	6	192	32.00	2	-	2	2	UESTC	China
4	Kumar, Mohit	5	81	16.20		1	2	2	IITRM	India
5	Chuan-Yuan, Wen	4	349	87.25	2	1	-	1	UAA	China
6	Ming-Lian, Zhang	4	349	87.25	2	1	-	1	UAA	China
7	Kang, Rui	4	48	12.00	-	1	-	1	BU	China
8	Garg, Harish	3	65	21.66		2	-	1	TU	India
9	Han, Chunming	3	11	3.66	-	-	-	3	CAS	China
10	Ling, Chunyan	3	11	3.66	-	-	-	2	NPU	China
11	Liu, Xiaoning	3	11	3.66	-	-	-	3	BIT	China
12	Lu, Zhenzhou	3	11	3.66	-	-	-	2	NPU	China
13	Quan, Quan	3	39	13.00	-	1	-	2	BU	China
14	Singh, S.B.	3	7	2.33	-	-	-	2	PUAT	India
15	Wen, Meilin	3	3	1.00	-	-	-	1	BU	China
16	Yadav, Shiv Prasad	3	36	12.00	-	1	-	2	IITR	India
17	Zhang, Hongwei	3	11	3.66	-	-	-	3	SJU	China
18	Zhao, Zhiyao	3	39	13.00	-	1	-	2	BU	China
19	Zu, Tianpei	3	3	1.00	-	-	-	1	BU	China

Source: Own elaboration based on WoS & Scopus 2020. Abbreviations: R: Ranking; TP: Total Publications; TC: Total citations; C/Y: Citations per year; U: University; C: Country; UESTC: University of Electronic Science and Technology; UAA: University of Aeronautics and Astronautics; IITRM: Institute of Infrastructure, Technology, Research, and Management; BU: Beihang University; TU: Thapar University; NPU: Northwestern Polytechnical University; UAT: Pant University of Agriculture and Technology; IITR: Indian Institute of Technology Roorkee; BIT: Beijing Institute of Technology; CAS: Chinese Academy of Sciences; SJU: Southwest Jiaotong University.

Figure 5. Relation Productivity/Impact; P: Publications, TC/TP: Citations per year

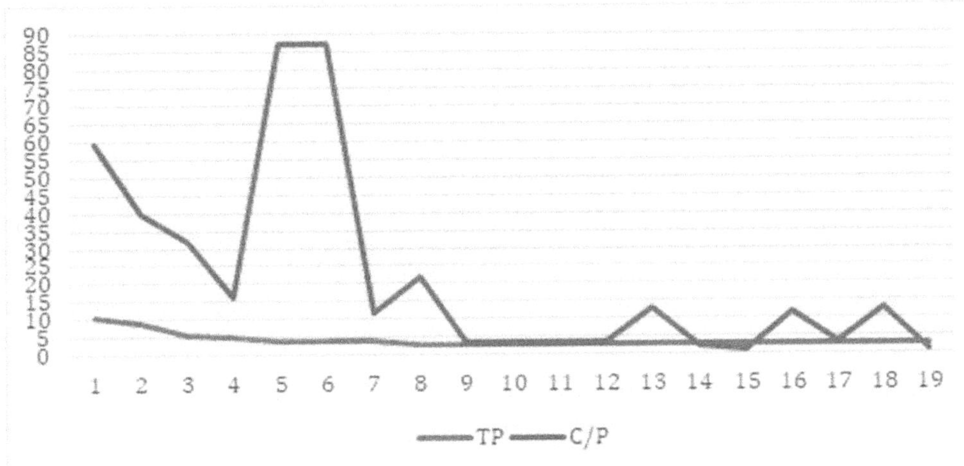

Figure 6a. Bibliometric mapping of authors with at least 25 bibliographic couplings, a threshold of 124, and the 3000 most representative connections.

five intervals considering those with 1 to 9 citations, 10 to 24 citations, 25 to 49 citations, 50 to 99 citations, and articles with 100 citations or more.

In Figure 8, the productivity and impact of the journal are presented. The three most productive journals are Fuzzy Sets and Systems, Microelectronics and Reliability & the Chinese Journal of Aeronautics. The three journals with the most significant impact are Reliability Engineering System Safety, Fuzzy Sets, and Systems & IEEE Transactions on Reliability.

Citations, and the way they connect with others were analyzed to determine how journals are structured in the field of fuzzy reliability theory. The first analysis is focused on bibliometric coupling (Bibcoup) (see Figure 9a, Figure 9b). Figure 8a shows the connection between journals and the influence of research in the field. In this analysis, six relevant clusters are identified, being Fuzzy Sets and Systems,

Figure 6b. Bibliometric mapping of authors with at least 25 bibliographic couplings, a threshold of 124, and the 1000 most representative connections.

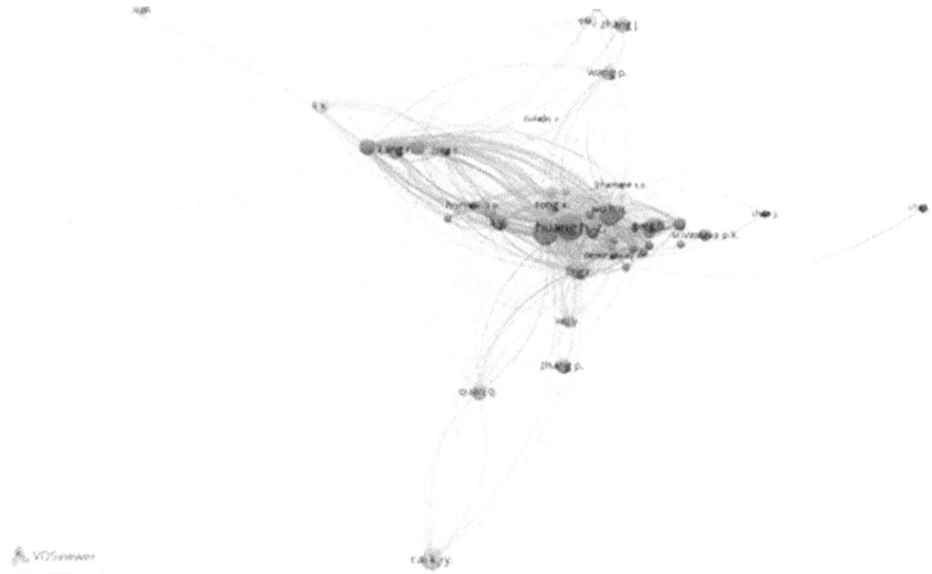

Microelectronics and Reliability, International Journal of System Science, IEEE Transactions on Reliability, International Journal of mathematical, Engineering and Management Sciences, and Journal of Multiple-Valued Logic and Soft Computing the most significant journals in the clusters. Figure 9b shows the connections between the journals in the core, which are not visible in Figure 9a.

The second analysis identifies co-citations between journals (see Figure 10a, Figure 10b). Figure 10a shows the influence of research in detail. Co-citation helps identify the possibility of documents B and C being cited in document A dealing with the same topic. In this analysis, fifteen relevant clusters are identified. Results show that articles published in the journals Fuzzy Sets and Systems, Reliability Engineering and System Safety, and IEEE Transactions on Reliability, are the most co-cited on the topic of fuzzy reliability theory. Figure 10b shows the connections among the journals in the core, which are not visible in Figure 10a.

The Most Influential Universities

Scientific publications depend on the number of researchers in a field and their productivity, as well as the number of higher education institutions working on the topic (Blanco-Mesa et al., 2017). It is within universities that researchers can pursue their interest in a scientific field. The WoS reported a total of 10 institutions, while

Table 5. The most cited journals

R	Journal	TP W	TC W	C/P W	TP S	TC S	C/P S	W &S	≥ 100	≥ 50	≥ 25	≥ 10	≥ 1
1	FSS	5	315	63.00	9	575	63.88	4	2	2	1	3	2
2	MR	2	124	62.00	4	160	40.00	2	1	-	1	-	2
3	CJA	1	4	4.00	2	55	27.50	1	-	-	1	1	-
4	CJME	1	6	6.00	1	8	8.00	1	-	-	-	-	1
5	IEEE TR	1	24	24.00	2	115	57.50	1	-	1	1	-	-
6	JH	1	1	1.00	1	1	1.00	1	-	-	-	-	1
7	JIFS	1	6	6.00	1	6	6.00	1	-	-	-	-	1
8	JMVLSC	1	5	5.00	-	-	-	-	-	-	-	-	1
9	MR	1	4	4.00	-	-	-	-	-	-	-	-	1
10	RESS	1	110	110.00	2	246	123.00	1	2	-	-	-	-
11	SC	1	42	42.00	2	42	21.00	1	-	-	1	-	-
12	SERRA	1	10	10.00	-	-	-		-	-	-	1	-
13	JMS	-	-	-	5	33	6.60		-	-	-	1	3
14	JBUAA	-	-	-	3	7	2.33		-	-	-	-	2
15	IJMEMS	-	-	-	3	14	4.66		-	-	-	-	2
16	IJSS	-	-	-	3	125	41.66		-	1	-	2	-
17	MESA	-	-	-	3	5	1.66		-	-	-	-	3
18	AMM	-	-	-	2	1	0.50		-	-	-	-	1
19	IJ	-	-	-	2	3	1.50		-	-	-	-	1
20	IJPE	-	-	-	2	1	0.50		-	-	-	-	1
21	IJRS	-	-	-	2	22	11.00		-	-	-	1	1
22	JME	-	-	-	2	70	35.00		-	-	1	1	-
23	JMVLSC	-	-	-	2	15	7.50		-	-	-	1	1
24	JRR	-	-	-	2	67	33.50		-	1	-	-	1
25	SCI	-	-	-	2	11	5.50		-	-	-	1	1

Source: Developed by authors, based on WoS & Scopus 2020. Abbreviations: R: Ranking; TPW: Total publications in WoS; TCW: Total citations in WoS; C/PW: Citations per year in WoS; TPS: Total publications in Scopus; TCS: Total citations in Scopus; C/PS: Citations per year in Scopus; W&S: Number of articles published by the Journal that are in both the Wos and Scopus; ≥ 100, ≥50 ≥ 25, ≥10 ≥ 1 = Number of journals with equal or more than 100, 50, 25, 10 and 1 citations. FSS: Fuzzy Sets and Systems; MR: Microelectronics and Reliability; CJA: Chinese Journal of Aeronautics; CJE: Chinese Journal of Mechanical Engineering; IEEE TR: IEEE Transactions on Reliability; JH: Journal of Hydroinformatics; JIFS: Journal of Intelligent Fuzzy Systems; JMVLSC: Journal of Multiple-Valued Logic and Soft Computing; MR: Microelectronics Reliability; RESS: Reliability Engineering System Safety; SC: Soft Computing; SERRA: Stochastic Environmental Research and Risk Assessment, JMS: Journal of Mechanical Strength; JBUAA: Journal of Beijing University of Aeronautics and Astronautics; IJMEMS: International Journal of Mathematical, Engineering and Management Sciences; IJSS: International Journal of Systems Science; MESA: Mathematics in Engineering, Science and Aerospace; AMM: Applied Mechanics and Materials; IJ: Information Japan; International Journal of Performability Engineering; IJRS: International Journal of Reliability and Safety; JME: Journal of Mechanical Engineering; Journal of Multiple-Valued Logic and Soft Computing; JRR. Journal of Risk and Reliability; SCI: Studies in Computational Intelligence.

Figure 7. Total of publications by journal in the WoS and Scopus

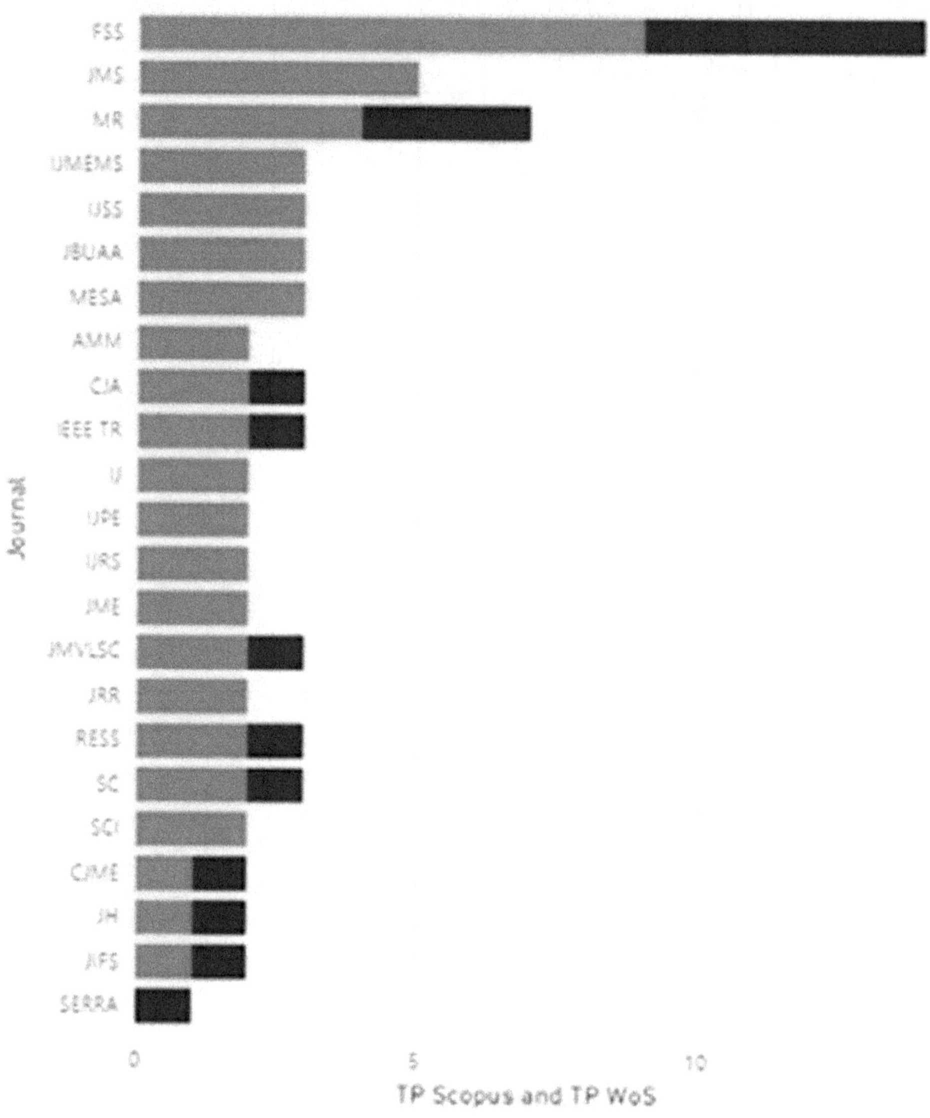

Scopus reported 117. All the institutions reported by both databases, with at least two articles and one citation, were considered.

Table 6 shows the top universities in terms of relevant publications in the area of fuzzy reliability theory. Universities are ranked considering the number of publications, the total number of citations was also considered, along with the

Figure 8. Relation Productivity / Impact. TP: Publications, C/P: Citations per year.

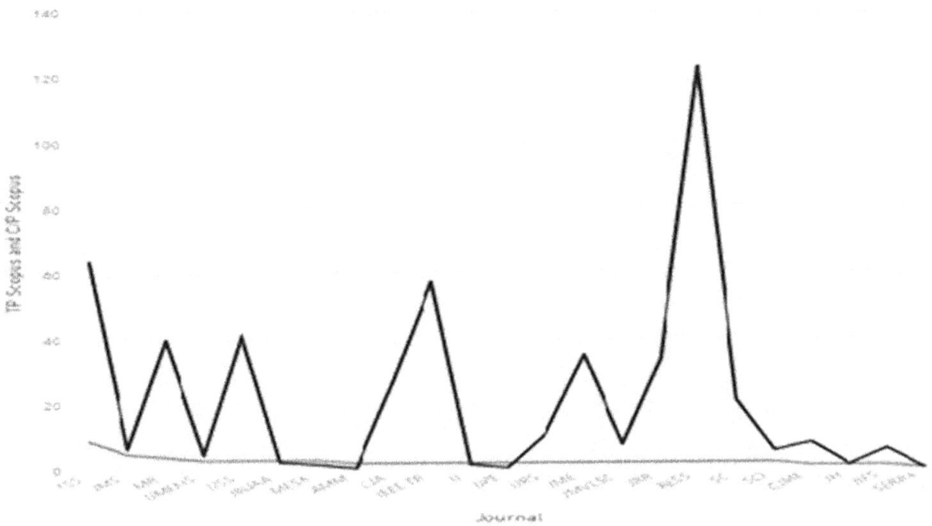

Figure 9a. Bibliometric mapping of journals. Bibliographic coupling with a threshold of 60 and considering the 1000 most influential connections.

Figure 9b. Bibliometric mapping of journals. Bibliographic coupling with a threshold of 60 and considering the 1000 most influential connections.

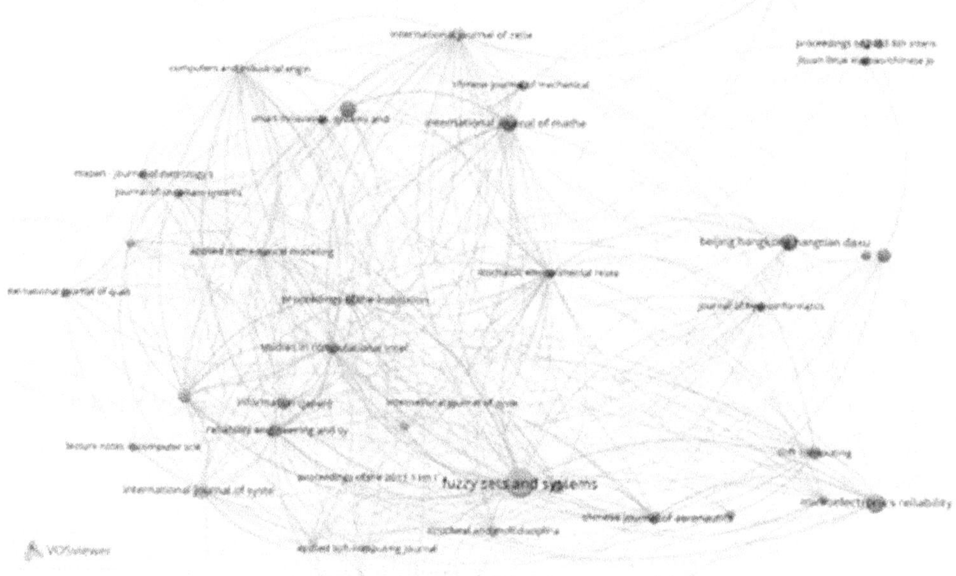

Figure 10a. Bibliometric mapping of journal co-citations with a threshold of 541 and considering the 1000 most influential connections.

Figure 10b. Bibliometric mapping of journal co-citations with a threshold of 541 and considering the 1000 most influential connections.

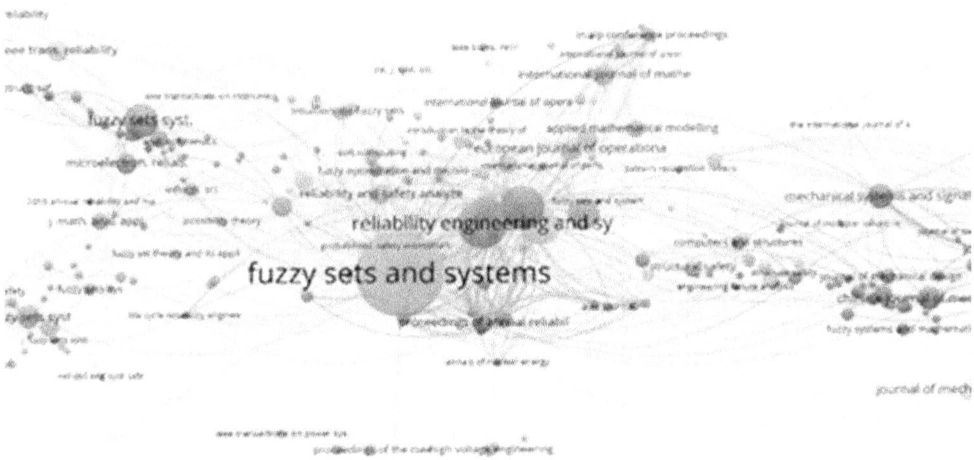

country, the number of publications reported by each institution, and the average citations per year.

The most productive universities (see Figure 11) are Beihang University, the University of Electronic Science and Technology of China, and the Indian Institute of Technology Roorkee. In contrast, those with the highest impact are the University of Alberta, the University of Alabama at Birmingham, and the Dalian University of Technology.

61.54% of the most relevant universities presented in Figure 11 are based in China and 23.08% in India.

A bibliometric Bibcoup analysis was performed to determine how universities are structured around research in the field of fuzzy reliability theory (see Figure 12) and the connection between citation and institutions (see Figure 13). In the first analysis, six significant clusters were identified. The most relevant institutions in the clusters are Beihang University, the School of Reliability and Systems Engineering, Dalian University of Technology, University of Electronic Science and Technology of China, the Complutense University, North Dakota State University, and the Indian Institute of Technology Roorkee. In a secondary analysis, ten clusters were identified. The universities leading these groups are the University of Aeronautics and Astronautics, the Indian Institute of Technology Roorkee, Beihang University, and the University of Electronic Science and Technology of China.

Table 6. The most influential universities

R	I	C	TP W	TC W	C/P W	TP S	TC S	C/P S	W & S	≥ 100	≥ 50	≥ 25	≥ 10	≥ 1
1	BU	China	9	421	47	20	743	37.1	9	3	1	4	2	7
2	BIASE	China	1	6	6	1	8	8.0	1	-	-	-	-	1
3	DUT	China	1	110	110	4	164	41.0	1	1	-	-	1	1
4	FTA	Russia	1	62	62	-	-	-	-	-	1	-	-	-
5	NITC	India	1	1	1	1	1	1.0	1	-	-	-	-	1
6	SJTU	China	1	6	6	1	8	8.0	1	-	-	-	-	1
7	SU	Iran	1	10	10	1	10	10.0	1	-	-	-	1	-
8	UAB	USA	1	42	42	1	42	42.0	1	-	-	1	-	-
9	UA	Canada	1	110	110	2	141	70.5		1	-	-	-	1
10	UESTC	China	1	5	5	8	194	24.2	1	-	2	-	2	4
11	IITR	India	-	-	-	6	91	15.1	-	-	-	2	1	3
12	NPU	China	-	-	-	4	11	2.7	-	-	-	-	-	2
13	WPCSE	China	-	-	-	3	11	3.6	-	-	-	-	-	3
14	MEC	China	-	-	-	3	23	7.6	-	-	-	-	1	1
15	GBPUAT	India	-	-	-	3	7	2.3	-	-	-	-	-	2
16	SPUC	China	-	-	-	3	4	1.3	-	-	-	-	-	2
17	GEDU	India	-	-	-	3	9	3.0	-	-	-	-	-	2
18	NITK	India	-	-	-	2	8	4.0	-	-	-	-	-	1
19	HUT	China	-	-	-	2	5	2.5	-	-	-	-	-	1
20	XJU	China	-	-	-	2	70	35.0	-	-	-	1	1	-
21	SCUT	China	-	-	-	2	1	0.5	-	-	-	-	-	1
22	XU	China	-	-	-	2	4	2.0	-	-	-	-	-	2
23	LTU	China	-	-	-	2	1	0.5	-	-	-	-	-	1
24	JIIT	India	-	-	-	2	4	2.0	-	-	-	-	-	2
25	SKLMSE	China	-	-	-	2	70	35.0	-	-	-	1	1	-
26	TUST	China	-	-	-	2	3	1.5	-	-	-	-	-	1

Source: Developed by authors, based on WoS & Scopus 2020. Abbreviations: R: Ranking; I: Institution; C: Country, TPW: Total publications in WoS; TCW: Total citations in WoS; C/PW: Citations per year in WoS; TPS: Total publications in Scopus; TCS: Total citations in Scopus; C/PS: Citations per year in Scopus; W&S: Number of articles published by higher education institutions reported in the Web of Science and Scopus; ≥ 100, ≥50 ≥ 25, ≥10 ≥ 1 = Number of journals with equal or more than 100, 50, 25, 10 and 1 citations. BU: Beihang University; BIASE: Beijing Institute of Aerospace Systems Engineering; DUT: Dalian University of Technology; FTA: Forest Tech Academy; NITC: National Institute of Technology Calicut; SJTU: Shanghai Jiao Tong University; SU: Shiraz University; UAB: University of Alabama at Birmingham; UA: University of Alberta; UESTC: University of Electronic Science and Technology of China; IITR: Indian Institute of Technology Roorkee; NPU: Northwestern Polytechnical University; WPCSE: Wuhan Polytechnic College of Software and Engineering; MEC: Ministry of Education China; GBPUAT: Govind Ballabh Pant University of Agriculture and Technology; SPUC: Southwest Petroleum University China; GEDU: Graphic Era Deemed to be University; NITK: National Institute of Technology of Kurukshetra; HUT: Hefei University of Technology; XJU: Xi'an Jiaotong University; SCUT: South China University of Technology; XU: Xidian University; LTU: Liaoning Technical University; JIIT: Jaypee Institute of Information Technology; SKLMSE: State Key Laboratory for Manufacturing Systems Engineering; TUST: Tianjin University of Science & Technology.

Figure 11. Productivity and impact ratio; TP: Publications, C/P: Citations per year

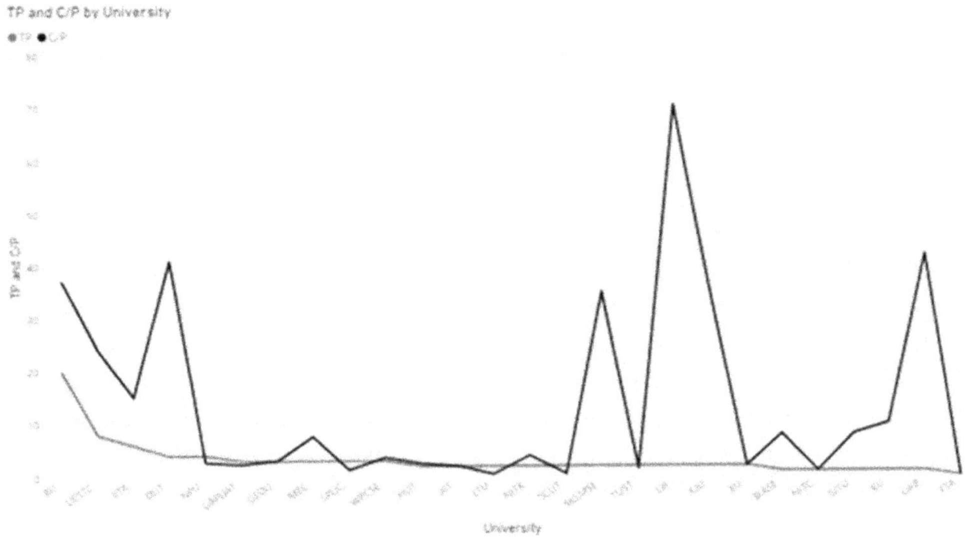

Figure 12. Bibliometric mapping of universities with more than 25 bibliographic couplings considering the 1000 most influential connections and a threshold of 34.

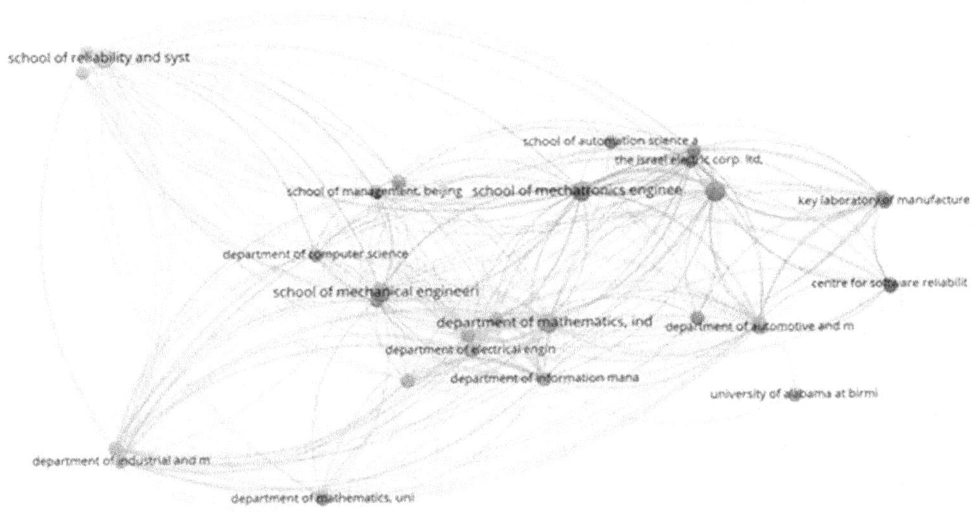

The Most Influential Countries

Analyzing countries in which publications are made is another fundamental criteria by analyzing research in a specific field. Table 7 and Table 8 show the countries

Figure 13. Bibliometric mapping of universities citations with a threshold of 75 and the 1000 most influential connections.

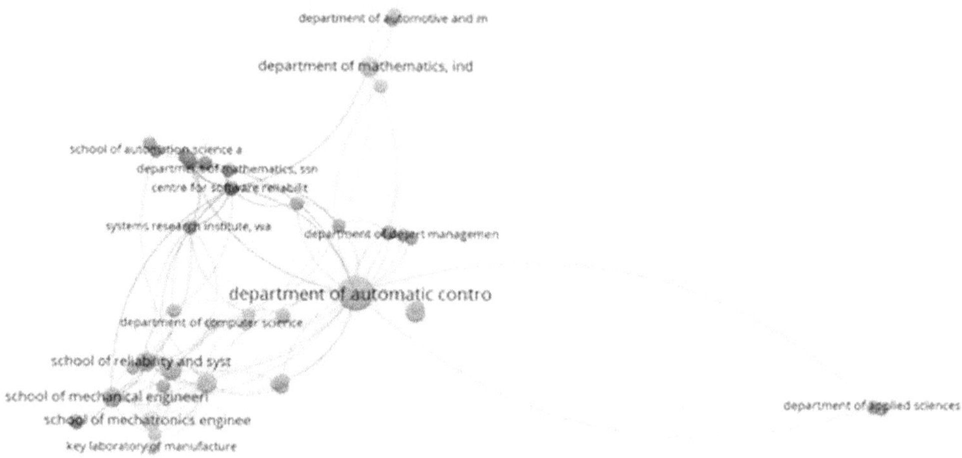

with the highest number of publications on the topic. The countries have been ranked according to the number of publications (TC) considering the WoS and Scopus. The total number of citations is also included (C/P) along with the number of publications that have between one and 9 citations, 10 and 24, 25 to 49, 50 to 99, and 100 citations and over. Additional information includes the number of country-based publications considering total population (P/P_0) and the average number of citations per country, based on the total population in that country (C/P_0).

The productivity/impact connection considering TP and C/P can be observed in Figure 14. China is the world leader in scientific publications on fuzzy reliability theory, followed by India and Canada. In terms of impact, the United Kingdom, Kuwait, and Russia are the leading countries, which can be corroborated with the bibliometric map, which establishes bibliographic coupling (see Figure 16). Nevertheless, when analyzing the productivity /impact connection, considering P/Po and $C/P_{0,}$ taking into account the total population of inhabitants per country, the countries with the highest productivity are Lithuania, Kuwait, and Taiwan, while those with the highest impact are Kuwait, Israel and UK (see Figure 15).

Overall Analysis of the Full Results

One of the most significant results is the correlation between authors, universities, and countries, as can be observed in Figure 17.

Table 7. The most productive and influential countries

R	Country	TP W	TC W	C/P W	TP S	TC S	C/P S	W& S	≥ 100	≥ 50	≥ 25	≥ 10	≥ 1
1	Peoples R. China	12	542	45.1	89	1268	14.2	12	4	3	5	6	40
2	Canada	1	110	110.0	4	141	35.2	1	1	-	-	-	1
3	India	1	1	1.0	22	198	9.0	1	-	-	4	2	13
4	Iran	1	10	10.0	1	10	10.0	1				1	
5	Russia	1	62	62.0	1	68	68.0	1	-	1	-	-	-
6	USA	1	42	42.0	2	61	30.5	1	-	-	1	1	-
7	Poland	-	-		3	10	3.3	-	-	-	-	1	-
8	Taiwan	-	-		3	82	27.3	-	-	1	1	-	1
9	France	-	-		2	131	65.5	-	-	1	1	-	-
10	Italy	-	-		2	65	32.5	-	-	-	1	1	-
11	United Kingdom	-	-		2	244	122.0	-	2	-	-	-	-
12	Israel	-	-		1	63	63.0	-	-	1	-	-	-
13	Kuwait	-	-		1	106	106.0	-	1	-	-	-	-
14	Lithuania	-	-		1	-	-	-	-	-	-	-	-
15	South Korea	-	-		1	5	5.0	-	-	-	-	-	1
16	Spain	-	-		1	19	19.0	-	-	-	-	1	-
17	Turkey	-	-		1	4	4.0	-	-	-	-	-	1

Source: Developed by authors, based on WoS & Scopus 2020. Abbreviations: R: Ranking; TPW: Total publications in WoS; TCW: Total citations in WoS; C/PW: Citations per year in WoS; TPS: Total publications in Scopus; TCS: Total citations in Scopus; C/PS: Citations per year in Scopus; W&S: Number of articles published by the countries reported in the WoS and Scopus; ≥ 100, ≥50 ≥ 25, ≥10 ≥ 1 = Number of journals with equal or more than 100, 50, 25, 10 and 1 citations.

Figure 17 shows the total publications by country (clear blue), by university, grouped by country (dark blue) and by author, grouped by country (red). There is a clear leadership of Asian countries, specifically China and India, which have been acknowledged as the pioneers in the study of fuzzy reliability theory, as observed in previous sections.

An interesting result shows that there is no correlation between productivity and impact. In this result, the authors, journals, and universities were analyzed. This means that the most published authors are not those with the highest citation impact. The same is true for journals and universities.

The keyword map shows the different areas involved in fuzzy reliability theory processes. The most relevant topics include fuzzy sets, reliability, failure analysis,

Table 8. The most productive and influential countries by million inhabitants

R	Country	TP W	TC W	C/P W	TP S	TC S	C/P S	Po	TP/Po	TC/Po
1	Peoples R. China	12	542	45.1	89	1268	14.2	1395.2	0.06	0.908
2	Canada	1	110	110.0	4	141	35.2	37.3	0.107	3.77
3	India	1	1	1.0	22	198	9.0	1375.8	0.015	0.14
4	Iran	1	10	10.0	1	10	10.0	81.8	0.012	0.12
5	Russia	1	62	62.0	1	68	68.0	147.0	0.006	0.46
6	USA	1	42	42.0	2	61	30.5	329.0	0.006	0.18
7	Poland	-	-		3	10	3.3	38.4	0.078	0.26
8	Taiwan	-	-		3	82	27.3	23.5	0.127	3.47
9	France	-	-		2	131	65.5	65.2	0.030	2.00
10	Italy	-	-		2	65	32.5	60.6	0.033	1.07
11	United Kingdom	-	-		2	244	122.0	66.6	0.030	3.66
12	Israel	-	-		1	63	63.0	8.9	0.111	7.03
13	Kuwait	-	-		1	106	106.0	4.1	0.242	25.66
14	Lithuania	-	-		1	-	-	2.7	0.358	0.00
15	South Korea	-	-		1	5	5.0	51.8	0.019	0.09
16	Spain	-	-		1	19	19.0	46.7	0.021	0.40
17	Turkey	-	-		1	4	4.0	81.8	0.012	0.04

Source: Developed by authors, based on WoS & Scopus 2020. Abbreviations: R: Ranking; TPW: Total publications in WoS; TCW: Total citations in WoS; C/PW: Citations per year in WoS; TPS: Total publications in Scopus; TCS: Total citations in Scopus; C/PS: Citations per year in Scopus; $P_{o:}$ Million inhabitants, P/P_o and C/P_o = Papers and cites per million inhabitants

Figure 14. Productivity and impact ratio; TP: Publications, TC/TP: Citations per year

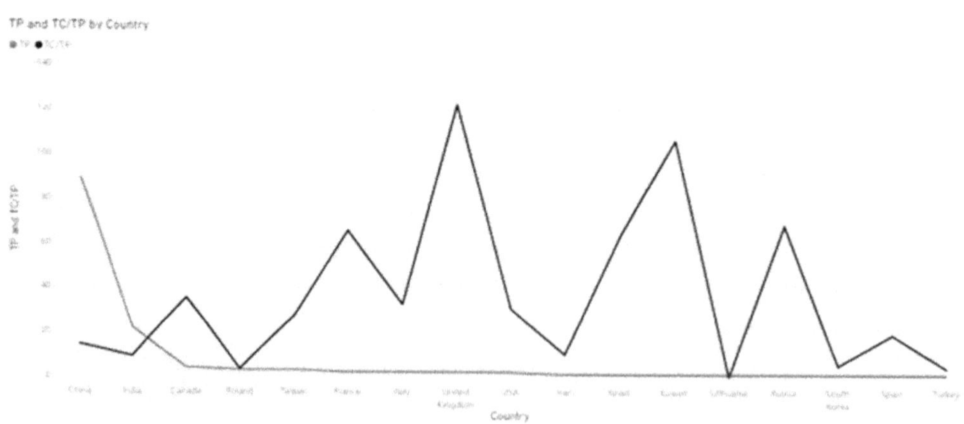

Figure 15. Productivity and impact ratio; TP/P_0: Publications per habitant, TC/P_0: Citations per habitant

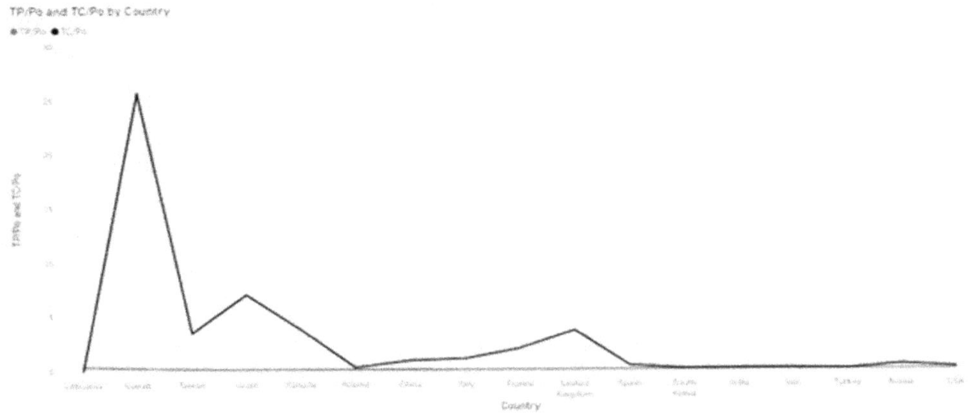

Figure 16. Bibliometric mapping bibliographic couplings of countries considering the 1000 most influential connections.

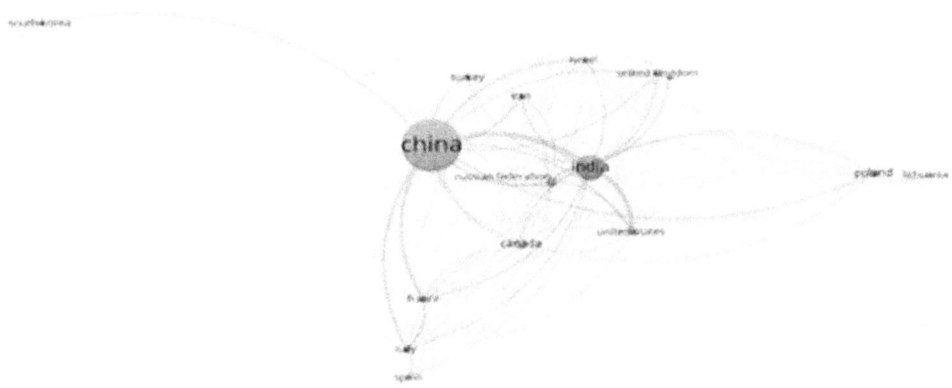

uncertainty analysis, system reliability, reliability theory, probability, membership functions, numerical methods, and others. These topics have a connection with high impact journals in which the main lines of research are computer science, artificial intelligence, mathematics, fuzzy logic, engineering, safety, risk, reliability, aerospace, and others.

Figure 17. Correlation between authors, universities, and countries

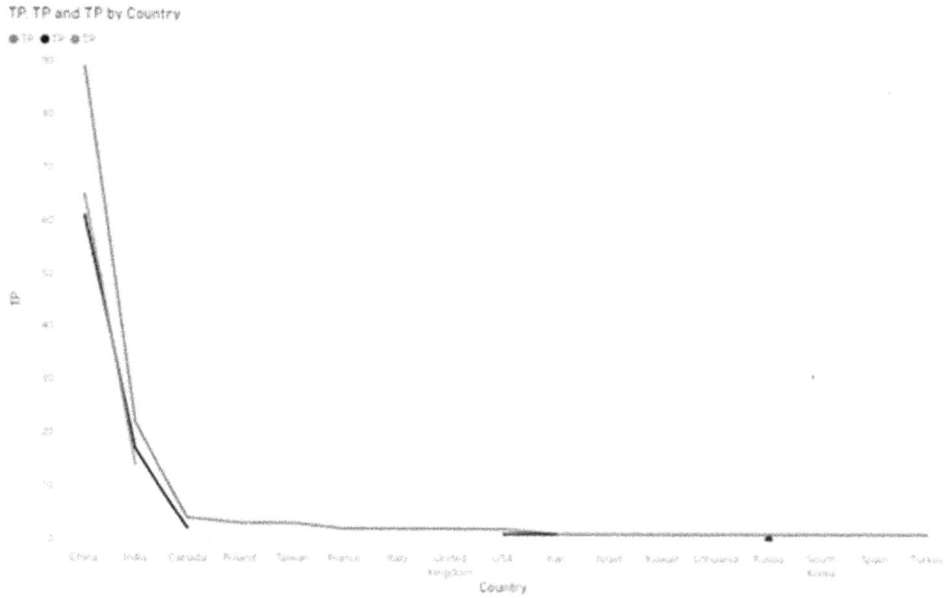

SOLUTIONS AND RECOMMENDATIONS

Bibliometric analysis is a powerful tool in analyzing the potential of a topic in terms of the interest that it creates in scientific fields. It presents interesting solutions to the problem of understanding how scientific research on the topic of fuzzy reliability theory is structured. Still, it does not come without some caveats, and the research here presented has some limitations; therefore, careful selection of the data to be analyzed is recommended. The publications selected for analysis came from the WoS and Scopus, still, other databases are available and if used, can provide different results.

Additional recommendations include the consideration of the formats in which the materials are presented, the depth and breadth of the documents, the geographical and language coverage as well as the depth of backfiles regarding the topics being analyzed.

FUTURE RESEARCH DIRECTIONS

The analysis and information presented in this chapter can be considered for future research efforts regarding Fuzzy Reliability Theory to be focused in more specific ways. For example, the keywords show the different areas of science that are most interested in the contributions of these topics. Therefore, directing research efforts in these areas is important. Also, identifying the countries, universities and leading researchers in the field allows to identify and promote collaborative research groups to enhance this area of knowledge.

CONCLUSION

A bibliometric analysis was conducted to identify how research regarding the concept of fuzzy reliability theory has evolved. The process that was performed was explained throughout the research. Different analyses were conducted to identify the number of published articles, the general structure of citations, and the most cited papers in the field of fuzzy reliability theory. Relevant publications in the field were identified along with most productive authors, countries, and institutions, as well as other valuable bibliometric indicators. This helped create a general context where science is leading in terms of fuzzy reliability theory. Visualizations were made to identify specific variable clusters and provide more detailed and graphic information about how it connects with other notions.

It can be concluded that the concepts that were analyzed are fairly recent since the first articles are dated from 1991. This is an area of sciences that has not been fully studied, with significant areas of opportunity for future research, deepening ongoing studies that would be of interest to researchers, universities, and countries from other regions of the world.

This research analyzed 117 publications connected to the concept of fuzzy reliability theory. The analysis was performed from different angles, combining bibliometric analysis with information collected from ISI Web of Knowledge, Scopus, and visual tools such as VOSviewer and Excel charts. Results show that this field has been receiving increased attention.

REFERENCES

Bamrungsetthapong, W., & Pongpullponsak, A. (2014). Parameter interval estimation of system reliability for repairable multistate series-parallel system with fuzzy data. *The Scientific World Journal*, *2014*, 1–10. doi:10.1155/2014/275374 PMID:24987728

Bhamare, S. S., Yedav, O. P., & Rathore, A. (2007). Evolution of reliability engineering discipline over the last six decades: A comprehensive review. *International Journal of Reliability and Safety*, *1*(4), 377–410. doi:10.1504/IJRS.2007.016256

Buckley, J. J. (2004). Uncertain probabilities III. The continuous case. *Soft Computing*, *8*(3), 200–206. doi:10.100700500-003-0263-5

Cai, K. Y. (1991). Fuzzy reliability theories. *Fuzzy Sets and Systems*, *40*(3), 510–511. doi:10.1016/0165-0114(91)90177-R

Cai, K. Y. (1996). System failure engineering and fuzzy methodology an introductory overview. *Fuzzy Sets and Systems*, *83*(2), 113–133. doi:10.1016/0165-0114(95)00385-1

Cai, K. Y. (1996). *Introduction on Fuzzy Reliability*. Kluwer Academic Publishers. doi:10.1007/978-1-4613-1403-5

Cai, K. Y., Wen, C. Y., & Zhang, M. L. (1991). Fuzzy variables as a basis for a theory of fuzzy reliability in the possibility context. *Fuzzy Sets and Systems*, *42*(2), 145–172. doi:10.1016/0165-0114(91)90143-E

Cai, K. Y., Wen, C. Y., & Zhang, M. L. (1991). Posbist reliability behavior of typical systems with 2 types of failure. *Fuzzy Sets and Systems*, *43*(1), 17–32. doi:10.1016/0165-0114(91)90018-L

Cai, K. Y., Wen, C. Y., & Zhang, M. L. (1993). Fuzzy states as a basis for a theory of fuzzy reliability. *Microelectronics and Reliability*, *33*(15), 2253–2263. doi:10.1016/0026-2714(93)90065-7

Cai, K. Y., Wen, C. Y., & Zhang, M. L. (1995). Posbist reliability behavior of Fault-Tolerant systems. *Microelectronics and Reliability*, *35*(1), 4956.

Cai, K. Y., Wen, C. Y., & Zhang, M. L. (1995). Mixture-Models in Profust Reliability Theory. *Microelectronics and Reliability*, *35*(6), 985–993. doi:10.1016/0026-2714(94)00058-V

Cappelle, B., & Kerre, E. E. (1993). On a possibilistic approach to reliability theory. In *1993 (2nd) International Symposium on Uncertainty Modeling and Analysis* (pp. 415-418). IEEE. 10.1109/ISUMA.1993.366736

Chaube, S., & Singh, S. B. (2016). Fuzzy reliability theory based on membership function. *International Journal of Mathematical. Engineering and Management Sciences*, *1*(1), 34–40.

Chen, K., Ma, C., Zheng, M., & Gao, F. (2015). Optimization and Mechanical Accuracy Reliability of a New Type of Forging Manipulator. *Chinese Journal of Mechanical Engineering*, *28*(2), 236–248. doi:10.3901/CJME.2014.1128.173

Cobo, M., López-Herrera, A., Herrera-Viedma, E., & Herrera, F. (2011). An approach for detection, quantifying and visualizing the evolution of a research field: A practical application to the fuzzy sets theory field. *Journal of Informetrics*, *5*(1), 146–166. doi:10.1016/j.joi.2010.10.002

Cutello, V., Montero, J., & Yáñez, J. (1996). Structure functions with fuzzy states. *Fuzzy Sets and Systems*, *83*(2), 189–202. doi:10.1016/0165-0114(95)00390-8

Dong, Y. G., Chen, X. Z., Zhao, X. D., & Quan, Z. W. (2005). General approach for fuzzy reliability analysis based on the fuzzy probability theory. *Journal of Computational Mechanics*, *22*(3), 281–286.

Elsevier. (2020). *Elsevier Scopus*. https://www.elsevier.com/es-mx/solutions/scopus

Fahimnia, B., Sarkis, J., & Davarzani, H. (2015). Green supply chain management: A review and bibliometric analysis. *International Journal of Production Economics*, *162*, 101–114. doi:10.1016/j.ijpe.2015.01.003

Gao, C. (2009). Fuzzy availability analysis of a repairable consecutive-2-out-of-3: F system. *IEEE International Conference on Grey Systems and Intelligent Services*, 434-437.

Garfield, E. (1972). Citation analysis as a tool in journal evaluation. *Science*, *178*(4060), 417–479. doi:10.1126cience.178.4060.471 PMID:5079701

Garfield, E. (Ed.). (1979). *SCI journal citation reports. A bibliometric analysis of science journals in the ISI data base*. Institute for Scientific Information.

Garg, H. (2014). Reliability, Availability and Maintainability Analysis of Industrial Systems Using PSO and Fuzzy Methodology. *Journal of Metrology Society of India*, *29*(2), 115–129. doi:10.100712647-013-0081-x

Garg, H., & Sharma, S. P. (2012). Behavior analysis of synthesis unit in fertilizer plant. *International Journal of Quality & Reliability Management*, *29*(2), 217–232. doi:10.1108/02656711211199928

Guan, J. C., & Wu, Y. Q. (2006). Repairable consecutive -k-out-of-n: F system with fuzzy sets. *Fuzzy Sets and Systems*, *157*(1), 121–142. doi:10.1016/j.fss.2005.05.025

He, Z., Cao, H., Zi, Y., & Li, B. (2014). Developments and thoughts on operational reliability assessment of mechanical equipment. *Jixie Gongcheng Xuebao*, *50*(2), 171–186. doi:10.3901/JME.2014.02.171

Hryniewicz, O. (2007). Fuzzy sets in the evaluation of reliability. *Studies in Computational Intelligence*, *40*, 363–386. doi:10.1007/978-3-540-37372-8_13

Huang, H., Sun, Z., Guo, D., Zhou, F., & Tian, Z. (2001). Fuzzy reliability analysis in the case of random stress and fuzzy strength. *Journal of Mechanical Strength, 23*(3), 305–307.

Huang, H. Z., & He, L. (2008). New approaches to system analysis and design: a review. In *Handbook of Performability Engineering* (pp. 477–498). Springer. doi:10.1007/978-1-84800-131-2_31

Huang, H. Z., Li, Y., & Liu, Y. (2010). Posbist Reliability Theory of k-out-of-n: G System. *Journal of Multiple-Valued Logic and Soft Computing*, *16*(1-2), 45–63.

Huang, H. Z., Tong, X., & Zuo, M. J. (2004). Posbits fault tree analysis of coherent systems. *Reliability Engineering & System Safety*, *84*(2), 141–148. doi:10.1016/j.ress.2003.11.002

Kacprzyk, J., & Onisawa, T. (1995). *Reliability and safety analyses under fuzziness*. Physica-Verlag.

Kang, R., Zhang, Q., Zeng, Z., Zio, E., & Li, X. (2016). Measuring reliability under epistemic uncertainty: Review on non-probabilistic reliability metric. *Chinese Journal of Aeronautics*, *29*(3), 571–579. doi:10.1016/j.cja.2016.04.004

Kerre, E., Onisawa, T., Cappelle, B., & Gazdik, I. (1998). Reliability. In R. Słowiński (Ed.), The Handbooks of Fuzzy Sets Series: Vol. 1. *Fuzzy Sets in Decision Analysis, Operations Research and Statistics*. Springer. doi:10.1007/978-1-4615-5645-9_12

Knezevic, J., & Odoom, E. R. (2001). Reliability modeling of repairable systems using Petri nets and fuzzy Lambda-Tau methodology. *Reliability Engineering & System Safety*, *73*(1), 1–17. doi:10.1016/S0951-8320(01)00017-5

Kumar, A., & Ram, M. (2018). System reliability analysis based on Weibull distribution and hesitant fuzzy set. *International Journal of Mathematical. Engineering and Management Sciences*, *3*(4), 513–521.

Kumar, K., & Kumar, P. (2011). Fuzzy availability modeling and analysis of biscuit manufacturing plant: A case study. *International Journal of Systems Assurance Engineering and Management*, *2*(3), 193–204. doi:10.100713198-011-0076-3

Kumar, M. (2014). Applying weakest-t-norm-based approximate intuitionistic fuzzy arithmetic operations on different types of intuitionistic fuzzy numbers to evaluate reliability of PCBE fault. *Applied Soft Computing*, *23*, 387–406. doi:10.1016/j.asoc.2014.06.036

Kumar, M., Prasad Yadav, S., & Kumar, S. (2013). Fuzzy system reliability evaluation using time-dependent intuitionistic fuzzy set. *International Journal of Systems Science*, *44*(1), 50–66. doi:10.1080/00207721.2011.581393

Kumar, M., & Yadav, S. P. (2012). A novel approach for analyzing fuzzy system reliability using different types of intuitionistic fuzzy failure rates of components. *ISA Transactions*, *51*(2), 115–129. doi:10.1016/j.isatra.2011.10.002 PMID:22134065

Li, Z., Zhang, Q., Fan, F., & Shen, S. (2020). A method for calculating uncomfortable rates of people due to vertical floor vibrations using corresponding assessment values based on fuzzy reliability theory. *Journal of Building Engineering*, *28*, 101061. doi:10.1016/j.jobe.2019.101061

Ling, C., Lu, Z., Feng, H., & Sun, B. (2019). Efficient numerical simulation methods for estimating fuzzy failure probability-based importance measure indices. *Structural and Multidisciplinary Optimization*, *59*(2), 577–593. doi:10.100700158-018-2085-1

Liu, Y., & Huang, H. Z. (2010). Reliability assessments for fuzzy multi-state systems. *International Journal of Systems Science*, *41*(4), 365–379. doi:10.1080/00207720903042939

Liu, Y., & Huang, H. Z. (2011). Optimal replacement policy for fuzzy multi-state elements. *Journal of Multiple-Valued Logic and Soft Computing*, *17*(1), 69–92.

Liu, Y., Huang, H. Z., & Levitin, G. (2008). Reliability and performance assessment for fuzzy multi-state elements. *Journal of Risk and Reliability*, *222*(4), 675–686. doi:10.1243/1748006XJRR180

Liu, Y., Huang, H. Z., & Ling, D. (2013). Reliability prediction for evolutionary product in the conceptual design phase using a neural network-based fuzzy synthetic assessment. *International Journal of Systems Science*, *44*(3), 545–555. doi:10.1080/00207721.2011.617887

Ma, C., Xiao, X., Yang, H., & Li, H. (2012). Uncertain evaluation measure system and model for power grid catastrophic accident. *Zhongguo Dianji Gongcheng Xuebao*, *32*(10), 118–125.

Martínez, M. A., Díaz, M., Lima, A. I., Herrera, M., & Herrera-Viedma, E. (2014). A bibliometric análisis of Spanish production of Social Work category according to the Journal Citation Report. *Cuadernos de Trabajo Social*, *27*(2), 429–438.

Mirakbari, M., & Ganji, A. (2010). Reliability analysis of a rangeland system: The application of Profust Theory. *Stochastic Environmental Research and Risk Assessment*, *24*(3), 399–409. doi:10.100700477-009-0329-8

Nair, S. J., & Sasikumar, K. (2019). Fuzzy reliability-based optimization of a hydropower reservoir. *Journal of Hydroinformatics*, *21*(2), 308–317. doi:10.2166/hydro.2019.078

Nikolaidis, E., Chen, S., Cudney, H., Haftka, R. T., & Rosca, R. (2004). Comparison of probability and possibility for design against catastrophic failure under uncertainty. *Journal of Mechanical Design*, *126*(3), 386–394. doi:10.1115/1.1701878

Pinto-López, I. N., Montaudon-Tomas, C. M., & De la Vega Meneses, J. G. (2019). Bibliometrics as a robust analytical tool for research in higher education. *Conference proceedings ICERI 2019 12th International Conference of Education Research and Innovation*, 1106-1118. 10.21125/iceri.2019.0339

Pinto-López, I. N., Montaudon-Tomas, C. M., & Yáñez-Moneda, A. L. (2020). Bibliometric Analysis as a tool for Research-Based Learning (RBL). *Conference proceedings INTED 2020 14th International Technology, Education, and Development Conference*, 7108-7116.

Reyes, A. D., Olate, Y., & Godoy, C. (2019). Análisis bibliométrico de la producción científica de la Revista Actividad Física y Ciencias durante el período 2009-2018. *Revista Ciencias de la Actividad Física UCM*, *20*(2), 1–25. doi:10.29035/rcaf.20.2.9

Rotshtein, A., & Shtovba, S. (1997). *Fuzzy Reliability Analysis of Algorithmic Processes*. Continent-PRIM.

Simon, C., & Weber, P. (2009). Evidential networks for reliability analysis and performance evaluation of systems with imprecise knowledge. *IEEE Transactions on Reliability*, *58*(1), 69–87. doi:10.1109/TR.2008.2011868

Sun, C., He, Z., Zhang, Z., Ning, X., & Zou, L. (2013). Operating reliability assessment for aero-engine based on condition monitoring information. *Jixie Gongcheng Xuebao*, *49*(6), 30–37. doi:10.3901/JME.2013.06.030

Utkin, L. V., & Gurov, S. V. (1996). A general formal approach for fuzzy reliability analysis in the possibility context. *Fuzzy Sets and Systems*, *83*(2), 203–213. doi:10.1016/0165-0114(95)00391-6

Van Eck, N. J., & Waltman, L. (2010). Software survey: VOSviewer, a computer program for bibliometric mapping. *Scientometrics*, *84*(2), 523–538. doi:10.100711192-009-0146-3 PMID:20585380

Wang, P., Zhang, J., Zhai, H., & Qiu, J. (2017). A new structural reliability index based on uncertainty theory. *Chinese Journal of Aeronautics*, *30*(4), 1451–1458. doi:10.1016/j.cja.2017.04.008

Wang, Y., Li, X., Zhang, B., & Li, M. (2013). Dynamic progressive failure evolution research on landslide under precipitation. *Journal of Hydraulic Engineering*, *44*(4), 416–425.

Wang, Y. X. (2010). Application of fuzzy mathematics to slope stability analysis. *Yantu Lixue*, *31*(9), 3000–3004.

Wang, Z., Li, Y., Huang, H. Z., & Liu, Y. (2012). Reliability analysis of structure for fuzzy safety state. *Intelligent Automation Soft Computing*, *18*(3), 215–224. doi:10.1080/10798587.2008.10643238

Wu, H. C. (2004). Fuzzy reliability estimation using Bayesian approach. *Computers & Industrial Engineering*, *46*(3), 467–493. doi:10.1016/j.cie.2004.01.009

Wu, H. C. (2006). Fuzzy Bayesian system reliability assessment based on exponential distribution. *Applied Mathematical Modelling*, *30*(6), 509–530. doi:10.1016/j.apm.2005.05.014

Xiaoning, L. (2008). Fuzzy reliability theory of determining testing pressure of steel pressure vessels. *China Petroleum Machinery, 4*. Available online: http://en.cnki.com.cn/Article_en/CJFDTotal-SYJI200804004.htm

Zadeh, L. A. (1965). Fuzzy sets. *Information and Control*, *8*(3), 338–353. doi:10.1016/S0019-9958(65)90241-X

Zhang, H., Chang, A., Wang, Q., Zhou, Q., & Liang, Y. (2016). Research and application of transformer condition evaluation model based on multi-source information. *High Voltage Apparatus*, *52*(2), 19–27.

Zhao, Z., Quan, Q., & Cai, K. Y. (2014). A modified profust-performance-reliability algorithm and its application to dynamic systems. *Journal of Intelligent & Fuzzy Systems*, *32*(1), 643–660. doi:10.3233/JIFS-152544

Zhao, Z., Quan, Q., & Cai, K. Y. (2014). A profust Reliability-Based Approach to Prognostics and Health Management. *IEEE Transactions on Reliability*, *63*(1), 26–41. doi:10.1109/TR.2014.2299111

Zhou, D., Zhang, X. F., Yang, Z., & Zhang, Y. M. (2015). Vibration reliability analysis on tractive transmission system of shearer. *Journal of the China Coal Society*, *40*(11), 2546–2551.

ADDITIONAL READING

Chraponska, A., Rzepecki, J., Isaac, C. W., Mazur, K., & Pawelczyk, M. (2020). Spectral analysis of macro-fiber composites measured vibrations of double-panel structure coupled with solenoids. *Sensors (Basel)*, *20*(12), 1–18. doi:10.339020123505 PMID:32575863

Li, Z., Zhang, Q., Fan, F., & Shen, S. (2020). A method for calculating uncomfortable rates of people due to vertical floor vibrations using corresponding assessment values based on fuzzy reliability theory. *Journal of Building Engineering*, *28*, 28. doi:10.1016/j.jobe.2019.101061

Ling, C., Lu, Z., Sun, B., & Wang, M. (2020). An efficient method combining active learning Kriging and Monte Carlo simulation for profust failure probability. *Fuzzy Sets and Systems*, *387*, 89–107. doi:10.1016/j.fss.2019.02.003

Nagar, P., Srivastava, A., & Srivastava, P. K. (2019). Optimization of species transportation via an exclusive fuzzy trapezoidal centroid approach. *Mathematics in Engineering. Science and Aerospace*, *10*(2), 271–280.

Nair, S. J., & Sasikumar, K. (2019). Fuzzy reliability-based optimization of a hydropower reservoir. *Journal of Hydroinformatics*, *21*(2), 308–317. doi:10.2166/hydro.2019.078

Ren, J., & Wu, C. (2020). Posbist Reliability Theory for Typical Systems with Multi components. *Mathematical Problems in Engineering*, *2020*, 1–18. doi:10.1155/2020/6509736

Song, Y., Peng, H., & Xia, Y. (2017). A fuzzy reliability analysis on the longitudinal vibration of remanufacturing sucker ord. *Acta Petrolei Sinica*, *38*(4), 468–474.

Zhang, X., Gao, H., Huang, H. Z., & Bchcra, D. (2018). An Equivalent Method for Fuzzy Reliability Analysis. Proceedings *Annual Reliability and Maintainability Symposium*. 10.1109/RAM.2018.8463085

Zu, T., Kang, R., Wen, M., & Yang, Y. (2018). An optimal model using data envelopment analysis for uncertainty metrics in reliability. *Soft Computing*, *22*(16), 5561–5568. doi:10.100700500-018-3329-0

KEY TERMS AND DEFINITIONS

Bibliometric: Use of statistical methods to analyze books, articles, and other publications.

Database: A structured set of data held in a computer, especially one that is accessible in various ways.

Fuzzy: Difficult to perceive clearly or understand and explain precisely, indistinct, or vague.

Fuzzy Sets: Mathematical means of representing vagueness and imprecise information.

Reliability: The degree to which the result of a measurement, calculation, or specification can be depended on to be accurate.

Theory: A supposition or a system of ideas intended to explain something, especially one based on general principles independent of the thing to be explained.

Trends: A general direction in which something is developing or changing.

Compilation of References

Rotshtein, A. P. (1995). Fuzzy Reliability Analysis of Labour (Man –Machine) Systems. In T. Onisawa & J. Kasprzyk (Eds.), *Reliability and Safety Analyses under Fuzzyness* (pp. 245–273). Physica-Verlag. doi:10.1007/978-3-7908-1898-7_17

Zadeh, L. A. (1973). Outline of a New Approach to the Analysis of Complex Systems and Decision Process. *IEEE Transactions on Systems, Man, and Cybernetics*, *SMS-3*(1), 28–44. doi:10.1109/TSMC.1973.5408575

Rotshtein, A. P., & Shtovba, S. D. (1998). Prediction of Reliability of Algorithmic Processes with Fuzzy Input Data. *Cybernetics and Systems Analysis*, *34*(4), 545–552. doi:10.1007/BF02666999

Rotshtein, A. P., & Shtovba, S. D. (2009). Modeling of the Human Operator Reliability with Aid of the Sugeno Fuzzy Knowledge Base. *Automation and Remote Control*, *70*(1), 163–178. doi:10.1134/S0005117909010123

Rotshtein, A. P. (2010). Algebra of Algorithms and Fuzzy Logic in System Reliability Analysis. *Journal of Computer and Systems Sciences International*, *49*(2), 254–264. doi:10.1134/S1064230710020115

Rotshtein, A. P. (2011). Fuzzy Algorithmic Reliability Analysis of Complex Systems. *Cybernetics and Systems Analysis*, *49*(6), 545–552. doi:10.100710559-011-9371-x

Katelnikov, D. I., & Rotshtein, A. P. (2010). Fuzzy Algorithmic Simulation of Reliability: Control and Correction Resources Optimization. *Journal of Computer and Systems Sciences International*, *49*(6), 967–971. doi:10.1134/S1064230710060146

Rotshtein, A., Katelnikov, D., & Pustylnik, L. (2012). Reliability Analysis and Optimization using Fuzzy Logic and Chaos Theory. *International Journal of Quality, Statistics and Reliability*. 9.

International Electrotechnical Commission (IEC). (2006). *Analysis Techniques for System Reliability: Procedure for Failure Mode Effect Analysis (FMEA)*. Available from: https://webstore.iec.ch/preview/info

Rotshtein, A. P., & Rakytyanska, H. B. (2001). Solution of a Diagnostics Problem on the Basis of Fuzzy Relation and Genetic Algorithm. *Cybernetics and Systems Analysis*, *37*(6), 918–925. doi:10.1023/A:1014594201670

Barlow, R., & Proschan, F. (1975). *Statistical Theory of Reliability and Life Testing*. Holt, Rinehart and Winston.

Rotshtein, A., Posner, M., & Rakytyanska, H. (2006). Cause and Effective Analysis based on Fuzzy Relation. *Reliability and Safety Engineering*, *91*(9), 1095–1101. doi:10.1016/j.ress.2005.11.041

Rotshtein, A. P., & Rakytyanska, H. B. (2008). Diagnoses Problem Solving using Fuzzy Relation. *IEEE Transactions on Fuzzy Systems*, *16*(3), 664–676. doi:10.1109/TFUZZ.2007.905908

Rotshtein, A., & Dreyfuss, M. (2018). Ranking of Elements in System Reliability Modeling: The Least Influence Method. *Systems Engineering*, *21*(5).

Rotshtein, A. P. (2017). Ranking of System's Elements Based on Fuzzy Relations and Transitive Closure. *Cybernetics and Systems Analysis*, (1), 68–75.

Bellman, R. E., & Zadeh, L. A. (1970). Decision Making in a Fuzzy Environment. *Management Science*, *17*(4), 141–164. doi:10.1287/mnsc.17.4.B141

Rotshtein, A. P. (2018). Selection of Human Performance Conditions Based on Fuzzy Perfection. *Journal of Computer and Systems Sciences International*, *49*(6), 927–971. doi:10.1134/S1064230718060096

Rotshtein, A. P. (2019). Reliability – Based Design of Human Performance Conditions using Fuzzy Perfection. *Cybernetics and Systems Analysis*, *55*(2), 68–79. doi:10.100710559-019-00128-5

Kosko, B. (1986). Fuzzy cognitive maps. *International Journal of Man-Machine Studies*, *24*(1), 65–75. doi:10.1016/S0020-7373(86)80040-2

Papageorgiu, E. I. (Ed.). (2014). *Fuzzy Cognitive Maps for Applied Sciences and Engineering From Fundamentals to Extensions and Learning Algorithms*. Springer. doi:10.1007/978-3-642-39739-4

Palaez, C. E., & Bowles, J. B. (1996). Using Fuzzy Cognitive Maps as a System Model for Failure Modes and Effect Analysis. *Intelligent Systems*, *88*, 177–199.

Rybinin, I.A. (1976). *Reliability of Engineering Systems, Principles and Analysis*. Moscow: Mir.

Salmeron, J., & Gutierrez, E. (2012). Fuzzy Grey Cognitive Maps in Reliability Engineering. *Applied Soft Computing*, *12*(12), 3818–3824. doi:10.1016/j.asoc.2012.02.003

Birnbaum, Z. W. (1969). On the Importance of Different Components in the Multicomponent System. In Multivariate Analysis (2nd ed., pp. 581-592). New York: Academic.

Axelrod, R. (1976). *Structure of Decision: The Cognitive Maps of Political Elites*. Princeton University Press.

Rotshtein, A., & Rakytyanska, H. (2012). *Fuzzy Evidence in Identification, Forecasting and Diagnosis*. Springer. doi:10.1007/978-3-642-25786-5

Stylios, C. D., & Groumpos, P. P. (2004). Modeling Complex Systems using Fuzzy Cognitive Maps, *IEEE Transaction on Systems, Man and Cybernetics. Systems and Humans, 34*(1), 155–162. doi:10.1109/TSMCA.2003.818878

Hong, J. S., & Lie, C. H. (1993). Joint Reliability Importance of Two Edges in Undirected Network. *IEEE Transactions on Reliability, 42*(1), 17–23. doi:10.1109/24.210266

Barnard, A. (2012). *Why you cannot predict electronic product reliability.* ARS.

Zadeh, L. A. (1965). Fuzzy Sets. *Information and Control, 8*(3), 338–353. doi:10.1016/S0019-9958(65)90241-X

Onisawa, T., & Kasprzyk, J. (Eds.). (1995). *Reliability and Safety Analysis under Fuzzyness.* Physica - Verlag. doi:10.1007/978-3-7908-1898-7

Cai, K. Y. (1996). *Introduction on Fuzzy Reliability.* Kluwer Academic Publisher. doi:10.1007/978-1-4613-1403-5

Rotshtein, A. P., & Shtovba, S. D. (1997). *Fuzzy Reliability of Algorithmic Processes.* Vinnitsa, Continent –Prim. (in Russian)

Utkin, L. V., & Shubinsky, I. B. (1998). *Non-conventional Methods of Reliability Analysis of Information Systems.* Lubavic. (in Russian)

Dubois, D., & Prade, H. (1978). Operations on Fuzzy Numbers. *International Journal of Systems Science, 9*(6), 613–623. doi:10.1080/00207727808941724

Ashraf, Z., & Muhuri, P. K., Danish Lohani, Q. M., & Nath, R. (2014). Fuzzy multi-objective reliability-redundancy allocation problem. *2014 IEEE International Conference on Fuzzy Systems (FUZZ-IEEE)*, 2580–2587. 10.1109/FUZZ-IEEE.2014.6891889

Ashraf, Z., Malhotra, D., Muhuri, P. K., & Danish Lohani, Q. M. (2017). Interval type-2 fuzzy demand based vendor managed inventory model. *IEEE International Conference on Fuzzy Systems.* 10.1109/FUZZ-IEEE.2017.8015738

Ashraf, Z., Muhuri, P. K., & Danish Lohani, Q. M. (2015). Particle swam optimization based reliability-redundancy allocation in a type-2 fuzzy environment. *2015 IEEE Congress on Evolutionary Computation, CEC 2015 - Proceedings*, 1212–1219. 10.1109/CEC.2015.7257027

Ashraf, Z., Roy, M. L., Muhuri, P. K., & Danish Lohani, Q. M. (2018). A Novel Image Steganography Approach Based on Interval Type-2 Fuzzy Similarity. *2018 IEEE International Conference on Fuzzy Systems (FUZZ-IEEE)*, 1–8. 10.1109/FUZZ-IEEE.2018.8491582

Ashraf, Z., Khan, M. S., & Lohani, Q. M. D. (2019). New bounded variation based similarity measures between Atanassov intuitionistic fuzzy sets for clustering and pattern recognition. *Applied Soft Computing, 85*, 105529. Advance online publication. doi:10.1016/j.asoc.2019.105529

Ashraf, Z., Muhuri, P. K., Lohani, Q. M. D., & Roy, M. L. (2019). Type-2 fuzzy reliability–redundancy allocation problem and its solution using particle-swarm optimization algorithm. *Granular Computing*, *4*(2), 145–166. doi:10.100741066-018-0106-5

Ashraf, Z., Roy, M. L., Muhuri, P. K., & Danish Lohani, Q. M. (2020). Interval type-2 fuzzy logic system based similarity evaluation for image steganography. *Heliyon*, *6*(5), e03771. doi:10.1016/j.heliyon.2020.e03771 PMID:32420466

Atanassov, K. (1983). *Intuitionistic fuzzy sets. VII.* ITKR's Session. Sofia, June 1,1983. Deposed in Central Science—Technology Library of Bulgaria Academy of Science—1697/84.

Atanassov, K. T. (1986). Intuitionistic fuzzy sets. *Fuzzy Sets and Systems*, *20*(1), 87–96. doi:10.1016/S0165-0114(86)80034-3

Atanassov, K. T. (1999). *Intuitionistic fuzzy sets*. Springer Physica-Verlag. doi:10.1007/978-3-7908-1870-3

Atanassov, K., & Gargov, G. (1989). Interval valued intuitionistic fuzzy sets. *Fuzzy Sets and Systems*, *31*(3), 343–349. doi:10.1016/0165-0114(89)90205-4

Attoh-Okine, N., & Ayyub, B. (2005). Applied Research in Uncertainty Modeling and Analysis. In *Applied Research in Uncertainty Modeling and Analysis* (1. Aufl., Vol. 20). Springer Netherlands. doi:10.1007/b101807

Axelrod, R. (1977). *Structure of decision: the cognitive maps of political elite*. Princeton University Press.

Bakri, E. L. A., & Boumhidi, I. (2018). Fuzzy model-based faults diagnosis of the wind turbine benchmark. *Procedia Computer Science, 127*, 464–470. doi:10.1016/j.procs.2018.01.144

Baloui Jamkhaneh, E. (2011). An evaluation of the systems reliability using fuzzy lifetime distribution, *Journal of Applied Mathematics. Islamic Azad University of Lahijan*, *7*, 73–80.

Baloui Jamkhaneh, E. (2012). Reliability estimation under the fuzzy environments,140804. *The Journal of Mathematics and Computer Science*, *5*(01), 28–39. doi:10.22436/jmcs.05.01.04

Baloui Jamkhaneh, E. (2017). System Reliability using Generalized Intuitionistic Fuzzy Exponential Lifetime Distribution. *International Journal of Soft Computing and Engineering*, *7*(4), 1–7.

Baloui, J. E. (2011). *An evaluation of the systems reliability using fuzzy lifetime distribution*. Academic Press.

Bamrungsetthapong, W., & Pongpullponsak, A. (2014). Parameter interval estimation of system reliability for repairable multistate series-parallel system with fuzzy data. *TheScientificWorldJournal*, *2014*, 1–10. doi:10.1155/2014/275374 PMID:24987728

Baziuk, P. A., Rivera, S. S., & Núñez Mc Leod, J. (2016). Fuzzy human reliability analysis: Applications and contributions review. *Advances in Fuzzy Systems*, *2016*, 2016. doi:10.1155/2016/4612086

Baziuk, P., Rivera, S., & Núñez McLeod, J. (2016). Fuzzy Human Reliability Analysis: Applications and Contributions Review. *Advances in Fuzzy Systems*, *2016*, 1–9. https://doi.org/10.1155/2016/4612086

Behzadian, M., Otaghsara, S. K., Yazdani, M., & Ignatius, J. (2012). A state-of the-art survey of TOPSIS applications. *Expert Systems with Applications*, *39*(17), 13051–13069. doi:10.1016/j.eswa.2012.05.056

Beielstein, T. (2002). Tuning PSO parameters through sensitivity analysis. *Technical Report, Reihe Computational Intelligence CI 124/02. 2002. Collaborative Research Cent*, (3). Advance online publication. doi:10.13140/2.1.1466.2080

Bevilacqua, M., & Braglia, M. (2000). The analytic hierarchy process applied to maintenance strategy selection. *Reliability Engineering & System Safety*, *70*(1), 71–83. doi:10.1016/S0951-8320(00)00047-8

Bhamare, S. S., Yedav, O. P., & Rathore, A. (2007). Evolution of reliability engineering discipline over the last six decades: A comprehensive review. *International Journal of Reliability and Safety*, *1*(4), 377–410. doi:10.1504/IJRS.2007.016256

Bian, T., Zheng, H., Yin, L., & Deng, Y. (2018). Failure mode and effects analysis based on D numbers and TOPSIS. *Quality and Reliability Engineering International*, *34*(4), 501–515. doi:10.1002/qre.2268

Bohra, K. S., & Singh, S. B. (2015). Evaluating fuzzy system reliability using intuitionistic fuzzy Rayleigh lifetime distribution. *Journal MESA, 6*(2), 245-254.

Bollinger, R. C. (1982). Direct computation for consecutive-k-out-of-n: F systems. *IEEE Transactions on Reliability*, *31*(5), 444–446. doi:10.1109/TR.1982.5221424

Boral, S., Howard, I., Chaturvedi, S. K., McKee, K., & Naikan, V. N. A. (2020). An integrated approach for fuzzy failure modes and effects analysis using fuzzy AHP and fuzzy MAIRCA. *Engineering Failure Analysis*, *108*, 104195. doi:10.1016/j.engfailanal.2019.104195

Boran, F. E., Genç, S., Kurt, M., & Akay, D. (2009). A multi-criteria intuitionistic fuzzy group decision making for supplier selection with TOPSIS method. *Expert Systems with Applications*, *36*(8), 11363–11368. doi:10.1016/j.eswa.2009.03.039

Bowles, J. B., & Peláez, C. E. (1995). Fuzzy logic prioritization of failures in a system failure mode, effects and criticality analysis. *Reliability Engineering & System Safety*, *50*(2), 203–213. doi:10.1016/0951-8320(95)00068-D

Buckley, J. J. (2004). Uncertain probabilities III. The continuous case. *Soft Computing*, *8*(3), 200–206. doi:10.100700500-003-0263-5

Bustince, H., Barrenechea, E., Pagola, M., Fernandez, J., Xu, Z., Bedregal, B., Montero, J., Hagras, H., Herrera, F., & De Baets, B. (2016). A Historical Account of Types of Fuzzy Sets and Their Relationships. *IEEE Transactions on Fuzzy Systems*, *24*(1), 179–194. doi:10.1109/TFUZZ.2015.2451692

Büyüközkan, G., & Güleryüz, S. (2016). Multi criteria group decision making approach for smart phone selection using intuitionistic fuzzy TOPSIS. *International Journal of Computational Intelligence Systems*, *9*(4), 709–725. doi:10.1080/18756891.2016.1204119

Cai, K. Y. (1991). Fuzzy reliability theories. *Fuzzy Sets and Systems*, *40*(3), 510–511. doi:10.1016/0165-0114(91)90177-R

Cai, K. Y. (1996). System failure engineering and fuzzy methodology an introductory overview. *Fuzzy Sets and Systems*, *83*(2), 113–133. doi:10.1016/0165-0114(95)00385-1

Cai, K. Y., Wen, C. Y., & Zhang, M. L. (1991). Fuzzy variables as a basis for a theory of fuzzy reliability in the possibility context. *Fuzzy Sets and Systems*, *42*(2), 145–172. doi:10.1016/0165-0114(91)90143-E

Cai, K. Y., Wen, C. Y., & Zhang, M. L. (1991). Posbist reliability behavior of typical systems with 2 types of failure. *Fuzzy Sets and Systems*, *43*(1), 17–32. doi:10.1016/0165-0114(91)90018-L

Cai, K. Y., Wen, C. Y., & Zhang, M. L. (1993). Fuzzy states as a basis for a theory of fuzzy reliability. *Microelectronics and Reliability*, *33*(15), 2253–2263. doi:10.1016/0026-2714(93)90065-7

Cai, K. Y., Wen, C. Y., & Zhang, M. L. (1995). Mixture-Models in Profust Reliability Theory. *Microelectronics and Reliability*, *35*(6), 985–993. doi:10.1016/0026-2714(94)00058-V

Cai, K. Y., Wen, C. Y., & Zhang, M. L. (1995). Posbist reliability behavior of Fault-Tolerant systems. *Microelectronics and Reliability*, *35*(1), 4956.

Cappelle, B., & Kerre, E. E. (1993). On a possibilistic approach to reliability theory. In *1993 (2nd) International Symposium on Uncertainty Modeling and Analysis* (pp. 415-418). IEEE. 10.1109/ISUMA.1993.366736

Chang, K. H., & Cheng, C. H. (2010). A risk assessment methodology using intuitionistic fuzzy set in FMEA. *International Journal of Systems Science*, *41*(12), 1457–1471. doi:10.1080/00207720903353633

Chang, K. H., & Cheng, C.-H. (2011). Evaluating the risk of failure using the fuzzy OWA and DEMATEL method. *Journal of Intelligent Manufacturing*, *22*(2), 113–129. doi:10.100710845-009-0266-x

Chaube, S., & Singh, S. B. (2016). Fuzzy reliability theory based on membership function. *International Journal of Mathematical. Engineering and Management Sciences*, *1*(1), 34–40.

Cheng, C. H., & Mon, D. L. (1993). Fuzzy system reliability analysis by interval of confidence. *Fuzzy Sets and Systems*, *56*(1), 29–35. doi:10.1016/0165-0114(93)90182-H

Chen, K., Ma, C., Zheng, M., & Gao, F. (2015). Optimization and Mechanical Accuracy Reliability of a New Type of Forging Manipulator. *Chinese Journal of Mechanical Engineering*, *28*(2), 236–248. doi:10.3901/CJME.2014.1128.173

Chen, S. H. (1985). Operations on fuzzy numbers with function principle. *Tamkang Journal of Management Sciences*, *6*(1), 13–26.

Chen, S. M. (1994). Fuzzy system reliability analysis using fuzzy number arithmetic operations. *Fuzzy Sets and Systems*, *64*(1), 31–38. doi:10.1016/0165-0114(94)90004-3

Chen, S. M. (1996). New method for fuzzy system reliability analysis. *Cybernetics and Systems: An International Journal*, *27*(4), 385–401. doi:10.1080/019697296126480

Chen, S. M., & Huang, Z. C. (2017). Multiattribute decision making based on interval-valued intuitionistic fuzzy values and linear programming methodology. *Information Sciences*, *381*, 341–351. doi:10.1016/j.ins.2016.11.010

Chern, M. S. (1992). On the computational complexity of reliability redundancy allocation in a series system. *Operations Research Letters*, *11*(5), 309–315. doi:10.1016/0167-6377(92)90008-Q

Chiozza, M. L., & Ponzetti, C. (2009). FMEA: A model for reducing medical errors. *Clinica Chimica Acta*, *404*(1), 75–78. doi:10.1016/j.cca.2009.03.015 PMID:19298799

Chongshan, G. (2009, November). Fuzzy availability analysis of a repairable consecutive-2-out-of-3: F System. In *2009 IEEE International Conference on Grey Systems and Intelligent Services (GSIS 2009)* (pp. 434-437). IEEE.

Chung, W. K. (1988). A k-out-of-N: G redundant system with dependent failure rates and common-cause failures. *Microelectronics and Reliability*, *28*(2), 201–203. doi:10.1016/0026-2714(88)90351-4

Clerc, M. (2006). Particle Swarm Optimization. Particle Swarm Optimization. doi:10.1002/9780470612163

Clerc, M., & Kennedy, J. (2002). The particle swarm - explosion, stability, and convergence in a multidimensional complex space. *IEEE Transactions on Evolutionary Computation*, *6*(1), 58–73. doi:10.1109/4235.985692

Cobo, M., López-Herrera, A., Herrera-Viedma, E., & Herrera, F. (2011). An approach for detection, quantifying and visualizing the evolution of a research field: A practical application to the fuzzy sets theory field. *Journal of Informetrics*, *5*(1), 146–166. doi:10.1016/j.joi.2010.10.002

Coello Coello, C. A. (2002). Theoretical and numerical constraint-handling techniques used with evolutionary algorithms: A survey of the state of the art. *Computer Methods in Applied Mechanics and Engineering*, *191*(11–12), 1245–1287. doi:10.1016/S0045-7825(01)00323-1

Cutello, V., Montero, J., & Yáñez, J. (1996). Structure functions with fuzzy states. *Fuzzy Sets and Systems*, *83*(2), 189–202. doi:10.1016/0165-0114(95)00390-8

da Costa Bueno, V., & do Carmo, I. M. (2007). Active redundancy allocation for a k-out-of-n: F system of dependent components. *European Journal of Operational Research*, *176*(2), 1041–1051. doi:10.1016/j.ejor.2005.09.012

Dag, M., Yalçın, Z. G., Oz, A., & Aydoğmuş, E. (2018). Risk Analysis by FMEA Method in Hotel. *Journal of Physical Chemistry and Functional Materials*, *1*(1), 63–71.

De Luca, A., & Termini, S. (1993). A definition of a nonprobabilistic entropy in the setting of fuzzy sets theory. In *Readings in Fuzzy Sets for Intelligent Systems* (pp. 197–202). Elsevier. doi:10.1016/B978-1-4832-1450-4.50020-1

Deb, K. (2001). *Multi-objective optimization using evolutionary algorithms*. John Wiley & Sons, Inc.

Deluca, A. T. S., & Termini, S. (1972). A definition of a non probabilistic entropy in the setting of fuzzy set theory. *Information and Control*, *20*(4), 301–312. doi:10.1016/S0019-9958(72)90199-4

Derrida, J. (2001). *Limited Inc*. Passagen.

Dincer, H., Yüksel, S., & Senel, S. (2018). Analyzing the Global Risks for the Financial Crisis after the Great Depression Using Comparative Hybrid Hesitant Fuzzy Decision-Making Models: Policy Recommendations for Sustainable Economic Growth. *Sustainability (Basel, Switzerland)*, *10*(9), 3126– doi:10.3390u10093126

Ding, Y., & Lisnianski, A. (2008). Fuzzy universal generating functions for multi-state system reliability assessment. *Fuzzy Sets and Systems*, *159*(3), 307–324. doi:10.1016/j.fss.2007.06.004

Ding, Y., Wang, P., Goel, L., Loh, P. C., & Wu, Q. (2010). Long-term reserve expansion of power systems with high wind power penetration using universal generating function methods. *IEEE Transactions on Power Systems*, *26*(2), 766–774.

Ding, Y., Zuo, M. J., Lisnianski, A., & Li, W. (2010). A framework for reliability approximation of multi-state weighted *k*-out-of-*n* systems. *IEEE Transactions on Reliability*, *59*(2), 297–308. doi:10.1109/TR.2010.2048659

Dong, Y. G., Chen, X. Z., Zhao, X. D., & Quan, Z. W. (2005). General approach for fuzzy reliability analysis based on the fuzzy probability theory. *Journal of Computational Mechanics*, *22*(3), 281–286.

Dubois, D., & Prade, H. (1988). *Possibility theory: An approach to computerized processing of uncertainty*. Plenum Press. doi:10.1007/978-1-4684-5287-7

Ebrahimipour, V., & Sheikhalishahi, M. (2011). Application of multi-objective particle swarm optimization to solve a fuzzy multi-objective reliability redundancy allocation problem. *2011 IEEE International Systems Conference*, 326–333. 10.1109/SYSCON.2011.5929085

Efe, B. (2019a). Website Evaluation Using Interval Type-2 Fuzzy-Number-Based TOPSIS Approach. In Multi-Criteria Decision-Making Models for Website Evaluation (pp. 166-185). IGI Global. doi:10.4018/978-1-5225-8238-0.ch009

Efe, Ö. F. (2020). Hybrid Multi-Criteria Models: Joint Health and Safety Unit Selection on Hybrid Multi-Criteria Decision Making. In Multi-Criteria Decision Analysis in Management (pp. 62-84). IGI Global.

Efe, B. (2019b). Analysis of operational safety risks in shipbuilding using failure mode and effect analysis approach. *Ocean Engineering*, *187*, 106214. doi:10.1016/j.oceaneng.2019.106214

Efe, B., & Efe, Ö. F. (2018). Intuitionistic Fuzzy Number Based Group Decision Making Approach For Personnel Selection. *Uludağ University Journal of The Faculty of Engineering*, *23*(3), 11–26. doi:10.17482/uumfd.338406

Efe, B., Kurt, M., & Efe, Ö. F. (2017). An Integrated Intuitionistic Fuzzy Set And Mathematical Programming Approach For An Occupational Health And Safety Policy. *Gazi University Journal of Science*, *30*(2), 73–95.

Efe, B., Yerlikaya, M. A., & Efe, Ö. F. (2016). İş Güvenliğinde Bulanık Promethee Yöntemiyle Hata Türleri ve Etkilerinin Analizi: Bir İnşaat Firmasında Uygulama. *Gümüşhane Üniversitesi Fen Bilimleri Enstitüsü Dergisi*, *6*(2), 126–137.

Elsevier. (2020). *Elsevier Scopus*. https://www.elsevier.com/es-mx/solutions/scopus

Engelbrecht, A. P. (2007). Computational Intelligence. *Computational Intelligence*. Advance online publication. doi:10.1002/9780470512517

Fahimnia, B., Sarkis, J., & Davarzani, H. (2015). Green supply chain management: A review and bibliometric analysis. *International Journal of Production Economics*, *162*, 101–114. doi:10.1016/j.ijpe.2015.01.003

Fattahi, R., & Khalilzadeh, M. (2018). Risk evaluation using a novel hybrid method based on FMEA, extended MULTIMOORA, and AHP methods under fuzzy environment. *Safety Science*, *102*, 290–300. doi:10.1016/j.ssci.2017.10.018

Gao, C. (2009). Fuzzy availability analysis of a repairable consecutive-2-out-of-3: F system. *IEEE International Conference on Grey Systems and Intelligent Services,* 434-437.

Garfield, E. (1972). Citation analysis as a tool in journal evaluation. *Science*, *178*(4060), 417–479. doi:10.1126cience.178.4060.471 PMID:5079701

Garfield, E. (Ed.). (1979). *SCI journal citation reports. A bibliometric analysis of science journals in the ISI data base*. Institute for Scientific Information.

Garg, H. (2014). Reliability, Availability and Maintainability Analysis of Industrial Systems Using PSO and Fuzzy Methodology. *Journal of Metrology Society of India*, *29*(2), 115–129. doi:10.100712647-013-0081-x

Garg, H., & Kumar, K. (2018). Group decision making approach based on possibility degree measure under linguistic interval-valued intuitionistic fuzzy set environment. *Journal of Industrial and Management Optimization*, *13*(5), 1–23. doi:10.3934/jimo.2018162

Garg, H., & Kumar, K. (2019). Linguistic interval-valued Atanassov intuitionistic fuzzy sets and their applications to group decision-making problems. *IEEE Transactions on Fuzzy Systems*, *27*(12), 2302–2311. doi:10.1109/TFUZZ.2019.2897961

Garg, H., & Kumar, K. (2020a). A novel exponential distance and its based TOPSIS method for interval-valued intuitionistic fuzzy sets using connection number of SPA theory. *Artificial Intelligence Review*, *53*(1), 595–624. doi:10.100710462-018-9668-5

Garg, H., & Kumar, K. (2020b). A novel possibility measure to interval-valued intuitionistic fuzzy set using connection number of set pair analysis and its applications. *Neural Computing & Applications*, *32*(8), 3337–3348. doi:10.100700521-019-04291-w

Garg, H., & Kumar, K. (2020c). Power Geometric Aggregation Operators Based on Connection Number of Set Pair Analysis Under Intuitionistic Fuzzy Environment. *Arabian Journal for Science and Engineering*, *45*(3), 2049–2063. doi:10.100713369-019-03961-0

Garg, H., Rani, M., Sharma, S. P., & Vishwakarma, Y. (2014). Intuitionistic fuzzy optimization technique for solving multi-objective reliability optimization problems in interval environment. *Expert Systems with Applications*, *41*(7), 3157–3167.

Garg, H., & Sharma, S. P. (2012). Behavior analysis of synthesis unit in fertilizer plant. *International Journal of Quality & Reliability Management*, *29*(2), 217–232. doi:10.1108/02656711211199928

Garg, H., & Sharma, S. P. (2013). Multi-objective reliability-redundancy allocation problem using particle swarm optimization. *Computers & Industrial Engineering*, *64*(1), 247–255. doi:10.1016/j.cie.2012.09.015

Gaxiola, F., Melin, P., Valdez, F., Castro, J. R., & Castillo, O. (2016). Optimization of type-2 fuzzy weights in backpropagation learning for neural networks using GAs and PSO. *Applied Soft Computing*, *38*, 860–871. doi:10.1016/j.asoc.2015.10.027

Geum, Y., Cho, Y., & Park, Y. (2011). A systematic approach for diagnosing service failure: Service-specific FMEA and grey relational analysis approach. *Mathematical and Computer Modelling*, *54*(11-12), 3126–3142. doi:10.1016/j.mcm.2011.07.042

Golberg, D. E. (1989). *Machine Learning and Its Applications* (G. Paliouras, V. Karkaletsis, & C. D. Spyropoulos, Eds.). doi:10.1007/3-540-44673-7

Griffith, W. (1986). On consecutive k-out-of-n failure system and their generalizations. In B. A. North-Holland (Ed.), *Reliability and Quality Control* (pp. 157–165). Elsevier.

Guan, J. C., & Wu, Y. Q. (2006). Repairable consecutive -k-out-of-n: F system with fuzzy sets. *Fuzzy Sets and Systems*, *157*(1), 121–142. doi:10.1016/j.fss.2005.05.025

Güler, E., Avci, S., & Aladağ, Z. (2019). Multi-Criteria Decision Approach With AHP And IF-TOPSIS Methods For R&D Project Selectıon Process. *Journal of Engineering Studies and Research*, *25*(3), 22–32. doi:10.29081/jesr.v25i3.328

Habib, A. S., Al-Seedy, R. O., & Radwan, T. (2007). Reliability evaluation of multi-state consecutive *k*-out-of-*r*-from-*n*: G system. *Applied Mathematical Modelling*, *31*(11), 2412–2423. doi:10.1016/j.apm.2006.09.006

Habib, A. S., Yuge, T., Al-Seedy, R. O., & Ammar, S. I. (2010). Reliability of a consecutive (r, s)-out-of-(m, n): F lattice system with conditions on the number of failed components in the system. *Applied Mathematical Modelling*, *34*(3), 531–538. doi:10.1016/j.apm.2009.06.022

He, Z., Cao, H., Zi, Y., & Li, B. (2014). Developments and thoughts on operational reliability assessment of mechanical equipment. *Jixie Gongcheng Xuebao*, *50*(2), 171–186. doi:10.3901/JME.2014.02.171

Hryniewicz, O. (2007). Fuzzy sets in the evaluation of reliability. *Studies in Computational Intelligence*, *40*, 363–386. doi:10.1007/978-3-540-37372-8_13

Huang, H. (1997). Fuzzy multi-objective optimization decision-making of reliability of series system. *Microelectronics and Reliability*, *37*(3), 447–449. doi:10.1016/S0026-2714(96)00040-6

Huang, H. Z., & He, L. (2008). New approaches to system analysis and design: a review. In *Handbook of Performability Engineering* (pp. 477–498). Springer. doi:10.1007/978-1-84800-131-2_31

Huang, H. Z., Li, Y., & Liu, Y. (2010). Posbist Reliability Theory of k-out-of-n: G System. *Journal of Multiple-Valued Logic and Soft Computing*, *16*(1-2), 45–63.

Huang, H. Z., Tong, X., & Zuo, M. J. (2004). Posbits fault tree analysis of coherent systems. *Reliability Engineering & System Safety*, *84*(2), 141–148. doi:10.1016/j.ress.2003.11.002

Huang, H., Sun, Z., Guo, D., Zhou, F., & Tian, Z. (2001). Fuzzy reliability analysis in the case of random stress and fuzzy strength. *Journal of Mechanical Strength*, *23*(3), 305–307.

Huang, J., Li, Z. S., & Liu, H. C. (2017). New approach for failure mode and effect analysis using linguistic distribution assessments and TODIM method. *Reliability Engineering & System Safety*, *167*, 302–309. doi:10.1016/j.ress.2017.06.014

Hwang, C. L., & Yoon, K. (1981). *Multiple attributes decision making methods and applications*. Springer. doi:10.1007/978-3-642-48318-9

Jahan, A. (2019). Analysis of failure modes and solutions in the process of open-heart surgery using fuzzy FMEA and fuzzy TOPSIS techniques. *Japan-Hospitals*, *18*(1), 31–42.

Jee, T. L., Tay, K. M., & Lim, C. P. (2015). A New Two-Stage Fuzzy Inference System-Based Approach to Prioritize Failures in Failure Mode and Effect Analysis. *IEEE Transactions on Reliability*, *64*(3), 869–877. doi:10.1109/TR.2015.2420300

Jiang, Q., & Chen, C. H. (2003). A numerical algorithm of fuzzy reliability. *Reliability Engineering & System Safety*, *80*(3), 299–307.

Ji, Z., Xia, Y., Sun, Q., & Cao, G. (2014). Interval-valued possibilistic fuzzy C-means clustering algorithm. *Fuzzy Sets and Systems*, *253*, 138–156. doi:10.1016/j.fss.2013.12.011

John, R. (1998). Type 2 Fuzzy Sets: An Appraisal of Theory and Applications. *International Journal of Uncertainty, Fuzziness and Knowledge-based Systems, 06*(06), 563–576. doi:10.1142/S0218488598000434

Joshi, D., & Kumar, S. (2018). Improved accuracy function for interval-valued intuitionistic fuzzy sets and its application to multi—Attributes group decision making. *Cybernetics and Systems, 49*(1), 64–76. doi:10.1080/01969722.2017.1412890

Kabiri Chimeh, M., Heywood, P., Pennisi, M., Pappalardo, F., & Richmond, P. (2019). Parallelisation strategies for agent based simulation of immune systems. *BMC Bioinformatics, 20*(S6), 579–14. https://doi.org/10.1186/s12859-019-3181-y

Kacprzyk, J., & Onisawa, T. (1995). *Reliability and safety analyses under fuzziness*. Physica-Verlag.

Kang, R., Zhang, Q., Zeng, Z., Zio, E., & Li, X. (2016). Measuring reliability under epistemic uncertainty: Review on non-probabilistic reliability metric. *Chinese Journal of Aeronautics, 29*(3), 571–579. doi:10.1016/j.cja.2016.04.004

Kaplan, H. (2014). *Data analytic studies for Turkey's energy forecast*. ProQuest LLC.

Karnik, N. N., & Mendel, J. M. (2001). Operations on type-2 fuzzy sets. *Fuzzy Sets and Systems, 122*(2), 327–348. doi:10.1016/S0165-0114(00)00079-8

Karpíšek, Z., Štěpánek, P., & Jurák, P. (2010). Weibull fuzzy probability distribution for reliability of concrete structures. *Engineering Mechanics, 17*(5-6), 363-372.

Kennedy, J. (2011). Particle Swarm Optimization. In C. Sammut & G. I. Webb (Eds.), *Encyclopedia of Machine Learning* (Vol. 4, pp. 760–766). doi:10.1007/978-0-387-30164-8_630

Kerre, E., Onisawa, T., Cappelle, B., & Gazdik, I. (1998). Reliability. In R. Słowiński (Ed.), The Handbooks of Fuzzy Sets Series: Vol. 1. *Fuzzy Sets in Decision Analysis, Operations Research and Statistics*. Springer. doi:10.1007/978-1-4615-5645-9_12

Kilic, B., & Ok, S. (2012). Otel işletmelerinde müşteri şikayetleri ve şikayetlerin değerlendirilmesi. *Journal of Yasar University, 25*(7), 4189–4202.

Klir, G. J., & Yuan, B. (2008). *Fuzzy sets and fuzzy logic, theory and applications*. Retrieved from http://digilib.uin-suka.ac.id/7049/

Knezevic, J., & Odoom, E. R. (2001). Reliability modeling of repairable systems using Petri nets and fuzzy Lambda-Tau methodology. *Reliability Engineering & System Safety, 73*(1), 1–17. doi:10.1016/S0951-8320(01)00017-5

Komal, & Sharma, S. P. (2014). Fuzzy reliability analysis of repairable industrial systems using soft-computing based hybridized techniques. *Applied Soft Computing Journal, 24*, 264–276. doi:10.1016/j.asoc.2014.06.054

Kumar, D., & Singh, S. B. (2014, November). Evaluating fuzzy reliability using rough intuitionistic fuzzy set. In 2014 Innovative Applications of Computational Intelligence on Power, Energy and Controls with their impact on Humanity (CIPECH) (pp. 138-142). IEEE.

Kumar, A., & Ram, M. (2018). Reliability of the system based on Weibull distribution and hesitant fuzzy set. *International Journal of Mathematical. Engineering and Management Sciences*, *3*(4), 513–521.

Kumar, A., & Ram, M. (2018). System Reliability Analysis Based On Weibull Distribution and Hesitant Fuzzy Set, *International Journal of Mathematical. Engineering and Management Sciences*, *3*(4), 513–521.

Kumar, A., & Ram, M. (2018). System reliability analysis based on weibull distribution and hesitant fuzzy set. *International Journal of Mathematical. Engineering and Management Sciences*, *3*(4), 513–521.

Kumar, A., & Ram, M. (2018). System reliability analysis based on Weibull distribution and hesitant fuzzy set. *International Journal of Mathematical. Engineering and Management Sciences*, *3*(4), 513–521.

Kumar, A., & Ram, M. (2019). Reliability Analysis for Environment Systems Using Dual Hesitant Fuzzy Set. In *Advanced Fuzzy Logic Approaches in Engineering Science* (pp. 162–173). IGI Global.

Kumar, A., & Singh, S. B. (2018). Signature reliability of linear multi-state sliding window system. *International Journal of Quality & Reliability Management*, *35*(10), 2403–2413. doi:10.1108/IJQRM-04-2017-0083

Kumar, A., Singh, S. B., & Ram, M. (2019). Reliability appraisal for consecutive-*k*-out-of-*n*:F system of non-identical components with intuitionistic fuzzy set. *Int. J. Operational Research.*, *36*(3), 362–374.

Kumar, A., Singh, S. B., & Ram, M. (2020). Systems reliability assessment using hesitant fuzzy set. *International Journal of Operational Research*, *38*(1), 1–18. doi:10.1504/IJOR.2020.106357

Kumar, A., Yadav, S. P., & Kumar, S. (2006). Fuzzy reliability of a marine power plant using interval valued vague sets. *International Journal of Applied Science and Engineering*, *4*(1), 71–82.

Kumar, D., & Kumari, A. (2019). An Evaluation of the System Reliability Using Fuzzy Lifetime Distribution Emphasizing Hexagonal Fuzzy. *IOSR Journal of Engineering*, *9*(8), 17–25.

Kumar, D., & Singh, S. B. (2013). Reliability analysis of complex repairable system with reboot delay. *International Journal of Information and Computation Technology*, *3*, 341–344.

Kumar, J., & Goel, M. (2017). Fuzzy reliability analysis of a pulping system in paper industry with general distributions for all random variables. *Cogent Mathematics*, *4*(1), 1285467. doi:10.1080/23311835.2017.1285467

Kumar, K., & Garg, H. (2018). {TOPSIS} method based on the connection number of set pair analysis under interval-valued intuitionistic fuzzy set environment. *Computational & Applied Mathematics*, *37*(2), 1319–1329. doi:10.100740314-016-0402-0

Kumar, K., & Kumar, P. (2011). Fuzzy availability modeling and analysis of biscuit manufacturing plant: A case study. *International Journal of Systems Assurance Engineering and Management*, *2*(3), 193–204. doi:10.100713198-011-0076-3

Kumar, M. (2014). Applying weakest-t-norm-based approximate intuitionistic fuzzy arithmetic operations on different types of intuitionistic fuzzy numbers to evaluate reliability of PCBE fault. *Applied Soft Computing*, *23*, 387–406. doi:10.1016/j.asoc.2014.06.036

Kumar, M., Prasad Yadav, S., & Kumar, S. (2013). Fuzzy system reliability evaluation using time-dependent intuitionistic fuzzy set. *International Journal of Systems Science*, *44*(1), 50–66. doi:10.1080/00207721.2011.581393

Kumar, M., & Yadav, S. P. (2012). A novel approach for analyzing fuzzy system reliability using different types of intuitionistic fuzzy failure rates of components. *ISA Transactions*, *51*(2), 115–129. doi:10.1016/j.isatra.2011.10.002 PMID:22134065

Kumar, M., Yadav, S. P., & Kumar, S. (2011). A new approach for analysing the fuzzy system reliability using intuitionistic fuzzy number. *International Journal of Industrial and Systems Engineering*, *8*(2), 135–156.

Kumar, P., & Singh, S. B. (2015). Fuzzy system reliability using intuitionistic fuzzy Weibull lifetime distribution. *International Journal of Reliability and Applications*, *16*(1), 15–26.

Kumar, P., & Singh, S. B. (2018). *Fuzzy reliability of systems using different types of level (λ, 1) interval-valued fuzzy numbers*. Reliability Engineering Theory and Applications, CRC Press Taylor & Francis Group. doi:10.1201/9781351130363-10

Kuo, W., & Prasad, R. (2004). *System Reliability Optimization: An Overview*. doi:10.1007/978-1-4419-9021-1_2

Kuo, W., & Prasad, V. R. (2000). An annotated overview of system-reliability optimization. *IEEE Transactions on Reliability*, *49*(2), 176–187. doi:10.1109/24.877336

Kutlu, A. C., & Ekmekçioğlu, M. (2012). Fuzzy failure modes and effects analysis by using fuzzy TOPSIS-based fuzzy AHP. *Expert Systems with Applications*, *39*(1), 61–67. doi:10.1016/j.eswa.2011.06.044

Lam, Y., & Ng, H. K. T. (2001). A general model for consecutive-k-out-of-n: F repairable system with exponential distribution and (k− 1)-step Markov dependence. *European Journal of Operational Research*, *129*(3), 663–682. doi:10.1016/S0377-2217(99)00474-9

Levitin, G., Lisnianski, A., Haim, H. B., & Elmakis, D. (2000, April). Genetic algorithm and universal generating function technique for solving problems of power system reliability optimization. In *DRPT2000. International Conference on Electric Utility Deregulation and Restructuring and Power Technologies. Proceedings (Cat. No. 00EX382)* (pp. 582-586). IEEE.

Levitin, G. (2002). Evaluating correct classification probability for weighted voting classifiers with plurality voting. *European Journal of Operational Research*, *141*(3), 596–607.

Levitin, G. (2004). A universal generating function approach for the analysis of multi-state systems with dependent elements. *Reliability Engineering & System Safety*, *84*(3), 285–292. doi:10.1016/j.ress.2003.12.002

Levitin, G. (2005). Reliability of linear multistate multiple sliding window systems. *Naval Research Logistics*, *52*(3), 212–223.

Levitin, G. (2005). *The universal generating function in reliability analysis and optimization* (Vol. 6). Springer.

Levitin, G. (2005). *Universal Generating Functions in Reliability Analysis and Optimization*. Springer –Verlag.

Levitin, G., & Lisnianski, A. (2001). A new approach to solving problems of multi-state system reliability optimization. *Quality and Reliability Engineering International*, *17*(2), 93–104.

Lincoln, D., & Ross, T. (2012). Evaluation of commercial facility demand response using genetically evolved fuzzy cognitive mapping. *Int. J. Distributed Energy Resources*, *8*(4), 315–342.

Ling, C., Lu, Z., Feng, H., & Sun, B. (2019). Efficient numerical simulation methods for estimating fuzzy failure probability-based importance measure indices. *Structural and Multidisciplinary Optimization*, *59*(2), 577–593. doi:10.100700158-018-2085-1

Lisnianski, A., & Ding, Y. (2009). Redundancy analysis for repairable multi-state system by using combined stochastic processes methods and universal generating function technique. *Reliability Engineering & System Safety*, *94*(11), 1788–1795.

Liu, H. C. (2016). FMEA using uncertainty theories and MCDM methods. In *FMEA using uncertainty theories and MCDM methods* (pp. 13–27). Springer.

Liu, H. C., Liu, L., & Lin, Q. L. (2013). Fuzzy failure mode and effects analysis using fuzzy evidential reasoning and belief rule-based methodology. *IEEE Transactions on Reliability*, *62*(1), 23–36. doi:10.1109/TR.2013.2241251

Liu, H. C., Liu, L., & Li, P. (2014). Failure mode and effects analysis using intuitionistic fuzzy hybrid weighted Euclidean distance operator. *International Journal of Systems Science*, *45*(10), 2012–2030. doi:10.1080/00207721.2012.760669

Liu, H. C., Liu, L., & Liu, N. (2013). Risk evaluation approaches in failure mode and effects analysis: A literature review. *Expert Systems with Applications*, *40*(2), 828–838. doi:10.1016/j.eswa.2012.08.010

Liu, H. C., Liu, L., Liu, N., & Mao, L. X. (2012). Risk evaluation in failure mode and effects analysis with extended VIKOR method under fuzzy environment. *Expert Systems with Applications*, *39*(17), 12926–12934. doi:10.1016/j.eswa.2012.05.031

Liu, H. C., You, J. X., Chen, S., & Chen, Y. Z. (2016). An integrated failure mode and effect analysis approach for accurate risk assessment under uncertainty. *IIE Transactions*, *48*(11), 1027–1042. doi:10.1080/0740817X.2016.1172742

Liu, H. C., You, J. X., & Duan, C. Y. (2019). An integrated approach for failure mode and effect analysis under interval-valued intuitionistic fuzzy environment. *International Journal of Production Economics*, *207*, 163–172. doi:10.1016/j.ijpe.2017.03.008

Liu, H. C., You, J. X., Lin, Q. L., & Li, H. (2015). Risk assessment in system FMEA combining fuzzy weighted average with fuzzy decision-making trial and evaluation laboratory. *International Journal of Computer Integrated Manufacturing*, *28*(7), 701–714. doi:10.1080/0951192X.2014.900865

Liu, H. C., You, J. X., Shan, M. M., & Shao, L. N. (2015). Failure mode and effects analysis using intuitionistic fuzzy hybrid TOPSIS approach. *Soft Computing*, *19*(4), 1085–1098. doi:10.100700500-014-1321-x

Liu, H. C., You, J. X., You, X. Y., & Shan, M. M. (2015). A novel approach for failure mode and effects analysis using combination weighting and fuzzy VIKOR method. *Applied Soft Computing*, *28*, 579–588. doi:10.1016/j.asoc.2014.11.036

Liu, H. C., Zhang, L. J., Ping, Y. J., & Wang, L. (2020). Failure mode and effects analysis for proactive healthcare risk evaluation: A systematic literature review. *Journal of Evaluation in Clinical Practice*, *26*(4), 1320–1337. doi:10.1111/jep.13317 PMID:31849153

Liu, H., You, J., Li, P., & Su, Q. (Article in Press). Failure Mode and Effect Analysis Under Uncertainty: An Integrated Multiple Criteria Decision Making Approach (2016). *IEEE Transactions on Reliability*.

Liu, M., & Ren, H. (2014). A new intuitionistic fuzzy entropy and application in multi-attribute decision making. *Information (Basel)*, *5*(4), 587–601. doi:10.3390/info5040587

Liu, Y., & Huang, H. Z. (2010). Reliability assessment for fuzzy multi-state systems. *International Journal of Systems Science*, *41*(4), 365–379. doi:10.1080/00207720903042939

Liu, Y., & Huang, H. Z. (2011). Optimal replacement policy for fuzzy multi-state elements. *Journal of Multiple-Valued Logic and Soft Computing*, *17*(1), 69–92.

Liu, Y., Huang, H. Z., & Levitin, G. (2008). Reliability and performance assessment for fuzzy multi-state elements. *Journal of Risk and Reliability*, *222*(4), 675–686. doi:10.1243/1748006XJRR180

Liu, Y., Huang, H. Z., & Ling, D. (2013). Reliability prediction for evolutionary product in the conceptual design phase using a neural network-based fuzzy synthetic assessment. *International Journal of Systems Science*, *44*(3), 545–555. doi:10.1080/00207721.2011.617887

Li, W., & Zuo, M. J. (2008). Reliability evaluation of multi-state weighted k-out-of-n systems. *Reliability Engineering & System Safety*, *93*(1), 160–167. doi:10.1016/j.ress.2006.11.009

Li, W., Zuo, M. J., & Ding, Y. (2008). Optimal design of binary weighted k-out-of-n systems. *International Journal of Reliability Quality and Safety Engineering*, *15*(05), 425–440. doi:10.1142/S0218539308003143

Li, Y. L., Wang, R., & Chin, K.-S. (2019). New failure mode and effect analysis approach considering consensus under interval-valued intuitionistic fuzzy environment. *Soft Computing*, *23*(22), 11611–11626. doi:10.100700500-018-03706-5

Li, Z., Zhang, Q., Fan, F., & Shen, S. (2020). A method for calculating uncomfortable rates of people due to vertical floor vibrations using corresponding assessment values based on fuzzy reliability theory. *Journal of Building Engineering*, *28*, 101061. doi:10.1016/j.jobe.2019.101061

Lo, H. W., & Liou, J. J. (2018). A novel multiple-criteria decision-making-based FMEA model for risk assessment. *Applied Soft Computing*, *73*, 684–696. doi:10.1016/j.asoc.2018.09.020

Ma, C., Xiao, X., Yang, H., & Li, H. (2012). Uncertain evaluation measure system and model for power grid catastrophic accident. *Zhongguo Dianji Gongcheng Xuebao*, *32*(10), 118–125.

MacKenzie, D. (2006). *An engine, not a camera: how financial models shape markets*. MIT Press.

Mahapatra, G. S., & Roy, T. K. (2006). Fuzzy multi-objective mathematical programming on reliability optimization model. *Applied Mathematics and Computation*, *174*(1), 643–659. doi:10.1016/j.amc.2005.04.105

Maheswaran, K., & Loganathan, T. (2013). A novel approach for prioritization of failure modes in FMEA using MCDM. *International Journal of Engineering Research and Applications*, *3*(4), 733–739.

Maity, S., De, S. K., & Mondal, S. P. (2020). A study of a backorder EOQ model for cloud-type intuitionistic dense fuzzy demand rate. *International Journal of Fuzzy Systems*, *22*(1), 201–211. doi:10.100740815-019-00756-1

Makri, F. S., & Psillakis, Z. M. (1996). Bounds for reliability of k-within two-dimensional consecutive-r-out-of-n failure systems. *Microelectronics and Reliability*, *36*(3), 341–345. doi:10.1016/0026-2714(95)00102-6

Malik, H., Sharma, R., & Mishra, S. (2020). Fuzzy reinforcement learning based intelligent classifier for power transformer faults. *ISA Transactions*, *101*, 390–398. https://doi.org/10.1016/j.isatra.2020.01.016

Malinowski, J., & Preuss, W. (1996). Lower and upper bounds for the reliability of connected-(r, s)-out-of-(m, n): F lattice systems. *IEEE Transactions on Reliability*, *45*(1), 156–160. doi:10.1109/24.488935

Mansnerus, E. (2012). Understanding and Governing Public Health Risks by Modeling. In Handbook of Risk Theory (pp. 213–237). Springer Netherlands; https://doi.org/10.1007/978-94-007-1433-5_9.

Marler, R. T., & Arora, J. S. (2010). The weighted sum method for multi-objective optimization: New insights. *Structural and Multidisciplinary Optimization*, *41*(6), 853–862. doi:10.100700158-009-0460-7

Martínez, M. A., Díaz, M., Lima, A. I., Herrera, M., & Herrera-Viedma, E. (2014). A bibliometric análisis of Spanish production of Social Work category according to the Journal Citation Report. *Cuadernos de Trabajo Social*, *27*(2), 429–438.

Memari, A., Dargi, A., Jokar, M. R. A., Ahmad, R., & Rahim, A. R. A. (2019). Sustainable supplier selection: A multi-criteria intuitionistic fuzzy TOPSIS method. *Journal of Manufacturing Systems*, *50*, 9–24. doi:10.1016/j.jmsy.2018.11.002

Menaka, G. (2017). Ranking of Octagonal intuitionistic fuzzy number. *IOSR Journal of Mathematics*, *13*(3), 63–71. doi:10.9790/5728-1303026371

Mendel, J. M. (2007). Advances in type-2 fuzzy sets and systems. *Information Sciences*, *177*(1), 84–110. doi:10.1016/j.ins.2006.05.003

Mendel, J. M. (2010). *Perceptual Computing*. doi:10.1002/9780470599655

Mendel, J. M., & John, R. I. B. (2002). Type-2 fuzzy sets made simple. *IEEE Transactions on Fuzzy Systems*, *10*(2), 117–127. doi:10.1109/91.995115

Mendel, J. M., & Xinwang Liu. (2013). Simplified interval type-2 fuzzy logic systems. *IEEE Transactions on Fuzzy Systems*, *21*(6), 1056–1069. doi:10.1109/TFUZZ.2013.2241771

Meng, F., & Chen, X. (2016). Entropy and similarity measure for Atannasov's interval-valued intuitionistic fuzzy sets and their application. *Fuzzy Optimization and Decision Making*, *15*(1), 75–101. doi:10.100710700-015-9215-7

Mirakbari, M., & Ganji, A. (2010). Reliability analysis of a rangeland system: The application of Profust Theory. *Stochastic Environmental Research and Risk Assessment*, *24*(3), 399–409. doi:10.100700477-009-0329-8

Mishra, A. R., Rani, P., Pardasani, K. R., Mardani, A., Stević, Ž., & Pamučar, D. (2020). A novel entropy and divergence measures with multi-criteria service quality assessment using interval-valued intuitionistic fuzzy TODIM method. *Soft Computing*, *24*(15), 1–21. doi:10.100700500-019-04627-7

Mohsen, O., & Fereshteh, N. (2017). An extended VIKOR method based on entropy measure for the failure modes risk assessment – A case study of the geothermal power plant (GPP). *Safety Science*, *92*, 160–172. doi:10.1016/j.ssci.2016.10.006

Mon, D.-L., & Cheng, C.-H. (1994). Fuzzy system reliability analysis for components with different membership functions. *Fuzzy Sets and Systems*, *64*(2), 145–157. doi:10.1016/0165-0114(94)90330-1

Mondal, S. P. (2016). Differential equation with interval valued fuzzy number and its applications. *International Journal of System Assurance Engineering and Management*, *7*(3), 370–386. doi:10.100713198-016-0474-7

Mondal, S. P. (2018). Interval valued intuitionistic fuzzy number and its application in differential equation. *Journal of Intelligent & Fuzzy Systems*, *34*(1), 677–687. doi:10.3233/JIFS-161898

Mondal, S. P., Goswami, A., & Kumar De, S. (2019). Nonlinear Triangular Intuitionistic Fuzzy Number and Its Application in Linear Integral Equation. *Advances in Fuzzy Systems*, *2019*, 2019. doi:10.1155/2019/4142382

Mondal, S. P., Mandal, M., & Bhattacharya, D. (2018). Non-linear interval-valued fuzzy numbers and their application in difference equations. *Granular Computing*, *3*(2), 177–189. doi:10.100741066-017-0063-4

Mondal, S. P., & Roy, T. K. (2015). Second order linear differential equations with generalized trapezoidal intuitionistic Fuzzy boundary value. *Journal of Linear and Topological Algebra*, *4*(2), 115–129.

Muhuri, P. K., Ashraf, Z., & Goel, S. (2020). A Novel Image Steganographic Method based on Integer Wavelet Transformation and Particle Swarm Optimization. *Applied Soft Computing*, *92*, 106257. Advance online publication. doi:10.1016/j.asoc.2020.106257

Muhuri, P. K., Ashraf, Z., & Lohani, Q. M. D. (2018). Multiobjective Reliability Redundancy Allocation Problem With Interval Type-2 Fuzzy Uncertainty. *IEEE Transactions on Fuzzy Systems*, *26*(3), 1339–1355. doi:10.1109/TFUZZ.2017.2722422

Nair, S. J., & Sasikumar, K. (2019). Fuzzy reliability-based optimization of a hydropower reservoir. *Journal of Hydroinformatics*, *21*(2), 308–317. doi:10.2166/hydro.2019.078

Nannapaneni, S., & Mahadevan, S. (2016). Reliability analysis under epistemic uncertainty. *Reliability Engineering & System Safety*, *155*, 9–20. doi:10.1016/j.ress.2016.06.005

Negi, S., Jaiswal, N., & Singh, S. B. (2018). Reliability Assessment of Non-Repairable *k*-Out-Of-*n* System using Belief Universal Generating Function. *International Journal of Performability Engineering*, *14*(5), 831–840. doi:10.23940/ijpe.18.05.p1.831840

Negi, S., & Singh, S. B. (2015). Reliability analysis of non-repairable complex system with weighted subsystems connected in series. *Applied Mathematics and Computation*, *262*, 79–89. doi:10.1016/j.amc.2015.03.119

Nguyen, H. (2016). A new interval-valued knowledge measure for interval-valued intuitionistic fuzzy sets and application in decision making. *Expert Systems with Applications*, *56*, 143–155. doi:10.1016/j.eswa.2016.03.007

Nikolaidis, E., Chen, S., Cudney, H., Haftka, R. T., & Rosca, R. (2004). Comparison of probability and possibility for design against catastrophic failure under uncertainty. *Journal of Mechanical Design*, *126*(3), 386–394. doi:10.1115/1.1701878

Noguchi, K., Sasaki, M., Yanagi, S., & Yuge, T. (1996). Reliability of connected-(*rs*)-out-of-(*mn*): F lattice systems. *IEICE Transaction on Reliability, J79-A*, 954–962.

Nyaoga, R., Magutu, P., & Wang, M. (2016). Application of Grey-TOPSIS approach to evaluate value chain performance of tea processing chains. *Decision Science Letters*, *5*(3), 431–446. doi:10.5267/j.dsl.2016.1.002

Olivas, F., Amador-Angulo, L., Perez, J., Caraveo, C., Valdez, F., & Castillo, O. (2017). Comparative study of type-2 fuzzy Particle swarm, Bee Colony and Bat Algorithms in optimization of fuzzy controllers. *Algorithms*, *10*(3), 101. Advance online publication. doi:10.3390/a10030101

Olivas, F., Valdez, F., Castillo, O., & Melin, P. (2016). Dynamic parameter adaptation in particle swarm optimization using interval type-2 fuzzy logic. *Soft Computing*, *20*(3), 1057–1070. doi:10.100700500-014-1567-3

Pak, A., Parham, G. A., & Saraj, M. (2014). Reliability estimation in Rayleigh distribution based on fuzzy lifetime data. *International Journal of System Assurance Engineering and Management*, *5*(4), 487–494. doi:10.100713198-013-0190-5

Pinto-López, I. N., Montaudon-Tomas, C. M., & De la Vega Meneses, J. G. (2019). Bibliometrics as a robust analytical tool for research in higher education. *Conference proceedings ICERI 2019 12th International Conference of Education Research and Innovation*, 1106-1118. 10.21125/iceri.2019.0339

Pinto-López, I. N., Montaudon-Tomas, C. M., & Yáñez-Moneda, A. L. (2020). Bibliometric Analysis as a tool for Research-Based Learning (RBL). *Conference proceedings INTED 2020 14th International Technology, Education, and Development Conference*, 7108-7116.

Poli, R., Kennedy, J., & Blackwell, T. (2007). Particle swarm optimization. *Swarm Intelligence*, *1*(1), 33–57. doi:10.100711721-007-0002-0

Priyadharsini, J., & Balasubramaniam, P. (2019). Multi-criteria decision making method based on interval-valued intuitionistic fuzzy sets. *The Journal of Analysis*, *27*(1), 259–276. doi:10.100741478-018-0122-5

Qin, J., Xi, Y., & Pedrycz, W. (2020). Failure mode and effects analysis (FMEA) for risk assessment based on interval type-2 fuzzy evidential reasoning method. *Applied Soft Computing*, *89*, 106134. doi:10.1016/j.asoc.2020.106134

Ramirez-Marquez, J. E., & Coit, D. W. (2004). A heuristic for solving the redundancy allocation problem for multi-state series-parallel systems. *Reliability Engineering & System Safety*, *83*(3), 341–349. doi:10.1016/j.ress.2003.10.010

Ram, M., Singh, S. B., & Singh, V. V. (2013). Stochastic Analysis of a Standby System with Waiting Repair Strategy. *IEEE Transactions on Systems, Man, and Cybernetics. Systems*, *43*(3), 698–707. doi:10.1109/TSMCA.2012.2217320

Rashid, T., Faizi, S., & Zafar, S. (2018). Distance Based Entropy Measure of Interval-Valued Intuitionistic Fuzzy Sets and Its Application in Multicriteria Decision Making. *Advances in Fuzzy Systems*. doi:10.1155/2018/3637897

Rau, H., Liao, C. P., Shiang, W. J., & Lin, C. H. (2007, June). Using fuzzy theory for packaging attribute deployment for new notebook computer introduction. In *International Conference on Industrial, Engineering and Other Applications of Applied Intelligent Systems* (pp. 250-259). Springer. 10.1007/978-3-540-73325-6_25

Rausand, M. (2014). Reliability of Safety-Critical Systems. Reliability of Safety-Critical Systems. doi:10.1002/9781118776353

Ravi, V., Reddy, P. J., & Zimmermann, H. J. (2000). Fuzzy global optimization of complex system reliability. *IEEE Transactions on Fuzzy Systems*, *8*(3), 241–248. doi:10.1109/91.855914

Ren, W., Wu, K., Gu, Q., & Hu, Y. (2020). Intelligent decision making for service providers selection in maintenance service network: An adaptive fuzzy-neuro approach. *Knowledge-Based Systems*, *190*, 105263. https://doi.org/10.1016/j.knosys.2019.105263

Reyes, A. D., Olate, Y., & Godoy, C. (2019). Análisis bibliométrico de la producción científica de la Revista Actividad Física y Ciencias durante el período 2009-2018. *Revista Ciencias de la Actividad Física UCM*, *20*(2), 1–25. doi:10.29035/rcaf.20.2.9

Ross, T. (2016). Fuzzy Logic with Engineering Applications. In *Fuzzy Logic with Engineering Applications*. John Wiley & Sons, Incorporated.

Rotondaro, R. G., & De Oliveira, C. L. (2001). *Using failure mode effect analysis (FMEA) to improve service quality service operations management*. Academic Press.

Rotshtein, A., & Shtovba, S. (1997). *Fuzzy Reliability Analysis of Algorithmic Processes*. Continent-PRIM.

Saaty, T. L. (1980). *The analytic Hierarchy Process*. McGraw-Hill.

Sahaya Sudha, A. & Vijayalakshmi, K. R. (2018). Arithmetic Operations of Hexagonal Intuitionistic Fuzzy Number Using Extension Principle. *International Journal of Mathematics and its Applications*, *6*(2A), 219-225.

Sahba, A., Sahba, R., Rad, P., & Jamshidi, M. (2019). *Optimized IoT Based Decision Making for Autonomous Vehicles in Intersections*. doi:10.1109/UEMCON47517.2019.8992978

Salvia, A. A., & Lasher, W. C. (1990). Two-dimensional consecutive-k-out-of-n: F models. *IEEE Transactions on Reliability*, *39*(3), 382–385. doi:10.1109/24.103023

Selva Kumara, K., & Lavanya, S. (2017). Solving Fuzzy Game Problem Using Octagonal Intuitionistic Fuzzy Numbers. *International Journal of Pure and Applied Mathematics*, *113*(4), 188-198.

Senel, M., Senel, B., & Havle, C. A. (2018). Risk analysis of ports in Maritime Industry in Turkey using FMEA based intuitionistic Fuzzy TOPSIS Approach. In *ITM Web of Conferences* (Vol. 22, p. 01018). EDP Sciences.

Seyed-Hosseini, S. M., Safaei, N., & Asgharpour, M. J. (2006). Reprioritization of failures in a system failure mode and effects analysis by decision making trial and evaluation laboratory technique. *Reliability Engineering & System Safety*, *91*(8), 872–881. doi:10.1016/j.ress.2005.09.005

Shackley, S. (2001). Epistemic Lifestyles in Climate Change Modeling. In *Changing the Atmosphere: Expert Knowledge and Environmental Governance* (pp. 107–133). Cambridge: MIT Press.

Shannon, C. E. (1948). A mathematical theory of communication. *The Bell System Technical Journal*, *27*(3), 379–423. doi:10.1002/j.1538-7305.1948.tb01338.x

Shaw, A. K., & Roy, T. K. (2011). Generalized trapezoidal fuzzy number with its arithmetic operations and its application in fuzzy system reliability analysis. *International Journal of Pure and Applied Sciences and Technology*, *5*(2), 60–76.

Shi, Y., & Eberhart, R. C. (1998). *Parameter selection in particle swarm optimization*. doi:10.1007/BFb0040810

Shtovba, S., Rotshtein, A., & Pankevich, O. (2002). Fuzzy rule based system for diagnosis of stone construction cracks of buildings. In *Advances in Computational Intelligence and Learning* (pp. 401–411). Springer. doi:10.1007/978-94-010-0324-7_28

Simon, C., & Weber, P. (2009). Evidential networks for reliability analysis and performance evaluation of systems with imprecise knowledge. *IEEE Transactions on Reliability*, *58*(1), 69–87. doi:10.1109/TR.2008.2011868

Simon, H., & Newell, A. (1971). Human problem solving: The state of the theory in 1970. *The American Psychologist*, *26*(2), 145–159. doi:10.1037/h0030806

Sisman, B. (2019). Risk Evaluating by Fuzzy AHP and Fuzzy VIKOR methods in failure mode and effects analysis for automotive sector. *Mehmet Akif Ersoy Üniversitesi Sosyal Bilimler Enstitüsü Dergisi*, *9*(18), 234–250.

Soltani, R. (2014). Reliability optimization of binary state non-repairable systems: A state of the art survey. *International Journal of Industrial Engineering Computations*, *5*(3), 339–364. doi:10.5267/j.ijiec.2014.5.001

Song, W., Ming, X., Wu, Z., & Zhu, B. (2013). Failure modes and effects analysis using integrated weight-based fuzzy TOPSIS. *International Journal of Computer Integrated Manufacturing*, *26*(12), 1172–1186. doi:10.1080/0951192X.2013.785027

Song, X., Zhai, Z., Zhu, P., Guo, Y., & Zhang, Y. (2017, July). *Reliability approximation evaluation of multi-stated weighted k-out-of-n systems. In 2017 Prognostics and System Health Management Conference*. PHM-Harbin.

Srichetta, P., & Thurachon, W. (2012). Applying fuzzy analytic hierarchy process to evaluate and select product of notebook computers. *International Journal of Modeling and Optimization*, *2*(2), 168–173. doi:10.7763/IJMO.2012.V2.105

Sun, C., He, Z., Zhang, Z., Ning, X., & Zou, L. (2013). Operating reliability assessment for aero-engine based on condition monitoring information. *Jixie Gongcheng Xuebao*, *49*(6), 30–37. doi:10.3901/JME.2013.06.030

Sun, H., & Liao, J. (1990). The reliability *(n; f; k)* system. *Journal of Electronics (China)*, *12*, 436–439.

Szmidt, E., & Kacprzyk, J. (2009). Amount of information and its reliability in the ranking of Atanassov's intuitionistic fuzzy alternatives. In *Recent advances in decision making* (pp. 7–19). Springer. doi:10.1007/978-3-642-02187-9_2

Tian, Z. P., Wang, J. Q., & Zhang, H. Y. (2018). An integrated approach for failure mode and effects analysis based on fuzzy best-worst, relative entropy, and VIKOR methods. *App. S. Comp.*, *72*, 636–646. doi:10.1016/j.asoc.2018.03.037

Tian, Z., Zuo, M. J., & Huang, H. (2008). Reliability-redundancy allocation for multi-state series-parallel systems. *IEEE Transactions on Reliability*, *57*(2), 303–310. doi:10.1109/TR.2008.920871

Tillman, F. A., Hwang, C. L., Fan, L. T., & Lai, K. C. (1970). Optimal reliability of a complex system. *IEEE Transactions on Reliability*, *19*(3), 95–100.

Tong, Y. L. (1985). Rearrangement inequality for the longest run with an application to network reliability. *Journal of Applied Probability*, *22*(2), 286–393. doi:10.2307/3213781

Tooranloo, H. S., & Sadat Ayatollah, A. (2016). A model for failure mode and effects analysis based on intuitionistic fuzzy approach. *Applied Soft Computing*, *49*, 238-247.

Tooranloo, H. S. (2016). A model for failure mode and effects analysis based on intuitionistic fuzzy approach. *Applied Soft Computing*, *49*, 238–247. doi:10.1016/j.asoc.2016.07.047

Torra, V., & Narukawa, Y. (2009). On hesitant fuzzy sets and decision. *The 18th IEEE international Conference on Fuzzy Systems*, 1378-1382.

Torra, V. (2010). Hesitant fuzzy sets. *International Journal of Intelligent Systems*, *25*, 529–539.

Trelea, I. C. (2003). The particle swarm optimization algorithm: Convergence analysis and parameter selection. *Information Processing Letters*, *85*(6), 317–325. doi:10.1016/S0020-0190(02)00447-7

Tzafestas, S. G. (1980). Optimization of system reliability: A survey of problems and techniques†. *International Journal of Systems Science*, *11*(4), 455–486. doi:10.1080/00207728008967030

Ushakov, I. (1986). Universal Generating Function. *Soviet Journal of Computer Systems*, *24*, 118–129.

Ushakov, I. A. (1986). A universal generating function. *Soviet Journal of Computer and Systems Sciences*, *24*(5), 118–129.

Utkin, L. V., & Gurov, S. V. (1996). A general formal approach for fuzzy reliability analysis in the possibility context. *Fuzzy Sets and Systems*, *83*(2), 203–213. doi:10.1016/0165-0114(95)00391-6

Utkin, L., & Coolen, F. (2007). Imprecise reliability: An introductory overview. In *Intelligence in Reliability Engineering* (Vol. 40, pp. 261–306). Springer Berlin Heidelberg. doi:10.1007/978-3-540-37372-8_10

Uyanik, C., Tuzkaya, G., Kalender, Z. T., & Oguztimur, S. (2020). An integrated DEMATEL–IF-TOPSIS methodology for logistics centers' location selection problem: An application for Istanbul Metropolitan area. *Transport*, *0*(0), 1–9. doi:10.3846/transport.2020.12210

Vahdani, B., Salimi, M., & Charkhchian, M. (2015). A new FMEA method by integrating fuzzy belief structure and TOPSIS to improve risk evaluation process. *International Journal of Advanced Manufacturing Technology*, *77*(1–4), 357–368. doi:10.100700170-014-6466-3

Valdez, F., Vazquez, J. C., Melin, P., & Castillo, O. (2017). Comparative study of the use of fuzzy logic in improving particle swarm optimization variants for mathematical functions using co-evolution. *Applied Soft Computing*, *52*, 1070–1083. doi:10.1016/j.asoc.2016.09.024

Van Eck, N. J., & Waltman, L. (2010). Software survey: VOSviewer, a computer program for bibliometric mapping. *Scientometrics*, *84*(2), 523–538. doi:10.100711192-009-0146-3 PMID:20585380

Verma, R., & Sharma, B. D. (2013). Exponential entropy on intuitionistic fuzzy sets. *Kybernetika*, *49*(1), 114–127.

Wang, C. Y., & Chen, S.-M. (2017). An improved multiattribute decision making method based on new score function of interval-valued intuitionistic fuzzy values and linear programming methodology. *Information Sciences*, *411*, 176–184. doi:10.1016/j.ins.2017.05.022

Wang, J. Q., & Wang, P. (2012). Intuitionistic linguistic fuzzy multi-criteria decision-making method based on intuitionistic fuzzy entropy. *Control and Decision*, *27*(11), 1694–1698.

Wang, L. E., Liu, H. C., & Quan, M. Y. (2016). Evaluating the risk of failure modes with a hybrid MCDM model under interval-valued intuitionistic fuzzy environments. *Computers & Industrial Engineering*, *102*, 175–185. doi:10.1016/j.cie.2016.11.003

Wang, P., Zhang, J., Zhai, H., & Qiu, J. (2017). A new structural reliability index based on uncertainty theory. *Chinese Journal of Aeronautics*, *30*(4), 1451–1458. doi:10.1016/j.cja.2017.04.008

Wang, Y. M., Chin, K. S., Poon, G. K. K., & Yang, J. B. (2009). Risk evaluation in failure mode and effects analysis using fuzzy weighted geometric mean. *Expert Systems with Applications*, *36*(2), 1195–1207. doi:10.1016/j.eswa.2007.11.028

Wang, Y. X. (2010). Application of fuzzy mathematics to slope stability analysis. *Yantu Lixue*, *31*(9), 3000–3004.

Wang, Y., Li, L., Huang, S., & Chang, Q. (2012). Reliability and covariance estimation of weighted k-out-of-n multi-state systems. *European Journal of Operational Research, 221*(1), 138–147. doi:10.1016/j.ejor.2012.02.037

Wang, Y., Li, X., Zhang, B., & Li, M. (2013). Dynamic progressive failure evolution research on landslide under precipitation. *Journal of Hydraulic Engineering*, *44*(4), 416–425.

Wang, Z., Li, Y., Huang, H. Z., & Liu, Y. (2012). Reliability analysis of structure for fuzzy safety state. *Intelligent Automation Soft Computing*, *18*(3), 215–224. doi:10.1080/10798587.2008.10643238

Wang, Z., Tianshi, C., Tang, K., & Yao, X. (2009). A multi-objective approach to redundancy allocation problem in parallel-series systems. *2009 IEEE Congress on Evolutionary Computation, CEC 2009*, 582–589. 10.1109/CEC.2009.4982998

Wei, C. P., Gao, Z. H., & Guo, T. T. (2012). An intuitionistic fuzzy entropy measure based on trigonometric function. *Control and Decision*, *27*(4), 571–574.

Wu, H. C. (2004). Fuzzy reliability estimation using Bayesian approach. *Computers & Industrial Engineering*, *46*(3), 467–493.

Wu, H. C. (2006). Fuzzy Bayesian system reliability assessment based on exponential distribution. *Applied Mathematical Modelling*, *30*(6), 509–530. doi:10.1016/j.apm.2005.05.014

Wu, J. S., & Chen, R. J. (1994). An algorithm for computing the reliability of weighted-k-out-of-n systems. *IEEE Transactions on Reliability*, *43*(2), 327–328. doi:10.1109/24.295016

Xia, M., & Xu, Z. S. (2011). Hesitant fuzzy information aggregation in decision making. *International Journal of Approximate Reasoning*, *52*(3), 395–407. doi:10.1016/j.ijar.2010.09.002

Xiaoning, L. (2008). Fuzzy reliability theory of determining testing pressure of steel pressure vessels. *China Petroleum Machinery, 4*. Available online: http://en.cnki.com.cn/Article_en/CJFDTotal-SYJI200804004.htm

Xu, Z. S. (2007). Models for multiple attribute decision making with intuitionistic fuzzy information. *International Journal of Uncertainty, Fuzziness and Knowledge-based Systems*, *15*(03), 285–297. doi:10.1142/S0218488507004686

Xu, Z., & Liao, H. (2013). Intuitionistic fuzzy analytic hierarchy process. *IEEE Transactions on Fuzzy Systems*, *22*(4), 749–761. doi:10.1109/TFUZZ.2013.2272585

Yamamoto, H., & Miyakawa, M. (1995). Reliability of a linear connected-(r, s)-out-of-(m, n): F lattice system. *IEEE Transactions on Reliability*, *44*(2), 333–336. doi:10.1109/24.387391

Yazdani, M., Abdi, M. R., Kumar, N., Keshavarz-Ghorabaee, M., & Chan, F. T. (2019). Improved decision model for evaluating risks in construction projects. *J. Cons. Eng. Man.*, *145*(5), 04019024. doi:10.1061/(ASCE)CO.1943-7862.0001640

Yazdi, M., Nedjati, A., Zarei, E., & Abbassi, R. (2020). A reliable risk analysis approach using an extension of best-worst method based on democratic-autocratic decision-making style. *Journal of Cleaner Production, 256*, 120418. doi:10.1016/j.jclepro.2020.120418

Ye, F. (2010). An extended {TOPSIS} method with interval-valued intuitionistic fuzzy numbers for virtual enterprise partner selection. *Expert Systems with Applications, 37*(10), 7050–7055. doi:10.1016/j.eswa.2010.03.013

Yeh, W.-C., Lin, Y.-C., Chung, Y. Y., & Chih, M. (2010). A Particle Swarm Optimization Approach Based on Monte Carlo Simulation for Solving the Complex Network Reliability Problem. *IEEE Transactions on Reliability, 59*(1), 212–221. doi:10.1109/TR.2009.2035796

Yingkui, G., & Jing, L. (2012). *Multi-state system reliability: A new and systematic review*. Procedia Engineeri.

Yuge, T., Dehare, M., & Yanagi, S. (2000). Reliability and availability of a repairable lattice system. *IEICE TRANSACTIONS on Fundamentals of Electronics. Communications and Computer Sciences, 83*(5), 782–787.

Zadeh, L. (1965). Fuzzy sets. *Information and Computation, 8*, 338–353.

Zadeh, L. (1978). Fuzzy sets as a basis for a theory of possibility. *Fuzzy Sets and Systems, 1*(1), 3–28. doi:10.1016/0165-0114(78)90029-5

Zadeh, L. A. (1975). The concept of a linguistic variable and its application to approximate reasoning-I. *Information Sciences, 8*(3), 199–249. doi:10.1016/0020-0255(75)90036-5

Zammori, F., & Gabbrielli, R. (2012). ANP/RPN: A multi criteria evaluation of the risk priority number. *Quality and Reliability Engineering International, 28*(1), 85–104. doi:10.1002/qre.1217

Zavadskas, E. K., Antucheviciene, J., Hajiagha, S. H. R., & Hashemi, S. S. (2014). Extension of weighted aggregated sum product assessment with interval-valued intuitionistic fuzzy numbers (WASPAS-IVIF). *Applied Soft Computing, 24*, 1013–1021. doi:10.1016/j.asoc.2014.08.031

Zhang, H., Chang, A., Wang, Q., Zhou, Q., & Liang, Y. (2016). Research and application of transformer condition evaluation model based on multi-source information. *High Voltage Apparatus, 52*(2), 19–27.

Zhang, Q. S., & Jiang, S. Y. (2008). A note on information entropy measures for vague sets and its applications. *Information Sciences, 178*(21), 4184–4191. doi:10.1016/j.ins.2008.07.003

Zhao, H., You, J. X., & Liu, H. C. (2017). Failure mode and effect analysis using MULTIMOORA method with continuous weighted entropy under interval-valued intuitionistic fuzzy environment. *Soft Computing, 21*(18), 5355–5367. doi:10.100700500-016-2118-x

Zhao, Z., Quan, Q., & Cai, K. Y. (2014). A modified profust-performance-reliability algorithm and its application to dynamic systems. *Journal of Intelligent & Fuzzy Systems, 32*(1), 643–660. doi:10.3233/JIFS-152544

Zhao, Z., Quan, Q., & Cai, K. Y. (2014). A profust Reliability-Based Approach to Prognostics and Health Management. *IEEE Transactions on Reliability*, *63*(1), 26–41. doi:10.1109/TR.2014.2299111

Zhou, D., Zhang, X. F., Yang, Z., & Zhang, Y. M. (2015). Vibration reliability analysis on tractive transmission system of shearer. *Journal of the China Coal Society*, *40*(11), 2546–2551.

Zhou, Q., & Thai, V. V. (2016). Fuzzy and grey theories in failure mode and effect analysis for tanker equipment failure prediction. *Safety Science*, *83*, 74–79. doi:10.1016/j.ssci.2015.11.013

Zhu, J., Shuai, B., Li, G., Chin, K. S., & Wang, R. (2020). Failure mode and effect analysis using regret theory and PROMETHEE under linguistic neutrosophic context. *Journal of Loss Prevention in the Process Industries*, *64*, 104048. doi:10.1016/j.jlp.2020.104048

Zimmermann, H. (2010). Fuzzy set theory. *Wiley Interdisciplinary Reviews: Computational Statistics*, *2*(3), 317–332. doi:10.1002/wics.82

Zio, E. (2009). Reliability engineering: Old problems and new challenges. *Reliability Engineering & System Safety*, *94*(2), 125–141. doi:10.1016/j.ress.2008.06.002

Ziv, G., Watson, E., Young, D., Howard, D., Larcom, S., & Tanentzap, A. (2018). The potential impact of Brexit on the energy, water and food nexus in the UK: A fuzzy cognitive mapping approach. *Applied Energy*, *210*, 487–498. https://doi.org/10.1016/j.apenergy.2017.08.033

Zuo, M. J., Lin, D., & Wu, Y. (2000). Reliability evaluation of combined k-out-of-n: F, consecutive-k-out-of-n: F and linear connected-(r, s)-out-of-(m, n): F system structures. *IEEE Transactions on Reliability*, *49*(1), 99–104. doi:10.1109/24.855542

Related References

To continue our tradition of advancing academic research, we have compiled a list of recommended IGI Global readings. These references will provide additional information and guidance to further enrich your knowledge and assist you with your own research and future publications.

Abed, S., Khir, T., & Ben Brahim, A. (2016). Thermodynamic and Energy Study of a Regenerator in Gas Turbine Cycle and Optimization of Performances. *International Journal of Energy Optimization and Engineering*, *5*(2), 25–44. doi:10.4018/IJEOE.2016040102

Abu Bakar, W. A., Abdullah, W. N., Ali, R., & Mokhtar, W. N. (2016). Polymolybdate Supported Nano Catalyst for Desulfurization of Diesel. In T. Saleh (Ed.), *Applying Nanotechnology to the Desulfurization Process in Petroleum Engineering* (pp. 263–280). Hershey, PA: IGI Global. doi:10.4018/978-1-4666-9545-0.ch009

Addo-Tenkorang, R., Helo, P., & Kantola, J. (2016). Engineer-To-Order Product Development: A Communication Network Analysis for Supply-Chain's Sustainable Competitive Advantage. In R. Addo-Tenkorang, J. Kantola, P. Helo, & A. Shamsuzzoha (Eds.), *Supply Chain Strategies and the Engineer-to-Order Approach* (pp. 43–59). Hershey, PA: IGI Global. doi:10.4018/978-1-5225-0021-6.ch003

Adebiyi, I. D., Popoola, P. A., & Pityana, S. (2016). Mitigation of Wear Damage by Laser Surface Alloying Technique. In E. Akinlabi, R. Mahamood, & S. Akinlabi (Eds.), *Advanced Manufacturing Techniques Using Laser Material Processing* (pp. 172–196). Hershey, PA: IGI Global. doi:10.4018/978-1-5225-0329-3.ch007

Ahmad, W. (2016). Sulfur in Petroleum: Petroleum Desulfurization Techniques. In T. Saleh (Ed.), *Applying Nanotechnology to the Desulfurization Process in Petroleum Engineering* (pp. 1–52). Hershey, PA: IGI Global. doi:10.4018/978-1-4666-9545-0.ch001

Ahmed, I., Ahmad, N., Mehmood, I., Haq, I. U., Hassan, M., & Khan, M. U. (2016). Applications of Nanotechnology in Transportation Engineering. In A. Khitab & W. Anwar (Eds.), *Advanced Research on Nanotechnology for Civil Engineering Applications* (pp. 180–207). Hershey, PA: IGI Global. doi:10.4018/978-1-5225-0344-6.ch006

Aikhuele, D. (2018). A Study of Product Development Engineering and Design Reliability Concerns. *International Journal of Applied Industrial Engineering*, *5*(1), 79–89. doi:10.4018/IJAIE.2018010105

Al-Najar, B. T., & Bououdina, M. (2016). Bioinspired Nanoparticles for Efficient Drug Delivery System. In M. Bououdina (Ed.), *Emerging Research on Bioinspired Materials Engineering* (pp. 69–103). Hershey, PA: IGI Global. doi:10.4018/978-1-4666-9811-6.ch003

Al-Shebeeb, O. A., Rangaswamy, S., Gopalakrishan, B., & Devaru, D. G. (2017). Evaluation and Indexing of Process Plans Based on Electrical Demand and Energy Consumption. *International Journal of Manufacturing, Materials, and Mechanical Engineering*, *7*(3), 1–19. doi:10.4018/IJMMME.2017070101

Alexakis, H., & Makris, N. (2016). Validation of the Discrete Element Method for the Limit Stability Analysis of Masonry Arches. In V. Sarhosis, K. Bagi, J. Lemos, & G. Milani (Eds.), *Computational Modeling of Masonry Structures Using the Discrete Element Method* (pp. 292–325). Hershey, PA: IGI Global. doi:10.4018/978-1-5225-0231-9.ch012

AlMegren, H. A., Gonzalez-Cortes, S., Huang, Y., Chen, H., Qian, Y., Alkinany, M., ... Xiao, T. (2016). Preparation of Deep Hydrodesulfurzation Catalysts for Diesel Fuel using Organic Matrix Decomposition Method. In H. Al-Megren & T. Xiao (Eds.), *Petrochemical Catalyst Materials, Processes, and Emerging Technologies* (pp. 216–253). Hershey, PA: IGI Global. doi:10.4018/978-1-4666-9975-5.ch009

Alshammari, A., Kalevaru, V. N., Bagabas, A., & Martin, A. (2016). Production of Ethylene and its Commercial Importance in the Global Market. In H. Al-Megren & T. Xiao (Eds.), *Petrochemical Catalyst Materials, Processes, and Emerging Technologies* (pp. 82–115). Hershey, PA: IGI Global. doi:10.4018/978-1-4666-9975-5.ch004

Amel, M. (2016). Synthesis, Characterizations, and Biological Effects Study of Some Quinoline Family. In M. Bououdina (Ed.), *Emerging Research on Bioinspired Materials Engineering* (pp. 160–196). Hershey, PA: IGI Global. doi:10.4018/978-1-4666-9811-6.ch006

Amna, T., Haasan, M. S., Khil, M., & Hwang, I. (2016). Impact of Electrospun Biomimetic Extracellular Environment on Proliferation and Intercellular Communication of Muscle Precursor Cells: An Overview – Intercellular Communication of Muscle Precursor Cells with Extracellular Environment. In M. Bououdina (Ed.), *Emerging Research on Bioinspired Materials Engineering* (pp. 247–265). Hershey, PA: IGI Global. doi:10.4018/978-1-4666-9811-6.ch009

Amuda, M. O., Lawal, T. F., & Akinlabi, E. T. (2017). Research Progress on Rheological Behavior of AA7075 Aluminum Alloy During Hot Deformation. *International Journal of Materials Forming and Machining Processes*, *4*(1), 53–96. doi:10.4018/IJMFMP.2017010104

An, M., & Qin, Y. (2016). Challenges of Railway Safety Risk Assessment and Maintenance Decision Making. In B. Rai (Ed.), *Handbook of Research on Emerging Innovations in Rail Transportation Engineering* (pp. 173–211). Hershey, PA: IGI Global. doi:10.4018/978-1-5225-0084-1.ch009

Anil, M., Ayyildiz-Tamis, D., Tasdemir, S., Sendemir-Urkmez, A., & Gulce-Iz, S. (2016). Bioinspired Materials and Biocompatibility. In M. Bououdina (Ed.), *Emerging Research on Bioinspired Materials Engineering* (pp. 294–322). Hershey, PA: IGI Global. doi:10.4018/978-1-4666-9811-6.ch011

Armutlu, H. (2018). Intelligent Biomedical Engineering Operations by Cloud Computing Technologies. In U. Kose, G. Guraksin, & O. Deperlioglu (Eds.), *Nature-Inspired Intelligent Techniques for Solving Biomedical Engineering Problems* (pp. 297–317). Hershey, PA: IGI Global. doi:10.4018/978-1-5225-4769-3.ch015

Arokiyaraj, S., Saravanan, M., Bharanidharan, R., Islam, V. I., Bououdina, M., & Vincent, S. (2016). Green Synthesis of Metallic Nanoparticles Using Plant Compounds and Their Applications: Metallic Nanoparticles Synthesis Using Plants. In M. Bououdina (Ed.), *Emerging Research on Bioinspired Materials Engineering* (pp. 1–34). Hershey, PA: IGI Global. doi:10.4018/978-1-4666-9811-6.ch001

Atik, M., Sadek, M., & Shahrour, I. (2017). Single-Run Adaptive Pushover Procedure for Shear Wall Structures. In V. Plevris, G. Kremmyda, & Y. Fahjan (Eds.), *Performance-Based Seismic Design of Concrete Structures and Infrastructures* (pp. 59–83). Hershey, PA: IGI Global. doi:10.4018/978-1-5225-2089-4.ch003

Aydin, A., Akyol, E., Gungor, M., Kaya, A., & Tasdelen, S. (2018). Geophysical Surveys in Engineering Geology Investigations With Field Examples. In N. Ceryan (Ed.), *Handbook of Research on Trends and Digital Advances in Engineering Geology* (pp. 257–280). Hershey, PA: IGI Global. doi:10.4018/978-1-5225-2709-1.ch007

Azevedo, N. M., Lemos, J. V., & Rocha de Almeida, J. (2016). Discrete Element Particle Modelling of Stone Masonry. In V. Sarhosis, K. Bagi, J. Lemos, & G. Milani (Eds.), *Computational Modeling of Masonry Structures Using the Discrete Element Method* (pp. 146–170). Hershey, PA: IGI Global. doi:10.4018/978-1-5225-0231-9.ch007

Bamufleh, H. S., Noureldin, M. M., & El-Halwagi, M. M. (2016). Sustainable Process Integration in the Petrochemical Industries. In H. Al-Megren & T. Xiao (Eds.), *Petrochemical Catalyst Materials, Processes, and Emerging Technologies* (pp. 150–163). Hershey, PA: IGI Global. doi:10.4018/978-1-4666-9975-5.ch006

Banerjee, S., Gautam, R. K., Gautam, P. K., Jaiswal, A., & Chattopadhyaya, M. C. (2016). Recent Trends and Advancement in Nanotechnology for Water and Wastewater Treatment: Nanotechnological Approach for Water Purification. In A. Khitab & W. Anwar (Eds.), *Advanced Research on Nanotechnology for Civil Engineering Applications* (pp. 208–252). Hershey, PA: IGI Global. doi:10.4018/978-1-5225-0344-6.ch007

Bas, T. G. (2017). Nutraceutical Industry with the Collaboration of Biotechnology and Nutrigenomics Engineering: The Significance of Intellectual Property in the Entrepreneurship and Scientific Research Ecosystems. In T. Bas & J. Zhao (Eds.), *Comparative Approaches to Biotechnology Development and Use in Developed and Emerging Nations* (pp. 1–17). Hershey, PA: IGI Global. doi:10.4018/978-1-5225-1040-6.ch001

Beale, R., & André, J. (2017). *Design Solutions and Innovations in Temporary Structures*. Hershey, PA: IGI Global. doi:10.4018/978-1-5225-2199-0

Behnam, B. (2017). Simulating Post-Earthquake Fire Loading in Conventional RC Structures. In P. Samui, S. Chakraborty, & D. Kim (Eds.), *Modeling and Simulation Techniques in Structural Engineering* (pp. 425–444). Hershey, PA: IGI Global. doi:10.4018/978-1-5225-0588-4.ch015

Ben Hamida, I., Salah, S. B., Msahli, F., & Mimouni, M. F. (2018). Distribution Network Reconfiguration Using SPEA2 for Power Loss Minimization and Reliability Improvement. *International Journal of Energy Optimization and Engineering*, *7*(1), 50–65. doi:10.4018/IJEOE.2018010103

Benjamin, S. R., de Lima, F., & Rathoure, A. K. (2016). Genetically Engineered Microorganisms for Bioremediation Processes: GEMs for Bioremediaton. In A. Rathoure & V. Dhatwalia (Eds.), *Toxicity and Waste Management Using Bioremediation* (pp. 113–140). Hershey, PA: IGI Global. doi:10.4018/978-1-4666-9734-8.ch006

Bhaskar, S. V., & Kudal, H. N. (2017). Effect of TiCN and AlCrN Coating on Tribological Behaviour of Plasma-nitrided AISI 4140 Steel. *International Journal of Surface Engineering and Interdisciplinary Materials Science*, *5*(2), 1–17. doi:10.4018/IJSEIMS.2017070101

Bhowmik, S., Sahoo, P., Acharyya, S. K., Dhar, S., & Chattopadhyay, J. (2016). Effect of Microstructure Degradation on Fracture Toughness of 20MnMoNi55 Steel in DBT Region. *International Journal of Manufacturing, Materials, and Mechanical Engineering*, *6*(3), 11–27. doi:10.4018/IJMMME.2016070102

Bhutto, A. W., Abro, R., Abbas, T., Yu, G., & Chen, X. (2016). Desulphurization of Fuel Oils Using Ionic Liquids. In H. Al-Megren & T. Xiao (Eds.), *Petrochemical Catalyst Materials, Processes, and Emerging Technologies* (pp. 254–284). Hershey, PA: IGI Global. doi:10.4018/978-1-4666-9975-5.ch010

Bhuyan, D. (2018). Designing of a Twin Tube Shock Absorber: A Study in Reverse Engineering. In K. Kumar & J. Davim (Eds.), *Design and Optimization of Mechanical Engineering Products* (pp. 83–104). Hershey, PA: IGI Global. doi:10.4018/978-1-5225-3401-3.ch005

Bouloudenine, M., & Bououdina, M. (2016). Toxic Effects of Engineered Nanoparticles on Living Cells. In M. Bououdina (Ed.), *Emerging Research on Bioinspired Materials Engineering* (pp. 35–68). Hershey, PA: IGI Global. doi:10.4018/978-1-4666-9811-6.ch002

Brunetti, A., Sellaro, M., Drioli, E., & Barbieri, G. (2016). Membrane Engineering and its Role in Oil Refining and Petrochemical Industry. In H. Al-Megren & T. Xiao (Eds.), *Petrochemical Catalyst Materials, Processes, and Emerging Technologies* (pp. 116–149). Hershey, PA: IGI Global. doi:10.4018/978-1-4666-9975-5.ch005

Bügler, M., & Borrmann, A. (2016). Simulation Based Construction Project Schedule Optimization: An Overview on the State-of-the-Art. In F. Miranda & C. Abreu (Eds.), *Handbook of Research on Computational Simulation and Modeling in Engineering* (pp. 482–507). Hershey, PA: IGI Global. doi:10.4018/978-1-4666-8823-0.ch016

Calderon, F. A., Giolo, E. G., Frau, C. D., Rengel, M. G., Rodriguez, H., Tornello, M., ... Gallucci, R. (2018). Seismic Microzonation and Site Effects Detection Through Microtremors Measures: A Review. In N. Ceryan (Ed.), *Handbook of Research on Trends and Digital Advances in Engineering Geology* (pp. 326–349). Hershey, PA: IGI Global. doi:10.4018/978-1-5225-2709-1.ch009

Carmona-Murillo, J., & Valenzuela-Valdés, J. F. (2016). Motivation on Problem Based Learning. In D. Fonseca & E. Redondo (Eds.), *Handbook of Research on Applied E-Learning in Engineering and Architecture Education* (pp. 179–203). Hershey, PA: IGI Global. doi:10.4018/978-1-4666-8803-2.ch009

Ceryan, N. (2016). A Review of Soft Computing Methods Application in Rock Mechanic Engineering. In P. Samui (Ed.), *Handbook of Research on Advanced Computational Techniques for Simulation-Based Engineering* (pp. 1–70). Hershey, PA: IGI Global. doi:10.4018/978-1-4666-9479-8.ch001

Ceryan, N., & Can, N. K. (2018). Prediction of The Uniaxial Compressive Strength of Rocks Materials. In N. Ceryan (Ed.), *Handbook of Research on Trends and Digital Advances in Engineering Geology* (pp. 31–96). Hershey, PA: IGI Global. doi:10.4018/978-1-5225-2709-1.ch002

Ceryan, S. (2018). Weathering Indices Used in Evaluation of the Weathering State of Rock Material. In N. Ceryan (Ed.), *Handbook of Research on Trends and Digital Advances in Engineering Geology* (pp. 132–186). Hershey, PA: IGI Global. doi:10.4018/978-1-5225-2709-1.ch004

Chandrasekaran, S., Silva, B., Patil, A., Oo, A. M., & Campbell, M. (2016). Evaluating Engineering Students' Perceptions: The Impact of Team-Based Learning Practices in Engineering Education. *International Journal of Quality Assurance in Engineering and Technology Education, 5*(4), 42–59. doi:10.4018/IJQAETE.2016100103

Chen, H., Padilla, R. V., & Besarati, S. (2017). Supercritical Fluids and Their Applications in Power Generation. In L. Chen & Y. Iwamoto (Eds.), *Advanced Applications of Supercritical Fluids in Energy Systems* (pp. 369–402). Hershey, PA: IGI Global. doi:10.4018/978-1-5225-2047-4.ch012

Chen, L. (2017). Principles, Experiments, and Numerical Studies of Supercritical Fluid Natural Circulation System. In L. Chen & Y. Iwamoto (Eds.), *Advanced Applications of Supercritical Fluids in Energy Systems* (pp. 136–187). Hershey, PA: IGI Global. doi:10.4018/978-1-5225-2047-4.ch005

Clementi, F., Di Sciascio, G., Di Sciascio, S., & Lenci, S. (2017). Influence of the Shear-Bending Interaction on the Global Capacity of Reinforced Concrete Frames: A Brief Overview of the New Perspectives. In V. Plevris, G. Kremmyda, & Y. Fahjan (Eds.), *Performance-Based Seismic Design of Concrete Structures and Infrastructures* (pp. 84–111). Hershey, PA: IGI Global. doi:10.4018/978-1-5225-2089-4.ch004

Cortés-Polo, D., Calle-Cancho, J., Carmona-Murillo, J., & González-Sánchez, J. (2017). Future Trends in Mobile-Fixed Integration for Next Generation Networks: Classification and Analysis. *International Journal of Vehicular Telematics and Infotainment Systems*, *1*(1), 33–53. doi:10.4018/IJVTIS.2017010103

Cui, X., Zeng, S., Li, Z., Zheng, Q., Yu, X., & Han, B. (2018). Advanced Composites for Civil Engineering Infrastructures. In K. Kumar & J. Davim (Eds.), *Composites and Advanced Materials for Industrial Applications* (pp. 212–248). Hershey, PA: IGI Global. doi:10.4018/978-1-5225-5216-1.ch010

Dalgıç, S., & Kuşku, İ. (2018). Geological and Geotechnical Investigations in Tunneling. In N. Ceryan (Ed.), *Handbook of Research on Trends and Digital Advances in Engineering Geology* (pp. 482–529). Hershey, PA: IGI Global. doi:10.4018/978-1-5225-2709-1.ch014

de la Varga, D., Soto, M., Arias, C. A., van Oirschot, D., Kilian, R., Pascual, A., & Álvarez, J. A. (2017). Constructed Wetlands for Industrial Wastewater Treatment and Removal of Nutrients. In Á. Val del Río, J. Campos Gómez, & A. Mosquera Corral (Eds.), *Technologies for the Treatment and Recovery of Nutrients from Industrial Wastewater* (pp. 202–230). Hershey, PA: IGI Global. doi:10.4018/978-1-5225-1037-6.ch008

del Valle-Zermeño, R., Chimenos, J. M., & Formosa, J. (2016). Flue Gas Desulfurization: Processes and Technologies. In T. Saleh (Ed.), *Applying Nanotechnology to the Desulfurization Process in Petroleum Engineering* (pp. 337–377). Hershey, PA: IGI Global. doi:10.4018/978-1-4666-9545-0.ch011

Delgado, J. M., Henriques, A. A., & Delgado, R. M. (2016). Structural Non-Linear Models and Simulation Techniques: An Efficient Combination for Safety Evaluation of RC Structures. In F. Miranda & C. Abreu (Eds.), *Handbook of Research on Computational Simulation and Modeling in Engineering* (pp. 540–584). Hershey, PA: IGI Global. doi:10.4018/978-1-4666-8823-0.ch018

Delgado, P. S., Arêde, A., Pouca, N. V., & Costa, A. (2016). Numerical Modeling of RC Bridges for Seismic Risk Analysis. In F. Miranda & C. Abreu (Eds.), *Handbook of Research on Computational Simulation and Modeling in Engineering* (pp. 457–481). Hershey, PA: IGI Global. doi:10.4018/978-1-4666-8823-0.ch015

Deng, Y., & Liu, S. (2016). Catalysis with Room Temperature Ionic Liquids Mediated Metal Nanoparticles. In H. Al-Megren & T. Xiao (Eds.), *Petrochemical Catalyst Materials, Processes, and Emerging Technologies* (pp. 285–329). Hershey, PA: IGI Global. doi:10.4018/978-1-4666-9975-5.ch011

Deperlioglu, O. (2018). Intelligent Techniques Inspired by Nature and Used in Biomedical Engineering. In U. Kose, G. Guraksin, & O. Deperlioglu (Eds.), *Nature-Inspired Intelligent Techniques for Solving Biomedical Engineering Problems* (pp. 51–77). Hershey, PA: IGI Global. doi:10.4018/978-1-5225-4769-3.ch003

Dias, G. L., Magalhães, R. R., Ferreira, D. D., & Vitoriano, F. A. (2016). The Use of a Robotic Arm for Displacement Measurements in a Cantilever beam. *International Journal of Manufacturing, Materials, and Mechanical Engineering*, *6*(3), 45–57. doi:10.4018/IJMMME.2016070104

Dimitratos, N., Villa, A., Chan-Thaw, C. E., Hammond, C., & Prati, L. (2016). Valorisation of Glycerol to Fine Chemicals and Fuels. In H. Al-Megren & T. Xiao (Eds.), *Petrochemical Catalyst Materials, Processes, and Emerging Technologies* (pp. 352–384). Hershey, PA: IGI Global. doi:10.4018/978-1-4666-9975-5.ch013

Dixit, A. (2018). Application of Silica-Gel-Reinforced Aluminium Composite on the Piston of Internal Combustion Engine: Comparative Study of Silica-Gel-Reinforced Aluminium Composite Piston With Aluminium Alloy Piston. In K. Kumar & J. Davim (Eds.), *Composites and Advanced Materials for Industrial Applications* (pp. 63–98). Hershey, PA: IGI Global. doi:10.4018/978-1-5225-5216-1.ch004

Drei, A., Milani, G., & Sincraian, G. (2016). Application of DEM to Historic Masonries, Two Case-Studies in Portugal and Italy: Aguas Livres Aqueduct and Arch-Tympana of a Church. In V. Sarhosis, K. Bagi, J. Lemos, & G. Milani (Eds.), *Computational Modeling of Masonry Structures Using the Discrete Element Method* (pp. 326–366). Hershey, PA: IGI Global. doi:10.4018/978-1-5225-0231-9.ch013

Dutta, S., Roy, P. K., & Nandi, D. (2016). Optimal Allocation of Static Synchronous Series Compensator Controllers using Chemical Reaction Optimization for Reactive Power Dispatch. *International Journal of Energy Optimization and Engineering*, *5*(3), 43–62. doi:10.4018/IJEOE.2016070103

Dutta, S., Roy, P. K., & Nandi, D. (2016). Quasi Oppositional Teaching-Learning based Optimization for Optimal Power Flow Incorporating FACTS. *International Journal of Energy Optimization and Engineering*, *5*(2), 64–84. doi:10.4018/IJEOE.2016040104

Eloy, S., Dias, M. S., Lopes, P. F., & Vilar, E. (2016). Digital Technologies in Architecture and Engineering: Exploring an Engaged Interaction within Curricula. In D. Fonseca & E. Redondo (Eds.), *Handbook of Research on Applied E-Learning in Engineering and Architecture Education* (pp. 368–402). Hershey, PA: IGI Global. doi:10.4018/978-1-4666-8803-2.ch017

Elsayed, A. M., Dakkama, H. J., Mahmoud, S., Al-Dadah, R., & Kaialy, W. (2017). Sustainable Cooling Research Using Activated Carbon Adsorbents and Their Environmental Impact. In T. Kobayashi (Ed.), *Applied Environmental Materials Science for Sustainability* (pp. 186–221). Hershey, PA: IGI Global. doi:10.4018/978-1-5225-1971-3.ch009

Ercanoglu, M., & Sonmez, H. (2018). General Trends and New Perspectives on Landslide Mapping and Assessment Methods. In N. Ceryan (Ed.), *Handbook of Research on Trends and Digital Advances in Engineering Geology* (pp. 350–379). Hershey, PA: IGI Global. doi:10.4018/978-1-5225-2709-1.ch010

Erinosho, M. F., Akinlabi, E. T., & Pityana, S. (2016). Enhancement of Surface Integrity of Titanium Alloy with Copper by Means of Laser Metal Deposition Process. In E. Akinlabi, R. Mahamood, & S. Akinlabi (Eds.), *Advanced Manufacturing Techniques Using Laser Material Processing* (pp. 60–91). Hershey, PA: IGI Global. doi:10.4018/978-1-5225-0329-3.ch004

Farag, H., & Kishida, M. (2016). Kinetic Models for Complex Parallel–Consecutive Reactions Assessment of Reaction Network and Product Selectivity. In H. Al-Megren & T. Xiao (Eds.), *Petrochemical Catalyst Materials, Processes, and Emerging Technologies* (pp. 330–351). Hershey, PA: IGI Global. doi:10.4018/978-1-4666-9975-5.ch012

Faroz, S. A., Pujari, N. N., Rastogi, R., & Ghosh, S. (2017). Risk Analysis of Structural Engineering Systems Using Bayesian Inference. In P. Samui, S. Chakraborty, & D. Kim (Eds.), *Modeling and Simulation Techniques in Structural Engineering* (pp. 390–424). Hershey, PA: IGI Global. doi:10.4018/978-1-5225-0588-4.ch014

Fernando, P. R., Hamigah, T., Disne, S., Wickramasingha, G. G., & Sutharshan, A. (2018). The Evaluation of Engineering Properties of Low Cost Concrete Blocks by Partial Doping of Sand with Sawdust: Low Cost Sawdust Concrete Block. *International Journal of Strategic Engineering*, *1*(2), 26–42. doi:10.4018/IJoSE.2018070103

Fragiadakis, M., Stefanou, I., & Psycharis, I. N. (2016). Vulnerability Assessment of Damaged Classical Multidrum Columns. In V. Sarhosis, K. Bagi, J. Lemos, & G. Milani (Eds.), *Computational Modeling of Masonry Structures Using the Discrete Element Method* (pp. 235–253). Hershey, PA: IGI Global. doi:10.4018/978-1-5225-0231-9.ch010

Gaines, T. W., Williams, K. R., & Wagener, K. B. (2016). ADMET: Functionalized Polyolefins. In H. Al-Megren & T. Xiao (Eds.), *Petrochemical Catalyst Materials, Processes, and Emerging Technologies* (pp. 1–21). Hershey, PA: IGI Global. doi:10.4018/978-1-4666-9975-5.ch001

Garg, H. (2016). Bi-Criteria Optimization for Finding the Optimal Replacement Interval for Maintaining the Performance of the Process Industries. In P. Vasant, G. Weber, & V. Dieu (Eds.), *Handbook of Research on Modern Optimization Algorithms and Applications in Engineering and Economics* (pp. 643–675). Hershey, PA: IGI Global. doi:10.4018/978-1-4666-9644-0.ch025

Gaspar, P. D., Dinho da Silva, P., Gonçalves, J. P., & Carneiro, R. (2016). Computational Modelling and Simulation to Assist the Improvement of Thermal Performance and Energy Efficiency in Industrial Engineering Systems: Application to Cold Stores. In F. Miranda & C. Abreu (Eds.), *Handbook of Research on Computational Simulation and Modeling in Engineering* (pp. 1–68). Hershey, PA: IGI Global. doi:10.4018/978-1-4666-8823-0.ch001

Ge, H., Tang, M., & Wen, X. (2016). Ni/ZnO Nano Sorbent for Reactive Adsorption Desulfurization of Refinery Oil Streams. In T. Saleh (Ed.), *Applying Nanotechnology to the Desulfurization Process in Petroleum Engineering* (pp. 216–239). Hershey, PA: IGI Global. doi:10.4018/978-1-4666-9545-0.ch007

Ghosh, S., Mitra, S., Ghosh, S., & Chakraborty, S. (2017). Seismic Reliability Analysis in the Framework of Metamodelling Based Monte Carlo Simulation. In P. Samui, S. Chakraborty, & D. Kim (Eds.), *Modeling and Simulation Techniques in Structural Engineering* (pp. 192–208). Hershey, PA: IGI Global. doi:10.4018/978-1-5225-0588-4.ch006

Gil, M., & Otero, B. (2017). Learning Engineering Skills through Creativity and Collaboration: A Game-Based Proposal. In R. Alexandre Peixoto de Queirós & M. Pinto (Eds.), *Gamification-Based E-Learning Strategies for Computer Programming Education* (pp. 14–29). Hershey, PA: IGI Global. doi:10.4018/978-1-5225-1034-5.ch002

Gill, J., Ayre, M., & Mills, J. (2017). Revisioning the Engineering Profession: How to Make It Happen! In M. Gray & K. Thomas (Eds.), *Strategies for Increasing Diversity in Engineering Majors and Careers* (pp. 156–175). Hershey, PA: IGI Global. doi:10.4018/978-1-5225-2212-6.ch008

Gopal, S., & Al-Hazmi, M. H. (2016). Advances in Catalytic Technologies for Selective Oxidation of Lower Alkanes. In H. Al-Megren & T. Xiao (Eds.), *Petrochemical Catalyst Materials, Processes, and Emerging Technologies* (pp. 22–52). Hershey, PA: IGI Global. doi:10.4018/978-1-4666-9975-5.ch002

Goyal, N., Ram, M., Bhardwaj, A., & Kumar, A. (2016). Thermal Power Plant Modelling with Fault Coverage Stochastically. *International Journal of Manufacturing, Materials, and Mechanical Engineering*, *6*(3), 28–44. doi:10.4018/IJMMME.2016070103

Goyal, N., Ram, M., & Kumar, P. (2017). Welding Process under Fault Coverage Approach for Reliability and MTTF. In M. Ram & J. Davim (Eds.), *Mathematical Concepts and Applications in Mechanical Engineering and Mechatronics* (pp. 222–245). Hershey, PA: IGI Global. doi:10.4018/978-1-5225-1639-2.ch011

Gray, M., & Lundy, C. (2017). Engineering Study Abroad: High Impact Strategy for Increasing Access. In M. Gray & K. Thomas (Eds.), *Strategies for Increasing Diversity in Engineering Majors and Careers* (pp. 42–59). Hershey, PA: IGI Global. doi:10.4018/978-1-5225-2212-6.ch003

Guha, D., Roy, P. K., & Banerjee, S. (2016). Application of Modified Biogeography Based Optimization in AGC of an Interconnected Multi-Unit Multi-Source AC-DC Linked Power System. *International Journal of Energy Optimization and Engineering*, *5*(3), 1–18. doi:10.4018/IJEOE.2016070101

Guha, D., Roy, P. K., & Banerjee, S. (2016). Grey Wolf Optimization to Solve Load Frequency Control of an Interconnected Power System: GWO Used to Solve LFC Problem. *International Journal of Energy Optimization and Engineering*, *5*(4), 62–83. doi:10.4018/IJEOE.2016100104

Gupta, A. K., Dey, A., & Mukhopadhyay, A. K. (2016). Micromechanical and Finite Element Modeling for Composites. In S. Datta & J. Davim (Eds.), *Computational Approaches to Materials Design: Theoretical and Practical Aspects* (pp. 101–162). Hershey, PA: IGI Global. doi:10.4018/978-1-5225-0290-6.ch005

Guraksin, G. E. (2018). Internet of Things and Nature-Inspired Intelligent Techniques for the Future of Biomedical Engineering. In U. Kose, G. Guraksin, & O. Deperlioglu (Eds.), *Nature-Inspired Intelligent Techniques for Solving Biomedical Engineering Problems* (pp. 263–282). Hershey, PA: IGI Global. doi:10.4018/978-1-5225-4769-3.ch013

Hansman, C. A. (2016). Developing Mentoring Programs in Engineering and Technology Education. *International Journal of Quality Assurance in Engineering and Technology Education*, *5*(2), 1–15. doi:10.4018/IJQAETE.2016040101

Hasan, U., Chegenizadeh, A., & Nikraz, H. (2016). Nanotechnology Future and Present in Construction Industry: Applications in Geotechnical Engineering. In A. Khitab & W. Anwar (Eds.), *Advanced Research on Nanotechnology for Civil Engineering Applications* (pp. 141–179). Hershey, PA: IGI Global. doi:10.4018/978-1-5225-0344-6.ch005

Hejazi, T., & Akbari, L. (2017). A Multiresponse Optimization Model for Statistical Design of Processes with Discrete Variables. In M. Ram & J. Davim (Eds.), *Mathematical Concepts and Applications in Mechanical Engineering and Mechatronics* (pp. 17–37). Hershey, PA: IGI Global. doi:10.4018/978-1-5225-1639-2.ch002

Hejazi, T., & Hejazi, A. (2017). Monte Carlo Simulation for Reliability-Based Design of Automotive Complex Subsystems. In M. Ram & J. Davim (Eds.), *Mathematical Concepts and Applications in Mechanical Engineering and Mechatronics* (pp. 177–200). Hershey, PA: IGI Global. doi:10.4018/978-1-5225-1639-2.ch009

Hejazi, T., & Poursabbagh, H. (2017). Reliability Analysis of Engineering Systems: An Accelerated Life Testing for Boiler Tubes. In M. Ram & J. Davim (Eds.), *Mathematical Concepts and Applications in Mechanical Engineering and Mechatronics* (pp. 154–176). Hershey, PA: IGI Global. doi:10.4018/978-1-5225-1639-2.ch008

Henao, J., & Sotelo, O. (2018). Surface Engineering at High Temperature: Thermal Cycling and Corrosion Resistance. In A. Pakseresht (Ed.), *Production, Properties, and Applications of High Temperature Coatings* (pp. 131–159). Hershey, PA: IGI Global. doi:10.4018/978-1-5225-4194-3.ch006

Huirache-Acuña, R., Alonso-Nuñez, G., Rivera-Muñoz, E. M., Gutierrez, O., & Pawelec, B. (2016). Trimetallic Sulfide Catalysts for Hydrodesulfurization. In T. Saleh (Ed.), *Applying Nanotechnology to the Desulfurization Process in Petroleum Engineering* (pp. 240–262). Hershey, PA: IGI Global. doi:10.4018/978-1-4666-9545-0.ch008

Ilori, O. O., Adetan, D. A., & Umoru, L. E. (2017). Effect of Cutting Parameters on the Surface Residual Stress of Face-Milled Pearlitic Ductile Iron. *International Journal of Materials Forming and Machining Processes*, *4*(1), 38–52. doi:10.4018/IJMFMP.2017010103

Imam, M. H., Tasadduq, I. A., Ahmad, A., Aldosari, F., & Khan, H. (2017). Automated Generation of Course Improvement Plans Using Expert System. *International Journal of Quality Assurance in Engineering and Technology Education*, *6*(1), 1–12. doi:10.4018/IJQAETE.2017010101

Injeti, S. K., & Kumar, T. V. (2018). A WDO Framework for Optimal Deployment of DGs and DSCs in a Radial Distribution System Under Daily Load Pattern to Improve Techno-Economic Benefits. *International Journal of Energy Optimization and Engineering*, *7*(2), 1–38. doi:10.4018/IJEOE.2018040101

Ishii, N., Anami, K., & Knisely, C. W. (2018). *Dynamic Stability of Hydraulic Gates and Engineering for Flood Prevention*. Hershey, PA: IGI Global. doi:10.4018/978-1-5225-3079-4

J., J., Chowdhury, S., Goyal, P., Samui, P., & Dalkiliç, Y. (2016). Determination of Bearing Capacity of Shallow Foundation Using Soft Computing. In P. Saxena, D. Singh, & M. Pant (Eds.), *Problem Solving and Uncertainty Modeling through Optimization and Soft Computing Applications* (pp. 292-328). Hershey, PA: IGI Global. doi:10.4018/978-1-4666-9885-7.ch014

Jagan, J., Gundlapalli, P., & Samui, P. (2016). Utilization of Classification Techniques for the Determination of Liquefaction Susceptibility of Soils. In S. Bhattacharyya, P. Banerjee, D. Majumdar, & P. Dutta (Eds.), *Handbook of Research on Advanced Hybrid Intelligent Techniques and Applications* (pp. 124–160). Hershey, PA: IGI Global. doi:10.4018/978-1-4666-9474-3.ch005

Jayapalan, S. (2018). A Review of Chemical Treatments on Natural Fibers-Based Hybrid Composites for Engineering Applications. In K. Kumar & J. Davim (Eds.), *Composites and Advanced Materials for Industrial Applications* (pp. 16–37). Hershey, PA: IGI Global. doi:10.4018/978-1-5225-5216-1.ch002

Jeet, K., & Dhir, R. (2016). Software Module Clustering Using Bio-Inspired Algorithms. In P. Vasant, G. Weber, & V. Dieu (Eds.), *Handbook of Research on Modern Optimization Algorithms and Applications in Engineering and Economics* (pp. 445–470). Hershey, PA: IGI Global. doi:10.4018/978-1-4666-9644-0.ch017

Joshi, S. D., & Talange, D. B. (2016). Fault Tolerant Control for a Fractional Order AUV System. *International Journal of Energy Optimization and Engineering*, *5*(2), 1–24. doi:10.4018/IJEOE.2016040101

Julião, D., Ribeiro, S., de Castro, B., Cunha-Silva, L., & Balula, S. S. (2016). Polyoxometalates-Based Nanocatalysts for Production of Sulfur-Free Diesel. In T. Saleh (Ed.), *Applying Nanotechnology to the Desulfurization Process in Petroleum Engineering* (pp. 426–458). Hershey, PA: IGI Global. doi:10.4018/978-1-4666-9545-0.ch014

Kamthan, P. (2016). On the Nature of Collaborations in Agile Software Engineering Course Projects. *International Journal of Quality Assurance in Engineering and Technology Education*, *5*(2), 42–59. doi:10.4018/IJQAETE.2016040104

Karaman, O., Celik, C., & Urkmez, A. S. (2016). Self-Assembled Biomimetic Scaffolds for Bone Tissue Engineering. In M. Bououdina (Ed.), *Emerging Research on Bioinspired Materials Engineering* (pp. 104–132). Hershey, PA: IGI Global. doi:10.4018/978-1-4666-9811-6.ch004

Karkalos, N. E., Markopoulos, A. P., & Dossis, M. F. (2017). Optimal Model Parameters of Inverse Kinematics Solution of a 3R Robotic Manipulator Using ANN Models. *International Journal of Manufacturing, Materials, and Mechanical Engineering*, *7*(3), 20–40. doi:10.4018/IJMMME.2017070102

Kesimal, A., Karaman, K., Cihangir, F., & Ercikdi, B. (2018). Excavatability Assessment of Rock Masses for Geotechnical Studies. In N. Ceryan (Ed.), *Handbook of Research on Trends and Digital Advances in Engineering Geology* (pp. 231–256). Hershey, PA: IGI Global. doi:10.4018/978-1-5225-2709-1.ch006

Khanh, D. V., Vasant, P. M., Elamvazuthi, I., & Dieu, V. N. (2016). Multi-Objective Optimization of Two-Stage Thermo-Electric Cooler Using Differential Evolution: MO Optimization of TEC Using DE. In F. Miranda & C. Abreu (Eds.), *Handbook of Research on Computational Simulation and Modeling in Engineering* (pp. 139–170). Hershey, PA: IGI Global. doi:10.4018/978-1-4666-8823-0.ch004

Kim, D., Hassan, M. K., Chang, S., & Bigdeli, Y. (2016). Nonlinear Vibration Control of 3D Irregular Structures Subjected to Seismic Loads. In P. Samui (Ed.), *Handbook of Research on Advanced Computational Techniques for Simulation-Based Engineering* (pp. 103–119). Hershey, PA: IGI Global. doi:10.4018/978-1-4666-9479-8.ch003

Knoflacher, H. (2017). The Role of Engineers and Their Tools in the Transport Sector after Paradigm Change: From Assumptions and Extrapolations to Science. In H. Knoflacher & E. Ocalir-Akunal (Eds.), *Engineering Tools and Solutions for Sustainable Transportation Planning* (pp. 1–29). Hershey, PA: IGI Global. doi:10.4018/978-1-5225-2116-7.ch001

Kose, U. (2018). Towards an Intelligent Biomedical Engineering With Nature-Inspired Artificial Intelligence Techniques. In U. Kose, G. Guraksin, & O. Deperlioglu (Eds.), *Nature-Inspired Intelligent Techniques for Solving Biomedical Engineering Problems* (pp. 1–26). Hershey, PA: IGI Global. doi:10.4018/978-1-5225-4769-3.ch001

Kostić, S. (2018). A Review on Enhanced Stability Analyses of Soil Slopes Using Statistical Design. In N. Ceryan (Ed.), *Handbook of Research on Trends and Digital Advances in Engineering Geology* (pp. 446–481). Hershey, PA: IGI Global. doi:10.4018/978-1-5225-2709-1.ch013

Kumar, A., Patil, P. P., & Prajapati, Y. K. (2018). *Advanced Numerical Simulations in Mechanical Engineering*. Hershey, PA: IGI Global. doi:10.4018/978-1-5225-3722-9

Kumar, G. R., Rajyalakshmi, G., & Manupati, V. K. (2017). Surface Micro Patterning of Aluminium Reinforced Composite through Laser Peening. *International Journal of Manufacturing, Materials, and Mechanical Engineering*, *7*(4), 15–27. doi:10.4018/IJMMME.2017100102

Kumari, N., & Kumar, K. (2018). Fabrication of Orthotic Calipers With Epoxy-Based Green Composite. In K. Kumar & J. Davim (Eds.), *Composites and Advanced Materials for Industrial Applications* (pp. 157–176). Hershey, PA: IGI Global. doi:10.4018/978-1-5225-5216-1.ch008

Kuppusamy, R. R. (2018). Development of Aerospace Composite Structures Through Vacuum-Enhanced Resin Transfer Moulding Technology (VERTMTy): Vacuum-Enhanced Resin Transfer Moulding. In K. Kumar & J. Davim (Eds.), *Composites and Advanced Materials for Industrial Applications* (pp. 99–111). Hershey, PA: IGI Global. doi:10.4018/978-1-5225-5216-1.ch005

Lemos, J. V. (2016). The Basis for Masonry Analysis with UDEC and 3DEC. In V. Sarhosis, K. Bagi, J. Lemos, & G. Milani (Eds.), *Computational Modeling of Masonry Structures Using the Discrete Element Method* (pp. 61–89). Hershey, PA: IGI Global. doi:10.4018/978-1-5225-0231-9.ch003

Loy, J., Howell, S., & Cooper, R. (2017). Engineering Teams: Supporting Diversity in Engineering Education. In M. Gray & K. Thomas (Eds.), *Strategies for Increasing Diversity in Engineering Majors and Careers* (pp. 106–129). Hershey, PA: IGI Global. doi:10.4018/978-1-5225-2212-6.ch006

Macher, G., Armengaud, E., Kreiner, C., Brenner, E., Schmittner, C., Ma, Z., ... Krammer, M. (2018). Integration of Security in the Development Lifecycle of Dependable Automotive CPS. In N. Druml, A. Genser, A. Krieg, M. Menghin, & A. Hoeller (Eds.), *Solutions for Cyber-Physical Systems Ubiquity* (pp. 383–423). Hershey, PA: IGI Global. doi:10.4018/978-1-5225-2845-6.ch015

Maghsoodlou, S., & Poreskandar, S. (2016). Controlling Electrospinning Jet Using Microscopic Model for Ideal Tissue Engineering Scaffolds. *International Journal of Chemoinformatics and Chemical Engineering*, *5*(2), 1–16. doi:10.4018/IJCCE.2016070101

Mahendramani, G., & Lakshmana Swamy, N. (2018). Effect of Weld Groove Area on Distortion of Butt Welded Joints in Submerged Arc Welding. *International Journal of Manufacturing, Materials, and Mechanical Engineering*, *8*(2), 33–44. doi:10.4018/IJMMME.2018040103

Maiti, S. (2016). Engineered Gellan Polysaccharides in the Design of Controlled Drug Delivery Systems. In M. Bououdina (Ed.), *Emerging Research on Bioinspired Materials Engineering* (pp. 266–293). Hershey, PA: IGI Global. doi:10.4018/978-1-4666-9811-6.ch010

Majumdar, J. D., Weisheit, A., & Manna, I. (2016). Laser Surface Processing for Tailoring of Properties by Optimization of Microstructure. In E. Akinlabi, R. Mahamood, & S. Akinlabi (Eds.), *Advanced Manufacturing Techniques Using Laser Material Processing* (pp. 121–171). Hershey, PA: IGI Global. doi:10.4018/978-1-5225-0329-3.ch006

Maldonado-Macías, A. A., García-Alcaraz, J. L., Hernández-Arellano, J. L., & Cortes-Robles, G. (2016). An Ergonomic Compatibility Perspective on the Selection of Advanced Manufacturing Technology: A Case Study for CNC Vertical Machining Centers. In G. Alor-Hernández, C. Sánchez-Ramírez, & J. García-Alcaraz (Eds.), *Handbook of Research on Managerial Strategies for Achieving Optimal Performance in Industrial Processes* (pp. 137–165). Hershey, PA: IGI Global. doi:10.4018/978-1-5225-0130-5.ch008

Mamaghani, I. H. (2016). Application of Discrete Finite Element Method for Analysis of Unreinforced Masonry Structures. In V. Sarhosis, K. Bagi, J. Lemos, & G. Milani (Eds.), *Computational Modeling of Masonry Structures Using the Discrete Element Method* (pp. 440–458). Hershey, PA: IGI Global. doi:10.4018/978-1-5225-0231-9.ch017

Mansor, M. R., Sapuan, S. M., Salim, M. A., Akop, M. Z., Musthafah, M. T., & Shaharuzaman, M. A. (2016). Concurrent Design of Green Composites. In D. Verma, S. Jain, X. Zhang, & P. Gope (Eds.), *Green Approaches to Biocomposite Materials Science and Engineering* (pp. 48–75). Hershey, PA: IGI Global. doi:10.4018/978-1-5225-0424-5.ch003

Mansouri, I., & Esmaeili, E. (2016). Nanotechnology Applications in the Construction Industry. In A. Khitab & W. Anwar (Eds.), *Advanced Research on Nanotechnology for Civil Engineering Applications* (pp. 111–140). Hershey, PA: IGI Global. doi:10.4018/978-1-5225-0344-6.ch004

Manzoor, A. (2016). MOOCs for Enhancing Engineering Education. In D. Fonseca & E. Redondo (Eds.), *Handbook of Research on Applied E-Learning in Engineering and Architecture Education* (pp. 204–223). Hershey, PA: IGI Global. doi:10.4018/978-1-4666-8803-2.ch010

Martin, A., Kalevaru, V. N., & Radnik, J. (2016). Palladium in Heterogeneous Oxidation Catalysis. In H. Al-Megren & T. Xiao (Eds.), *Petrochemical Catalyst Materials, Processes, and Emerging Technologies* (pp. 53–81). Hershey, PA: IGI Global. doi:10.4018/978-1-4666-9975-5.ch003

Melnyczuk, J. M., & Palchoudhury, S. (2016). Introduction to Bio-Inspired Hydrogel and Their Application: Hydrogels. In M. Bououdina (Ed.), *Emerging Research on Bioinspired Materials Engineering* (pp. 133–159). Hershey, PA: IGI Global. doi:10.4018/978-1-4666-9811-6.ch005

Mitra-Kirtley, S., Mullins, O. C., & Pomerantz, A. E. (2016). Sulfur and Nitrogen Chemical Speciation in Crude Oils and Related Carbonaceous Materials. In T. Saleh (Ed.), *Applying Nanotechnology to the Desulfurization Process in Petroleum Engineering* (pp. 53–83). Hershey, PA: IGI Global. doi:10.4018/978-1-4666-9545-0.ch002

Moalosi, R., Uziak, J., & Oladiran, M. T. (2016). Using Blended Learning Approach to Deliver Courses in An Engineering Programme. *International Journal of Quality Assurance in Engineering and Technology Education*, *5*(1), 23–39. doi:10.4018/IJQAETE.2016010103

Mohammadzadeh, S., & Kim, Y. (2017). Nonlinear System Identification of Smart Buildings. In P. Samui, S. Chakraborty, & D. Kim (Eds.), *Modeling and Simulation Techniques in Structural Engineering* (pp. 328–347). Hershey, PA: IGI Global. doi:10.4018/978-1-5225-0588-4.ch011

Mohanty, I., & Bhattacherjee, D. (2016). Artificial Neural Network and Its Application in Steel Industry. In S. Datta & J. Davim (Eds.), *Computational Approaches to Materials Design: Theoretical and Practical Aspects* (pp. 267–300). Hershey, PA: IGI Global. doi:10.4018/978-1-5225-0290-6.ch010

Mohebkhah, A., & Sarhosis, V. (2016). Discrete Element Modeling of Masonry-Infilled Frames. In V. Sarhosis, K. Bagi, J. Lemos, & G. Milani (Eds.), *Computational Modeling of Masonry Structures Using the Discrete Element Method* (pp. 200–234). Hershey, PA: IGI Global. doi:10.4018/978-1-5225-0231-9.ch009

Molina, G. J., Aktaruzzaman, F., Soloiu, V., & Rahman, M. (2017). Design and Testing of a Jet-Impingement Instrument to Study Surface-Modification Effects by Nanofluids. *International Journal of Surface Engineering and Interdisciplinary Materials Science*, *5*(2), 43–61. doi:10.4018/IJSEIMS.2017070104

Montalvan-Sorrosa, D., de los Cobos-Vasconcelos, D., & Gonzalez-Sanchez, A. (2016). Nanotechnology Applied to the Biodesulfurization of Fossil Fuels and Spent Caustic Streams. In T. Saleh (Ed.), *Applying Nanotechnology to the Desulfurization Process in Petroleum Engineering* (pp. 378–389). Hershey, PA: IGI Global. doi:10.4018/978-1-4666-9545-0.ch012

Montillet, J., Yu, K., Bonenberg, L. K., & Roberts, G. W. (2016). Optimization Algorithms in Local and Global Positioning. In P. Vasant, G. Weber, & V. Dieu (Eds.), *Handbook of Research on Modern Optimization Algorithms and Applications in Engineering and Economics* (pp. 1–53). Hershey, PA: IGI Global. doi:10.4018/978-1-4666-9644-0.ch001

Moreira, F., & Ferreira, M. J. (2016). Teaching and Learning Requirements Engineering Based on Mobile Devices and Cloud: A Case Study. In D. Fonseca & E. Redondo (Eds.), *Handbook of Research on Applied E-Learning in Engineering and Architecture Education* (pp. 237–262). Hershey, PA: IGI Global. doi:10.4018/978-1-4666-8803-2.ch012

Mukherjee, A., Saeed, R. A., Dutta, S., & Naskar, M. K. (2017). Fault Tracking Framework for Software-Defined Networking (SDN). In C. Singhal & S. De (Eds.), *Resource Allocation in Next-Generation Broadband Wireless Access Networks* (pp. 247–272). Hershey, PA: IGI Global. doi:10.4018/978-1-5225-2023-8.ch011

Mukhopadhyay, A., Barman, T. K., & Sahoo, P. (2018). Electroless Nickel Coatings for High Temperature Applications. In K. Kumar & J. Davim (Eds.), *Composites and Advanced Materials for Industrial Applications* (pp. 297–331). Hershey, PA: IGI Global. doi:10.4018/978-1-5225-5216-1.ch013

Náprstek, J., & Fischer, C. (2017). Dynamic Stability and Post-Critical Processes of Slender Auto-Parametric Systems. In V. Plevris, G. Kremmyda, & Y. Fahjan (Eds.), *Performance-Based Seismic Design of Concrete Structures and Infrastructures* (pp. 128–171). Hershey, PA: IGI Global. doi:10.4018/978-1-5225-2089-4.ch006

Nautiyal, L., Shivach, P., & Ram, M. (2018). Optimal Designs by Means of Genetic Algorithms. In M. Ram & J. Davim (Eds.), *Soft Computing Techniques and Applications in Mechanical Engineering* (pp. 151–161). Hershey, PA: IGI Global. doi:10.4018/978-1-5225-3035-0.ch007

Nazir, R. (2017). Advanced Nanomaterials for Water Engineering and Treatment: Nano-Metal Oxides and Their Nanocomposites. In T. Saleh (Ed.), *Advanced Nanomaterials for Water Engineering, Treatment, and Hydraulics* (pp. 84–126). Hershey, PA: IGI Global. doi:10.4018/978-1-5225-2136-5.ch005

Nogueira, A. F., Ribeiro, J. C., Fernández de Vega, F., & Zenha-Rela, M. A. (2018). Evolutionary Approaches to Test Data Generation for Object-Oriented Software: Overview of Techniques and Tools. In M. Khosrow-Pour, D.B.A. (Ed.), Incorporating Nature-Inspired Paradigms in Computational Applications (pp. 162-194). Hershey, PA: IGI Global. doi:10.4018/978-1-5225-5020-4.ch006

Nunes, J. F., Moreira, P. M., & Tavares, J. M. (2016). Human Motion Analysis and Simulation Tools: A Survey. In F. Miranda & C. Abreu (Eds.), *Handbook of Research on Computational Simulation and Modeling in Engineering* (pp. 359–388). Hershey, PA: IGI Global. doi:10.4018/978-1-4666-8823-0.ch012

Ogunlaja, A. S., & Tshentu, Z. R. (2016). Molecularly Imprinted Polymer Nanofibers for Adsorptive Desulfurization. In T. Saleh (Ed.), *Applying Nanotechnology to the Desulfurization Process in Petroleum Engineering* (pp. 281–336). Hershey, PA: IGI Global. doi:10.4018/978-1-4666-9545-0.ch010

Ong, P., & Kohshelan, S. (2016). Performances of Adaptive Cuckoo Search Algorithm in Engineering Optimization. In P. Vasant, G. Weber, & V. Dieu (Eds.), *Handbook of Research on Modern Optimization Algorithms and Applications in Engineering and Economics* (pp. 676–699). Hershey, PA: IGI Global. doi:10.4018/978-1-4666-9644-0.ch026

Osho, M. B. (2018). Industrial Enzyme Technology: Potential Applications. In S. Bharati & P. Chaurasia (Eds.), *Research Advancements in Pharmaceutical, Nutritional, and Industrial Enzymology* (pp. 375–394). Hershey, PA: IGI Global. doi:10.4018/978-1-5225-5237-6.ch017

Padmaja, P., & Marutheswar, G. (2017). Certain Investigation on Secured Data Transmission in Wireless Sensor Networks. *International Journal of Mobile Computing and Multimedia Communications*, *8*(1), 48–61. doi:10.4018/IJMCMC.2017010104

Paixão, S. M., Silva, T. P., Arez, B. F., & Alves, L. (2016). Advances in the Reduction of the Costs Inherent to Fossil Fuels' Biodesulfurization towards Its Potential Industrial Application. In T. Saleh (Ed.), *Applying Nanotechnology to the Desulfurization Process in Petroleum Engineering* (pp. 390–425). Hershey, PA: IGI Global. doi:10.4018/978-1-4666-9545-0.ch013

Palmer, S., & Hall, W. (2017). An Evaluation of Group Work in First-Year Engineering Design Education. In R. Tucker (Ed.), *Collaboration and Student Engagement in Design Education* (pp. 145–168). Hershey, PA: IGI Global. doi:10.4018/978-1-5225-0726-0.ch007

Panneer, R. (2017). Effect of Composition of Fibers on Properties of Hybrid Composites. *International Journal of Manufacturing, Materials, and Mechanical Engineering*, *7*(4), 28–43. doi:10.4018/IJMMME.2017100103

Parker, J. (2016). Hubble's Expanding Universe: A Model for Quality in Technology Infused engineering and Technology Education. *International Journal of Quality Assurance in Engineering and Technology Education*, *5*(2), 16–29. doi:10.4018/IJQAETE.2016040102

Paul, S., & Roy, P. (2018). Optimal Design of Power System Stabilizer Using a Novel Evolutionary Algorithm. *International Journal of Energy Optimization and Engineering*, *7*(3), 24–46. doi:10.4018/IJEOE.2018070102

Pavaloiu, A. (2018). Artificial Intelligence Ethics in Biomedical-Engineering-Oriented Problems. In U. Kose, G. Guraksin, & O. Deperlioglu (Eds.), *Nature-Inspired Intelligent Techniques for Solving Biomedical Engineering Problems* (pp. 219–231). Hershey, PA: IGI Global. doi:10.4018/978-1-5225-4769-3.ch010

Peña, F. (2016). A Semi-Discrete Approach for the Numerical Simulation of Freestanding Blocks. In V. Sarhosis, K. Bagi, J. Lemos, & G. Milani (Eds.), *Computational Modeling of Masonry Structures Using the Discrete Element Method* (pp. 416–439). Hershey, PA: IGI Global. doi:10.4018/978-1-5225-0231-9.ch016

Penchovsky, R., & Traykovska, M. (2016). Synthetic Approaches to Biology: Engineering Gene Control Circuits, Synthesizing, and Editing Genomes. In M. Bououdina (Ed.), *Emerging Research on Bioinspired Materials Engineering* (pp. 323–351). Hershey, PA: IGI Global. doi:10.4018/978-1-4666-9811-6.ch012

Pieroni, A., & Iazeolla, G. (2016). Engineering QoS and Energy Saving in the Delivery of ICT Services. In P. Vasant & N. Voropai (Eds.), *Sustaining Power Resources through Energy Optimization and Engineering* (pp. 208–226). Hershey, PA: IGI Global. doi:10.4018/978-1-4666-9755-3.ch009

Pioro, I., Mahdi, M., & Popov, R. (2017). Application of Supercritical Pressures in Power Engineering. In L. Chen & Y. Iwamoto (Eds.), *Advanced Applications of Supercritical Fluids in Energy Systems* (pp. 404–457). Hershey, PA: IGI Global. doi:10.4018/978-1-5225-2047-4.ch013

Plaksina, T., & Gildin, E. (2017). Rigorous Integrated Evolutionary Workflow for Optimal Exploitation of Unconventional Gas Assets. *International Journal of Energy Optimization and Engineering*, *6*(1), 101–122. doi:10.4018/IJEOE.2017010106

Puppala, A. J., Bheemasetti, T. V., Zou, H., Yu, X., Pedarla, A., & Cai, G. (2016). Spatial Variability Analysis of Soil Properties using Geostatistics. In P. Samui (Ed.), *Handbook of Research on Advanced Computational Techniques for Simulation-Based Engineering* (pp. 195–226). Hershey, PA: IGI Global. doi:10.4018/978-1-4666-9479-8.ch008

Ramdani, N., & Azibi, M. (2018). Polymer Composite Materials for Microelectronics Packaging Applications: Composites for Microelectronics Packaging. In K. Kumar & J. Davim (Eds.), *Composites and Advanced Materials for Industrial Applications* (pp. 177–211). Hershey, PA: IGI Global. doi:10.4018/978-1-5225-5216-1.ch009

Ramesh, M., Garg, R., & Subrahmanyam, G. V. (2017). Investigation of Influence of Quenching and Annealing on the Plane Fracture Toughness and Brittle to Ductile Transition Temperature of the Zinc Coated Structural Steel Materials. *International Journal of Surface Engineering and Interdisciplinary Materials Science*, *5*(2), 33–42. doi:10.4018/IJSEIMS.2017070103

Razavi, A. M., & Ahmad, R. (2016). Agile Software Development Challenges in Implementation and Adoption: Focusing on Large and Distributed Settings – Past Experiences, Emergent Topics. In I. Ghani, D. Jawawi, S. Dorairaj, & A. Sidky (Eds.), *Emerging Innovations in Agile Software Development* (pp. 175–207). Hershey, PA: IGI Global. doi:10.4018/978-1-4666-9858-1.ch010

Reccia, E., Cecchi, A., & Milani, G. (2016). FEM/DEM Approach for the Analysis of Masonry Arch Bridges. In V. Sarhosis, K. Bagi, J. Lemos, & G. Milani (Eds.), *Computational Modeling of Masonry Structures Using the Discrete Element Method* (pp. 367–392). Hershey, PA: IGI Global. doi:10.4018/978-1-5225-0231-9.ch014

Ro, H. K., & McIntosh, K. (2016). Constructing Conducive Environment for Women of Color in Engineering Undergraduate Education. In U. Thomas & J. Drake (Eds.), *Critical Research on Sexism and Racism in STEM Fields* (pp. 23–48). Hershey, PA: IGI Global. doi:10.4018/978-1-5225-0174-9.ch002

Rodulfo-Baechler, S. M. (2016). Dual Role of Perovskite Hollow Fiber Membrane in the Methane Oxidation Reactions. In H. Al-Megren & T. Xiao (Eds.), *Petrochemical Catalyst Materials, Processes, and Emerging Technologies* (pp. 385–430). Hershey, PA: IGI Global. doi:10.4018/978-1-4666-9975-5.ch014

Rudolf, S., Biryuk, V. V., & Volov, V. (2018). Vortex Effect, Vortex Power: Technology of Vortex Power Engineering. In V. Kharchenko & P. Vasant (Eds.), *Handbook of Research on Renewable Energy and Electric Resources for Sustainable Rural Development* (pp. 500–533). Hershey, PA: IGI Global. doi:10.4018/978-1-5225-3867-7.ch021

Sah, A., Bhadula, S. J., Dumka, A., & Rawat, S. (2018). A Software Engineering Perspective for Development of Enterprise Applications. In A. Elçi (Ed.), *Handbook of Research on Contemporary Perspectives on Web-Based Systems* (pp. 1–23). Hershey, PA: IGI Global. doi:10.4018/978-1-5225-5384-7.ch001

Sahoo, P., & Roy, S. (2017). Tribological Behavior of Electroless Ni-P, Ni-P-W and Ni-P-Cu Coatings: A Comparison. *International Journal of Surface Engineering and Interdisciplinary Materials Science*, *5*(1), 1–15. doi:10.4018/IJSEIMS.2017010101

Sahoo, S. (2018). Laminated Composite Hypar Shells as Roofing Units: Static and Dynamic Behavior. In K. Kumar & J. Davim (Eds.), *Composites and Advanced Materials for Industrial Applications* (pp. 249–269). Hershey, PA: IGI Global. doi:10.4018/978-1-5225-5216-1.ch011

Sahu, H., & Hungyo, M. (2018). Introduction to SDN and NFV. In A. Dumka (Ed.), *Innovations in Software-Defined Networking and Network Functions Virtualization* (pp. 1–25). Hershey, PA: IGI Global. doi:10.4018/978-1-5225-3640-6.ch001

Saikia, P., Bharadwaj, S. K., & Miah, A. T. (2016). Peroxovanadates and Its Bio-Mimicking Relation with Vanadium Haloperoxidases. In M. Bououdina (Ed.), *Emerging Research on Bioinspired Materials Engineering* (pp. 197–219). Hershey, PA: IGI Global. doi:10.4018/978-1-4666-9811-6.ch007

Saladino, R., Botta, G., & Crucianelli, M. (2016). Advances in Nanotechnology Transition Metal Catalysts in Oxidative Desulfurization (ODS) Processes: Nanotechnology Applied to ODS Processing. In T. Saleh (Ed.), *Applying Nanotechnology to the Desulfurization Process in Petroleum Engineering* (pp. 180–215). Hershey, PA: IGI Global. doi:10.4018/978-1-4666-9545-0.ch006

Saleh, T. A., Danmaliki, G. I., & Shuaib, T. D. (2016). Nanocomposites and Hybrid Materials for Adsorptive Desulfurization. In T. Saleh (Ed.), *Applying Nanotechnology to the Desulfurization Process in Petroleum Engineering* (pp. 129–153). Hershey, PA: IGI Global. doi:10.4018/978-1-4666-9545-0.ch004

Saleh, T. A., Shuaib, T. D., Danmaliki, G. I., & Al-Daous, M. A. (2016). Carbon-Based Nanomaterials for Desulfurization: Classification, Preparation, and Evaluation. In T. Saleh (Ed.), *Applying Nanotechnology to the Desulfurization Process in Petroleum Engineering* (pp. 154–179). Hershey, PA: IGI Global. doi:10.4018/978-1-4666-9545-0.ch005

Salem, A. M., & Shmelova, T. (2018). Intelligent Expert Decision Support Systems: Methodologies, Applications, and Challenges. In T. Shmelova, Y. Sikirda, N. Rizun, A. Salem, & Y. Kovalyov (Eds.), *Socio-Technical Decision Support in Air Navigation Systems: Emerging Research and Opportunities* (pp. 215–242). Hershey, PA: IGI Global. doi:10.4018/978-1-5225-3108-1.ch007

Samal, M. (2017). FE Analysis and Experimental Investigation of Cracked and Un-Cracked Thin-Walled Tubular Components to Evaluate Mechanical and Fracture Properties. In P. Samui, S. Chakraborty, & D. Kim (Eds.), *Modeling and Simulation Techniques in Structural Engineering* (pp. 266–293). Hershey, PA: IGI Global. doi:10.4018/978-1-5225-0588-4.ch009

Samal, M., & Balakrishnan, K. (2017). Experiments on a Ring Tension Setup and FE Analysis to Evaluate Transverse Mechanical Properties of Tubular Components. In P. Samui, S. Chakraborty, & D. Kim (Eds.), *Modeling and Simulation Techniques in Structural Engineering* (pp. 91–115). Hershey, PA: IGI Global. doi:10.4018/978-1-5225-0588-4.ch004

Santhanakumar, M., Adalarasan, R., & Rajmohan, M. (2016). An Investigation in Abrasive Waterjet Cutting of Al6061/SiC/Al2O3 Composite Using Principal Component Based Response Surface Methodology. *International Journal of Manufacturing, Materials, and Mechanical Engineering, 6*(4), 30–47. doi:10.4018/IJMMME.2016100103

Sareen, N., & Bhattacharya, S. (2016). Cleaner Energy Fuels: Hydrodesulfurization and Beyond. In T. Saleh (Ed.), *Applying Nanotechnology to the Desulfurization Process in Petroleum Engineering* (pp. 84–128). Hershey, PA: IGI Global. doi:10.4018/978-1-4666-9545-0.ch003

Sarhosis, V. (2016). Micro-Modeling Options for Masonry. In V. Sarhosis, K. Bagi, J. Lemos, & G. Milani (Eds.), *Computational Modeling of Masonry Structures Using the Discrete Element Method* (pp. 28–60). Hershey, PA: IGI Global. doi:10.4018/978-1-5225-0231-9.ch002

Sarhosis, V., Oliveira, D. V., & Lourenco, P. B. (2016). On the Mechanical Behavior of Masonry. In V. Sarhosis, K. Bagi, J. Lemos, & G. Milani (Eds.), *Computational Modeling of Masonry Structures Using the Discrete Element Method* (pp. 1–27). Hershey, PA: IGI Global. doi:10.4018/978-1-5225-0231-9.ch001

Satyam, N. (2016). Liquefaction Modelling of Granular Soils using Discrete Element Method. In P. Samui (Ed.), *Handbook of Research on Advanced Computational Techniques for Simulation-Based Engineering* (pp. 381–441). Hershey, PA: IGI Global. doi:10.4018/978-1-4666-9479-8.ch015

Sawant, S. (2018). Deep Learning and Biomedical Engineering. In U. Kose, G. Guraksin, & O. Deperlioglu (Eds.), *Nature-Inspired Intelligent Techniques for Solving Biomedical Engineering Problems* (pp. 283–296). Hershey, PA: IGI Global. doi:10.4018/978-1-5225-4769-3.ch014

Sezgin, H., & Berkalp, O. B. (2018). Textile-Reinforced Composites for the Automotive Industry. In K. Kumar & J. Davim (Eds.), *Composites and Advanced Materials for Industrial Applications* (pp. 129–156). Hershey, PA: IGI Global. doi:10.4018/978-1-5225-5216-1.ch007

Shah, M. Z., Gazder, U., Bhatti, M. S., & Hussain, M. (2018). Comparative Performance Evaluation of Effects of Modifier in Asphaltic Concrete Mix. *International Journal of Strategic Engineering*, *1*(2), 13–25. doi:10.4018/IJoSE.2018070102

Shah, V. S., Shah, H. R., & Samui, P. (2016). Application of Meta-Models (MPMR and ELM) for Determining OMC, MDD and Soaked CBR Value of Soil. In S. Bhattacharyya, P. Banerjee, D. Majumdar, & P. Dutta (Eds.), *Handbook of Research on Advanced Hybrid Intelligent Techniques and Applications* (pp. 454–482). Hershey, PA: IGI Global. doi:10.4018/978-1-4666-9474-3.ch015

Sharma, N., & Kumar, K. (2018). Fabrication of Porous NiTi Alloy Using Organic Binders. In K. Kumar & J. Davim (Eds.), *Composites and Advanced Materials for Industrial Applications* (pp. 38–62). Hershey, PA: IGI Global. doi:10.4018/978-1-5225-5216-1.ch003

Sharma, T. K. (2016). Application of Shuffled Frog Leaping Algorithm in Software Project Scheduling. In P. Saxena, D. Singh, & M. Pant (Eds.), *Problem Solving and Uncertainty Modeling through Optimization and Soft Computing Applications* (pp. 225–238). Hershey, PA: IGI Global. doi:10.4018/978-1-4666-9885-7.ch011

Shivach, P., Nautiyal, L., & Ram, M. (2018). Applying Multi-Objective Optimization Algorithms to Mechanical Engineering. In M. Ram & J. Davim (Eds.), *Soft Computing Techniques and Applications in Mechanical Engineering* (pp. 287–301). Hershey, PA: IGI Global. doi:10.4018/978-1-5225-3035-0.ch014

Shmelova, T. (2018). Stochastic Methods for Estimation and Problem Solving in Engineering: Stochastic Methods of Decision Making in Aviation. In S. Kadry (Ed.), *Stochastic Methods for Estimation and Problem Solving in Engineering* (pp. 139–160). Hershey, PA: IGI Global. doi:10.4018/978-1-5225-5045-7.ch006

Shukla, R., Anapagaddi, R., Singh, A. K., Allen, J. K., Panchal, J. H., & Mistree, F. (2016). Integrated Computational Materials Engineering for Determining the Set Points of Unit Operations for Production of a Steel Product Mix. In S. Datta & J. Davim (Eds.), *Computational Approaches to Materials Design: Theoretical and Practical Aspects* (pp. 163–191). Hershey, PA: IGI Global. doi:10.4018/978-1-5225-0290-6.ch006

Siero González, L. R., & Romo Vázquez, A. (2017). Didactic Sequences Teaching Mathematics for Engineers With Focus on Differential Equations. In M. Ramírez-Montoya (Ed.), *Handbook of Research on Driving STEM Learning With Educational Technologies* (pp. 129–151). Hershey, PA: IGI Global. doi:10.4018/978-1-5225-2026-9.ch007

Singh, R., & Dutta, S. (2018). Visible Light Active Nanocomposites for Photocatalytic Applications. In K. Kumar & J. Davim (Eds.), *Composites and Advanced Materials for Industrial Applications* (pp. 270–296). Hershey, PA: IGI Global. doi:10.4018/978-1-5225-5216-1.ch012

Singh, R., & Lou, H. H. (2016). Safety and Efficiency Enhancement in LNG Terminals. In H. Al-Megren & T. Xiao (Eds.), *Petrochemical Catalyst Materials, Processes, and Emerging Technologies* (pp. 164–176). Hershey, PA: IGI Global. doi:10.4018/978-1-4666-9975-5.ch007

Sözbilir, H., Özkaymak, Ç., Uzel, B., & Sümer, Ö. (2018). Criteria for Surface Rupture Microzonation of Active Faults for Earthquake Hazards in Urban Areas. In N. Ceryan (Ed.), *Handbook of Research on Trends and Digital Advances in Engineering Geology* (pp. 187–230). Hershey, PA: IGI Global. doi:10.4018/978-1-5225-2709-1.ch005

Stanciu, I. (2018). Stochastic Methods in Microsystems Engineering. In S. Kadry (Ed.), *Stochastic Methods for Estimation and Problem Solving in Engineering* (pp. 161–176). Hershey, PA: IGI Global. doi:10.4018/978-1-5225-5045-7.ch007

Strebkov, D., Nekrasov, A., Trubnikov, V., & Nekrasov, A. (2018). Single-Wire Resonant Electric Power Systems for Renewable-Based Electric Grid. In V. Kharchenko & P. Vasant (Eds.), *Handbook of Research on Renewable Energy and Electric Resources for Sustainable Rural Development* (pp. 449–474). Hershey, PA: IGI Global. doi:10.4018/978-1-5225-3867-7.ch019

Subburaman, D., Jagan, J., Dalkiliç, Y., & Samui, P. (2016). Reliability Analysis of Slope Using MPMR, GRNN and GPR. In F. Miranda & C. Abreu (Eds.), *Handbook of Research on Computational Simulation and Modeling in Engineering* (pp. 208–224). Hershey, PA: IGI Global. doi:10.4018/978-1-4666-8823-0.ch007

Sun, J., Wan, S., Lin, J., & Wang, Y. (2016). Advances in Catalytic Conversion of Syngas to Ethanol and Higher Alcohols. In H. Al-Megren & T. Xiao (Eds.), *Petrochemical Catalyst Materials, Processes, and Emerging Technologies* (pp. 177–215). Hershey, PA: IGI Global. doi:10.4018/978-1-4666-9975-5.ch008

Tüdeş, Ş., Kumlu, K. B., & Ceryan, S. (2018). Integration Between Urban Planning and Natural Hazards For Resilient City. In N. Ceryan (Ed.), *Handbook of Research on Trends and Digital Advances in Engineering Geology* (pp. 591–630). Hershey, PA: IGI Global. doi:10.4018/978-1-5225-2709-1.ch017

Tyukhov, I., Rezk, H., & Vasant, P. (2016). Modern Optimization Algorithms and Applications in Solar Photovoltaic Engineering. In P. Vasant & N. Voropai (Eds.), *Sustaining Power Resources through Energy Optimization and Engineering* (pp. 390–445). Hershey, PA: IGI Global. doi:10.4018/978-1-4666-9755-3.ch016

Ulamis, K. (2018). Soil Liquefaction Assessment by Anisotropic Cyclic Triaxial Test. In N. Ceryan (Ed.), *Handbook of Research on Trends and Digital Advances in Engineering Geology* (pp. 631–664). Hershey, PA: IGI Global. doi:10.4018/978-1-5225-2709-1.ch018

Umar, M. A., Tenuche, S. S., Yusuf, S. A., Abdulsalami, A. O., & Kufena, A. M. (2016). Usability Engineering in Agile Software Development Processes. In I. Ghani, D. Jawawi, S. Dorairaj, & A. Sidky (Eds.), *Emerging Innovations in Agile Software Development* (pp. 208–221). Hershey, PA: IGI Global. doi:10.4018/978-1-4666-9858-1.ch011

Üzüm, O., & Çakır, Ö. A. (2016). A Bio-Inspired Phenomena in Cementitious Materials: Self-Healing. In M. Bououdina (Ed.), *Emerging Research on Bioinspired Materials Engineering* (pp. 220–246). Hershey, PA: IGI Global. doi:10.4018/978-1-4666-9811-6.ch008

Valente, M., & Milani, G. (2017). Seismic Assessment and Retrofitting of an Under-Designed RC Frame Through a Displacement-Based Approach. In V. Plevris, G. Kremmyda, & Y. Fahjan (Eds.), *Performance-Based Seismic Design of Concrete Structures and Infrastructures* (pp. 36–58). Hershey, PA: IGI Global. doi:10.4018/978-1-5225-2089-4.ch002

Vasant, P. (2018). A General Medical Diagnosis System Formed by Artificial Neural Networks and Swarm Intelligence Techniques. In U. Kose, G. Guraksin, & O. Deperlioglu (Eds.), *Nature-Inspired Intelligent Techniques for Solving Biomedical Engineering Problems* (pp. 130–145). Hershey, PA: IGI Global. doi:10.4018/978-1-5225-4769-3.ch006

Vergara, D., Lorenzo, M., & Rubio, M. (2016). On the Use of Virtual Environments in Engineering Education. *International Journal of Quality Assurance in Engineering and Technology Education*, *5*(2), 30–41. doi:10.4018/IJQAETE.2016040103

Verrollot, J., Tolonen, A., Harkonen, J., & Haapasalo, H. J. (2018). Challenges and Enablers for Rapid Product Development. *International Journal of Applied Industrial Engineering*, *5*(1), 25–49. doi:10.4018/IJAIE.2018010102

Wagner, C., & Ryan, C. (2016). Physical and Digital Integration Strategies of Electronic Device Supply Chains and Their Applicability to ETO Supply Chains. In R. Addo-Tenkorang, J. Kantola, P. Helo, & A. Shamsuzzoha (Eds.), *Supply Chain Strategies and the Engineer-to-Order Approach* (pp. 224–245). Hershey, PA: IGI Global. doi:10.4018/978-1-5225-0021-6.ch011

Wang, Z., Wu, P., Lan, L., & Ji, S. (2016). Preparation, Characterization and Desulfurization of the Supported Nickel Phosphide Catalysts. In H. Al-Megren & T. Xiao (Eds.), *Petrochemical Catalyst Materials, Processes, and Emerging Technologies* (pp. 431–458). Hershey, PA: IGI Global. doi:10.4018/978-1-4666-9975-5.ch015

Yardimci, A. G., & Karpuz, C. (2018). Fuzzy Rock Mass Rating: Soft-Computing-Aided Preliminary Stability Analysis of Weak Rock Slopes. In N. Ceryan (Ed.), *Handbook of Research on Trends and Digital Advances in Engineering Geology* (pp. 97–131). Hershey, PA: IGI Global. doi:10.4018/978-1-5225-2709-1.ch003

Zhang, L., Ding, S., Sun, S., Han, B., Yu, X., & Ou, J. (2016). Nano-Scale Behavior and Nano-Modification of Cement and Concrete Materials. In A. Khitab & W. Anwar (Eds.), *Advanced Research on Nanotechnology for Civil Engineering Applications* (pp. 28–79). Hershey, PA: IGI Global. doi:10.4018/978-1-5225-0344-6.ch002

Zindani, D., & Kumar, K. (2018). Industrial Applications of Polymer Composite Materials. In K. Kumar & J. Davim (Eds.), *Composites and Advanced Materials for Industrial Applications* (pp. 1–15). Hershey, PA: IGI Global. doi:10.4018/978-1-5225-5216-1.ch001

Zindani, D., Maity, S. R., & Bhowmik, S. (2018). A Decision-Making Approach for Material Selection of Polymeric Composite Bumper Beam. In K. Kumar & J. Davim (Eds.), *Composites and Advanced Materials for Industrial Applications* (pp. 112–128). Hershey, PA: IGI Global. doi:10.4018/978-1-5225-5216-1.ch006

About the Contributors

Akshay Kumar received the B.Sc. and M.Sc. degree in science from Chaudhary Charan Singh University, Meerut, India, in 2010 and 2012, and the Ph.D. degree major in Mathematics and minor in Computer science from G. B. Pant University of Agriculture and Technology, Pantnagar, India, in 2017. He is currently working as an Assistant Professor in the Department of Mathematics, Graphic Era Hill University, Dehradun in India. He is a regular Reviewer for international journals including Elsevier, Springer, Emerald, IGI Global Publisher and many other publishers. He has published 20 research papers in IEEE, Emerald, inderscience, International Journal of Mathematical, Engineering and Management Sciences, IGI Global Publisher and many other national and international journals of repute and also presented his works at national and international conferences. His fields of research are reliability theory, fuzzy reliability and applied Mathematics.

Mangey Ram received the Ph.D. degree major in Mathematics and minor in Computer Science from G. B. Pant University of Agriculture and Technology, Pantnagar, India. He has been a Faculty Member for around twelve years and has taught several core courses in pure and applied mathematics at undergraduate, post-graduate, and doctorate levels. He is currently the Research Professor at Graphic Era (Deemed to be University), Dehradun, India. Before joining the Graphic Era, he was a Deputy Manager (Probationary Officer) with Syndicate Bank for a short period. He is Editor-in-Chief of International Journal of Mathematical, Engineering and Management Sciences, Journal of Reliability and Statistical Studies, Editorin-Chief of six Book Series with Elsevier, CRC Press-A Taylor and Frances Group, Walter De Gruyter Publisher Germany, River Publishers and the Guest Editor & Member of the editorial board of various journals. He has published 225 plus research publications (journal articles/books/book chapters/conference articles) in IEEE, Taylor & Francis, Springer, Elsevier, Emerald, World Scientific and many other national and international journals and conferences. Also, he has published more than 50 plus books (authored/edited) with international publishers like Elsevier, Springer Nature, CRC Press-A Taylor and Frances Group, Walter De Gruyter Publisher Germany,

River Publisher. His fields of research are reliability theory and applied mathematics. Dr. Ram is a Senior Member of the IEEE, Senior Life Member of Operational Research Society of India, Society for Reliability Engineering, Quality and Operations Management in India, Indian Society of Industrial and Applied Mathematics. He has been a member of the organizing committee of a number of international and national conferences, seminars, and workshops. He has been conferred with "Young Scientist Award" by the Uttarakhand State Council for Science and Technology, Dehradun, in 2009. He has been awarded the "Best Faculty Award" in 2011; "Research Excellence Award" in 2015; and recently "Outstanding Researcher Award" in 2018 for his significant contribution in academics and research at Graphic Era Deemed to be University, Dehradun, India.

Om Prakash Yadav received the B.S. degree in mechanical engineering from the Malaviya National Institute of Technology, Jaipur, India, the M.S. degree in industrial engineering from the National Institute of Industrial Engineering, Mumbai, India, and the Ph.D. degree in industrial engineering from Wayne State University, Detroit, MI, USA, in 1987, 1992, and 2002 respectively. He is currently a Professor and a Duin Endowed Fellow with the Department of Industrial and Manufacturing Engineering, North Dakota State University, Fargo, ND, USA. Dr. Yadav has served as department chair for three years and also is a founding Director of Center for Quality, Reliability and Maintainability Engineering, which is supported by local member companies. The research work of his group has been published in Reliability Engineering and Systems Safety, Journal of Risk and Reliability, Quality and Reliability Engineering International, International Journal of Production Research, Engineering Management Journal, and IEEE Transaction of Systems, Man, and Cybernetics Systems. He has authored/coauthored more than 150 research papers in the area of reliability, risk assessment, design optimization, and operations management. His research interests include reliability modeling and analysis, risk assessment, design optimization and robust design, and manufacturing systems analysis. Dr. Yadav is currently serving as an Editor-in-Chief for the International Journal of Reliability and Safety and on the Editorial Board of several international journals. He is a recipient of the IISE William A.J. Golomski Best Paper Awards in 2015 and 2018. He is currently a member of IISE, ASQ, SRE, and INFORMS.

* * *

Reeta Bhardwaj has been published more than 15 papers in an international journal of repute and published one book on Applied Mathematics.

Kamlesh Bisht was born in 1990 in Uttarakhand, India. He received the M.Sc. degree from Kumaun University, Nainital, India in 2013. He is currently pursuing the Ph.D. degree in mathematics at G.B. Pant University of Ag. & Technology. He has published 4 research papers on fuzzy time series forecasting in international journals and conferences.

Md. Gulzarul Hasan is a Ph.D. in Operations research from Aligarh Muslim University, Aligarh, India. He has published several research papers in Mathematical programming and it's applications in specific to Manpower planning. He proposed several models by incorporating fuzzy uncertainty to best deal with real-life problems. Presently he is working as an Assistant professor at Madanapalle Institute of Technology and Science, Madanapalle, AP, India.

Nihan Kabadayi is an Assistant Professor in the School of Business at Istanbul University (Istanbul, Turkey) where she has been a faculty member since 2005. She received her Ph.D. in Supply Chain Management from Istanbul University in 2013. Her research interest includes supply chain management, healthcare operations management, quality management, and multi-criteria decision making.

Denis Katelnikov is an Associate Professor at Vinnytsia National Technical University. Ph.D. in Computer Science.

Deepak Kumar is currently Assistant Professor in the Department of Mathematics, D.S.B. Campus, Kumaun University, Nainital-263002, Uttarakhand, India. He has around 10 years of teaching and research experience to Undergraduate and Post Graduate students at kumaun University. He is a regular reviewer of many International/ National Journals. He has been a member of organizing committee of many international and national conferences. He has been conferred with Inspire fellowship award by DST, Govt. of India. He has published his research works at national and international journals of repute. His area of research includes reliability theory and fuzzy set theory.

Kamal Kumar is an Assistant professor at Amity University Haryana, Gurugram, India. Prior to joining this university Dr Kumar received his doctoral degree at Thapar Institute of Engineering & Technology Patiala, Punjab, India in 2020. He has obtained his Master degree in Mathematics during 2012 - 2014 and a Gold Medalist during this program. He has more than three years of teaching experience in different subjects of Engineering Mathematics and statistics. His current research interest is on Aggregation operation, multi criteria decision making, Uncertainty theory, and intuitionistic fuzzy set theory. He has published 21 international articles in the different reputed SCI journals. His Google citation is over 600 with h-index 10 and i10index 10. He has been a reviewer for some prestigious journals of Elsevier, Springer, Taylor & Francis.

Anita Kumari is currently Assistant Professor in the Department of Mathematics, D.S.B. Campus, Kumaun University, Nainital-263002, Uttarakhand, India. She has around 10 years of teaching and research experience to Undergraduate and Post Graduate students at kumaun University. She is Life Member of International Academy of Physical Sciences, The Tensor Society. She has been a member of organizing committee of many international and national conferences. She has published his research works at national and international journals of repute. His area of research includes Riemannian geometry and fixed-point theory.

Cynthia Montaudon is a full-time professor and a business consultant at the Business School at UPAEP. Since September 2018, she became head of the Observatory on Competitiveness and New Ways of Working, which deals with numerous issues related to social problems and creates awareness about current and future social needs. She has obtained a Post Doctorate Certificate In Organizational Leadership from Regent University in Virginia, USA; a Ph.D. in Strategic Planning and Technology Management from UPAEP, in Puebla, Mexico; a Ph.D. in Business from the University of Lincoln in Lincolnshire, UK, and three masters degrees: one in Quality Engineering from the Universidad Iberoamericana in Puebla, Mexico, another in Communication and media from the Jean Moulin, Lyon II University in Lyon, France, and the last one in Business Administration from the Tec de Monterrey in Mexico.

Ingrid Pinto López is a university professor, researcher and consultor at Business School of UPAEP University, recognized by the National System of Researchers (SNI Conacyt). Ph.D. in Strategic Planning and Technology Management. Currently coordinator of the competitiveness observatory and new forms of work, coordinator of the international arbitration of the Latin American Association of Accounting and Administration Schools ALAFEC, member of Barcelona Economics Network and the Illustrious Latin American Academy of doctors.

Amit Sharma has published more than 28 research paper in an international journal of repute. He has published 8 book chapter and 10 papers presented in international conferences. He also wrote 1 books on Applied mathematics and 1 book on Integral calculus. He has more than 12 years of teaching and research experience.

S. B. Singh is currently Professor in the Department of Mathematics, Statistics and Computer Science, G. B. Pant University of Agriculture and Technology, Pantnagar, India. He has around 25 years of teaching and research experience to Undergraduate and Post Graduate students at different Engineering Colleges and University. Prof. Singh is a member of Indian Mathematical Society, Operations Research Society of India, ISST and National Society for Prevention of Blindness in India and Indian Science Congress Association. He is a regular reviewer of many books and International/ National Journals. He has been a member of organizing committee of many international and national conferences and workshops. He is an Editor of the 'Journal of Reliability and Statistical Studies'. He has authored and co-authored eight books on different courses of Applied/ Engineering Mathematics. He has been conferred with five national awards and the Best Teacher Award. He has published his research works at national and international journals of repute. His area of research is reliability theory.

Index

H

I

M

O

P

Q

R

S

T

U

V

W

www.ingramcontent.com/pod-product-compliance
Lightning Source LLC
LaVergne TN
LVHW081338300926
849161LV00008B/2013

* 9 7 8 1 7 9 9 8 7 5 6 4 2 *